Mineral Admixtures in Cement and Concrete

Mineral Admixtures in Cement and Concrete

Jayant D. Bapat

CRC Press
Taylor & Francis Group
Boca Raton London New York

CRC Press is an imprint of the
Taylor & Francis Group, an **informa** business

CRC Press
Taylor & Francis Group
6000 Broken Sound Parkway NW, Suite 300
Boca Raton, FL 33487-2742

First issued in paperback 2017

ISBN 13: 978-1-138-07644-0 (pbk)
ISBN 13: 978-1-4398-1792-6 (hbk)

Library of Congress Cataloging-in-Publication Data

Bapat, Jayant D.
    Mineral admixtures in cement and concrete / Jayant D. Bapat.
        p. cm.
    Includes bibliographical references and index.
    ISBN 978-1-4398-1792-6 (hardback)
    1. Cement--Additives. 2. Concrete--Additives. I. Title.

TP884.A3B37 2012
693'.5--dc23
                                                                2012021573

I dedicate this book to my spiritual Guru

Sri Narasinhasaraswati Swamimaharaj

Alandi (Devachi)

Pune, Maharashtra, India

Contents

List of Abbreviations

NOTE: The concentration or content, expressed as fraction or percentage, is on mass-to-mass basis, and mass expressed as ton(s) is metric ton(s), unless specified otherwise.

A	aluminum oxide (Al_2O_3) or alumina
AAR	alkali–aggregate reaction
AASHTO	American Association of State Highway and Transportation Officials
ACFB	atmospheric (pressure) circulating fluidized bed
ACI	American Concrete Institute
ACR	alkali-carbonate reaction
AEA	air-entraining agent/admixture
Af_t	ettringite or trisulfatehydrate ($C_6AS_3H_{32}$)
Af_m	monosulfatehydrate or sulfate-Af_m
AMBT	accelerated mortar bar test
APC	air pollution control
aq.	aqueous
AS_2	$Al_2O_3 \cdot 2SiO_2$ or metakaolin
A-S	aluminosilicate (glass)
ASCE	American Society of Civil Engineers
ASR	alkali–silica reaction
ASTM	American Society for Testing and Materials
BA	bottom ash
BE	backscattered electron
BET	Brunauer–Emmett–Taylor (method)
BFS	blast furnace slag
BRE	Building Research Establishment (UK)
BSEN	British Standard European Norm
b_i	binder intensity index
C	calcium oxide (CaO)
C_3A	tricalciumaluminate or aluminate
CC	chloride conductivity
CCA	corn cob ash
C&D	construction and demolition
CDM	Clean Development Mechanism
CEB	Euro-International Concrete Committee (Comité Euro-International du Béton)
CER	Certified Emission Reduction
CFB	circulating fluidized bed
CH	calcium hydroxide ($Ca(OH)_2$) or portlandite

CLSM	controlled low-strength materials
CPT	concrete prism test
C_2S	dicalciumsilicate or belite
C_3S	tricalciumsilicate or alite
$C\hat{S}$	calcium sulfate ($CaSO_4$)
C/S	CaO/SiO_2 (molar ratio)
CSA	Canadian Standards Association
CSF	condensed silica fume
C-S-H	calcium silicate hydrate
CTL	chloride threshold level
CWS	calcined wastepaper sludge
c_i	CO_2 intensity index
DEF	delayed ettringite formation
DI	durability index
DIN	Deutsches Institut fur Normung (German Institute for Standardisation)
EAF	electric-arc furnace
EAFD	electric-arc furnace dust
EDF	expected durability factor
EDS/EDX	energy dispersive x-ray spectroscopy
EPA	Environmental Protection Agency, USA
ESA	external sulfate attack
ESP	electrostatic precipitator
F	F_2O_3
FA	fly ash
FAS	ferric oxide + alumina + silica
FBC	fluidized bed combustor
FDOT	Florida Department of Transportation
FGD	flue gas desulfurization
FHWA	Federal Highway Administration
FI	foam index
fib	International Federation of Structural Concrete
FIP	Fédération Internationale de la Précontrainte (International Federation for Prestressing)
GE	General Electric
GGBS	ground granulated blast furnace slag
GHG	greenhouse gas
H	water (H_2O)
HAP	hazardous air pollutants
HPC	high-performance concrete
HRM	high-reactivity metakaolin
HRWRA	high-range water-reducing admixture
HVFAC	high-volume fly ash concrete
IISI	International Iron and Steel Institute
IS	Indian Standard

ISA	internal sulfate attack
Iss.	issue
ITZ	interfacial transition zone
LA	lagoon ash
LOI	loss on ignition
M	MgO
MAS-NMR	magic angle spinning nuclear magnetic resonance
MH	magnesium hydroxide ($Mg(OH)_2$)
MK	metakaolin or metakaolinite
m/m	mass/mass
M-S-H	magnesium silicate hydrate
MSW	municipal solid waste
MSWA	municipal solid waste incineration waste ash
NIST	National Institute of Standards and Technology
NMR	nuclear magnetic resonance
NN	neural network
OPI	oxygen permeability index
PAH	polycyclic aromatic hydrocarbons
PAI	pozzolanic activity index
PBR	Pilling and Bedworth ratio
PC	Portland cement
PCA	Portland Cement Association
PCDD/F	polychlorinated dibenzo-dioxins/furans
PCFB	pressurized circulating fluidized bed
PEL	permissible exposure limit
PORA	palm oil residue ash
PPC	Portland pozzolana cement
PSC	Portland slag cement
QA/QC	quality assurance/quality control
QXRD	quantitative x-ray diffraction
RC	reinforced concrete
RCA	recycled concrete aggregate
RCPT	rapid chloride permeability test
RH	rice husk
RHA	rice husk ash
RILEM	Reunion Internationale des Laboratoires et Experts des Materiaux, Systemes de Construction et Ouvrages (French: International Union of Laboratories and Experts in Construction Materials, Systems and Structures)
RIM	repair index method
RMC	ready mixed concrete
RP	reaction product
RPC	reactive powder concrete
RPI	repair performance indicator
RRSB	Rosin–Rammler–Sperling–Bennett

RTA	Road Transport Authority
S	silica (SiO_2)
$\hat{S}$	SO_3
SBA	sugarcane bagasse ash
SCC	self-compacting concrete
SCE	saturated calomel electrode
SCM	supplementary cementitious material
SEF	secondary ettringite formation
SEM	scanning electron microscope
SISD	simplified index of structural damage
SF	silica fume
SLS	serviceability limit state
SO_4^{2-}	sulfate ion
SP	superplasticizer
t	metric ton(s)
t/a	metric tons per annum
TAPPI	Technical Association of the Pulp and Paper Industry
TCLP	toxicity characteristic leaching procedure
TEM	transmission electron microscopy
TERI	The Energy Resources Institute
TG	thermogravimetry
TLV	threshold limit value
TSA	thaumasite form of sulfate attack
TPWT	two-point workability test
UBC	unburned carbon
UFFA	ultrafine fly ash
ULS	ultimate limit state
UN	United Nations
UNFCCC	United Nations Framework Convention on Climate Change
Vol.	volume
WBCSD	World Business Council for Sustainable Development
w/c or w/b	water-to-cement or water-to-binder ratio
WSA	wheat straw ash
WSI	water sorptivity index
WWA	wood waste ash

Preface

Since the beginning of my career in 1975, I have been fortunate to be engaged in teaching, research, training, and consultancy in the area of cement, concrete, and construction. The behavior of a structural system throughout its life is the primary concern of an engineer. This book has been written keeping in mind the requirements of structural engineers and of those who are engaged in manufacturing concrete. It should also serve as a reference book for civil engineering courses, both at undergraduate and postgraduate levels. The focus of the book is making good, that is, workable and durable, concrete. The physical, mineralogical, and chemical characteristics of mineral admixtures are discussed keeping this aspect in mind. As the book has been written for civil engineers, references to complex issues on microstructure or chemistry have been presented to an extent that help better understanding of the application of mineral admixtures in practice.

It is now an established fact that durable concrete means concrete with fewer microcracks (10–100 μm). Microcracks in concrete allow ingress of external deteriorating agents such as water, carbon dioxide, chlorides, sulfates, and so on, leading to the deterioration, distress, and destruction of the structure. They can be reduced by using pozzolanic or cementitious materials, collectively called mineral admixtures, to replace cement in concrete. The term includes all siliceous and aluminous materials, which, in finely divided form and in the presence of water, react chemically with the calcium hydroxide generated during cement hydration to form additional compounds possessing cementitious properties. They may be naturally occurring materials, industrial and agricultural wastes or by-products, or materials that require less energy to manufacture. The mineral admixtures covered under the scope of this book are pulverized fuel ash (PFA), blast furnace slag (BFS), silica fume (SF), rice husk ash (RHA), metakaolin (MK), and some new compounds currently under investigation. The action of mineral admixtures can be explained in a simplified manner as follows:

$$\text{Cement} + \text{water} = \text{hydrated paste} + \text{Ca(OH)}_2$$

$$\text{Mineral admixture} + \text{water} = \text{slurry}$$

$$\text{Mineral admixture} + \text{Ca(OH)}_2 + \text{H}_2\text{O} = \text{Calcium-silicate-hydrate or C-S-H}$$
$$\text{through secondary hydration}$$

C-S-H is a principal strength-giving compound in the hardened concrete. The formation of additional cementitious compounds during secondary hydration leads to a reduction in temperature rise and refinement of pore structure in the hardened concrete. Calcium hydroxide is considered as a weak link in the concrete structure. The consumption of calcium hydroxide to form strength-giving phases, principally C-S-H, during hydration leads to improved durability of the structure in terms of its resistance to deterioration through carbonation, corrosion, sulfate attack, alkali–silica reaction, and so on. Besides the chemical (pozzolanic or cementitious) reaction, the mineral admixtures also act physically. The finely divided particles act as fillers. This is particularly significant in the interfacial zone, where they produce denser packing at the cement paste–aggregate particle interface, reduce the amount of bleeding, and produce a more homogeneous microstructure and a narrower transition zone. The overall effect is the enhancement in the strength and durability or the service life of concrete structures.

The knowledge and experience that I gained through academic and industrial research activities and also through sustained and wide-scale interactions with practicing civil engineers over the past four decades revealed that better understanding of the material aspects of the constituents, in particular mineral admixtures, will help civil engineers make better, durable concrete. Today, there is a greater appreciation of this aspect in the civil engineering community.

In India and in many other countries, the syllabi of university courses related to structures and concrete do not place sufficient emphasis on the materials aspect. The availability of such work will perhaps make it easier for academicians to include this aspect in the syllabi.

Another equally important issue is related to the sustainable development of the cement and construction industry. It is known that the production of a ton of Portland cement expels an almost equal mass of carbon dioxide in the atmosphere. The cement industry contributes about 5% of the total anthropogenic carbon dioxide emissions globally. Thus, replacement of Portland cement by mineral admixtures leads to sustainability as the mineral admixtures are mostly industrial and agricultural wastes. The volume of these wastes currently produced worldwide exceeds their utilization. The "factor 4" concept as envisaged in the global goals of sustainability fixed by the Intergovernmental Panel on Climate Change (IPCC) aims at reducing CO_2 emissions in developed countries in 2050 by a factor of 4 from their 1990 levels, after they are first reduced by a factor of 2 by 2020. Against that goal, the global scenario on sustainability of the cement industry predicted by the World Business Council for Sustainable Development (WBCSD) envisages reduction in CO_2 emission only by a factor of 2 by 2050 from their 1990 levels. The technological shift will require not only changes in concrete raw materials and mix design, but also new building techniques, using less materials to obtain the desired structural properties. The understanding of the materials aspects of the mineral admixtures and their impact on the hydration, strength, and durability of concrete will make a positive contribution, encouraging greater and more fruitful utilization of these and even other wastes in cement and concrete, and lead to the sustainable growth of both the cement and construction industry, on the one hand, and the waste-producing industries, on the other.

Dr. Jayant D. Bapat
Pune, Maharashtra, India

Acknowledgments

At the outset, I confess that it is impossible to carry out a work of this nature without the active support of family, friends, colleagues, peers, and above all the Almighty. While I humbly acknowledge all, I mention only some of them. The list is definitely not complete; hence, kindly excuse me for the omissions.

My sincere gratitude and thanks go to Joseph Clements, editor; Kari Budyk, senior project coordinator, Editorial project development; and Robert Sims, project editor, Production division, CRC Press/Taylor & Francis Group, for their valuable suggestions and guidance. I also thank Andrea Dale and the entire editorial staff at CRC Press for their kind support throughout the course of work. My special thanks go to Deepa Kalaichelvan, project manager, SPi Content Solutions, SPi Global, and the team for providing valuable support that made the book's journey through the publishing process swift and smooth.

I thankfully acknowledge inclusion of the published work from the following researchers and organizations: published material from the U.S. Federal Government Departments and Organizations, namely, the National Institute of Standards and Technology, MD; Virginia Department of Transportation, VA; Naval Facilities Engineering Command, Washington Navy Yard; Office of the Infrastructure Research and Development, Federal Highway Administration; Department of the Air Force, FL; Professor S. Tamulevicious, editor in chief, *Materials Science (Medziagotyra)*, Lithuania; HPC Bridge Views; International Center for Aggregates Research (ACAR); Theodore W. Bremmer, PhD, PEng, professor emeritus and honorary research professor, Department of Civil Engineering, University of New Brunswick, New Brunswick, Canada; Portland Cement Association, IL; and Dr. Moises Frias Rojas, Instituto de Ciencias de la Construccion Eduardo Torroja (CSIC), Madrid, Spain.

Dr. Jayant D. Bapat
Pune, Maharashtra, India

Author

Dr. Jayant D. Bapat, BTech, ME, PhD (IIT, Delhi), currently works as an independent professional in sustainable engineering and higher education. He previously worked as principal and professor at the engineering colleges affiliated to the University of Pune, Maharashtra, India. He also held senior positions at the National Council for Cement and Building Materials (NCB), New Delhi, and Walchandnagar Industries Ltd. (WIL), Walchandnagar, Pune. He has about four decades of experience in teaching, research, and consultancy in the cement, concrete, and construction industry. He has a number of publications to his credit and is a reviewer of technical papers for international journals. His research areas of interest are cement manufacturing, durability of concrete, and utilization of industrial and agricultural wastes in building materials.

1

Pulverized Fuel Ash

1.1 Introduction

The "pulverized fuel ash" (PFA) or the so-called "fly ash" (FA), used as a mineral admixture in cement and concrete, is a product of the pulverized coal firing system, through conventional boilers, mostly used in the thermal power plants. While carbon burns in oxidizing surroundings, the inorganic mineral matter gets sintered and liquefied at high temperature. The melt flows down the walls of the furnace and about 25% gets collected as "bottom ash" (BA). It is crushed before disposal. The rest, PFA or FA, gets entrained in the up-flowing hot gas in the form of fine particles, which get trapped in the economizer, air-preheater, mechanical separator, and, finally, battery of electrostatic precipitators (ESP). As a general practice in many countries, PFA and BA are mixed with water and transported to ash ponds/lagoons. The ash thus deposited in lagoons is called "lagoon ash" (LA) or "pond ash. It causes problems besides occupying huge stretches of agricultural land. Notwithstanding the greater utilization of PFA (and BA) in recent times in cement and concrete, in bricks, and for land filling, a large quantity of ash still lies unutilized. In order to get a measure of the enormity of the problem, the estimate on coal ash production in 2020, in the major producing countries or regions, is given in Table 1.1 [1]. The data in Table 1.1 should be viewed along with the estimated cement demand and carbon dioxide (green house gas) generation by cement industry for that period. The principal sources of green house gas generation in the cement industry are the manufacturing process, fossil fuels, transport, and power. Table 1.1 gives an estimate of possible reduction in CO_2 emission using the mineral admixtures, namely, the PFA and the blast furnace slag. As per several estimates, the cement industry contributes about 5% of the global generation of carbon dioxide. The cement industry's sustainable program developed by the World Business Council for Sustainable Development (WBCSD) prepared an "Agenda for Action" for a 5 year period from 2002 to 2007 [2], endorsed by the leading cement manufacturers of the world.

TABLE 1.1

Estimated Cement Demand, CO_2 Generation, and Potential for CO_2 Reduction Using Mineral Admixtures, in Major Regions in 2020

		Cement		Mineral Admixtures		
Sl No	Country/ Region	Demand (10^6 Ton/ Annum)	CO_2 Generation Potential[a] (10^6 Ton/ Annum)	PFA Generation (10^6 Ton/ Annum)	BFS Generation (10^6 Ton/ Annum)	CO_2 Reduction Potential[a] (10^6 Ton/ Annum)
i	United States	106	92	29	16	43
ii	Europe	318	277	31	31	20
iii	Japan	88	77	4	15	22
iv	China	1154	1004	62	20	7
v	Southeast Asia	294	256	17	3	7
vi	India	215	187	16	4	9
vii	Former Soviet Union	175	152	15	13	16
viii	Latin America	341	297	11	7	5
ix	Africa	288	251	7	2	3
x	Middle East	188	164	3	1	2

[a] The estimate of CO_2 generation potential is based on a global average of 0.87 kg CO_2 per kg cement and the reduction potential is based on 100% use of available PFA and BFS as replacement of cement.

The agenda addressed the issues of (a) climate protection, (b) fuels and raw materials use, and (c) emission reduction besides other issues. The comprehensive study on the sustainability of cement industry, carried out by Bettelle and the WBCSD [1], envisaged the following scenario, insofar as the CO_2 emissions are concerned:

- Considering 1990 as the base, the global demand for cement in 2020 shall increase 1.15–1.8 times regionwise. The demand in developing countries (China, India, and others) shall far exceed that in developed countries.

- At the global level, the cement industry will be required to reduce the CO_2 generation by 30%–40% in 2020 and by about 50% in 2050, above the 1990 measure.

- In order to achieve the desired level of CO_2 reduction, the cement industry will have to develop alternative cement formulations and new technologies to improve energy efficiency; use alternative building materials such as PFA and the blast furnace slag; use alternative, low carbon fuels; and adopt CO_2 capture and sequestration techniques.

Notwithstanding the WBCSD's agenda of action and the aforementioned observations, cement companies are not expecting the emergence of major environmentally friendly cement manufacturing technologies in the foreseeable future. That leads us to the conclusion that the answer to the problem of greenhouse gas emissions, on account of cement manufacturing, lies in reducing the output of clinker (raw cement before grinding, in the form of coarse particles obtained from the manufacturing process) and overcoming the loss in clinker production by the use of PFA and other supplementary cementitious or pozzolanic materials (mineral admixtures) in cement and concrete.

The addition of PFA to cement and concrete improves performance in terms of long term strength and the durability. The production of blended cement (cement with mineral admixtures) also leads to substantial reduction in the energy consumption, as illustrated in Table 1.2 [3,4]. The reduction in energy consumption leads to the corresponding reduction in the consumption of fossil fuels required for the generation of thermal and electrical energy. In the last few years, the emphasis in construction industry has shifted from high-strength to high-performance cement and concrete. This change has come on account of the fact that majority of the cement sales presently go toward the repairs of the structures, most of which are built using the so-called high-strength Portland cement (PC). The knowledge and experience gained through laboratory experiments and field trials worldwide have revealed that the strength and durability properties of cement/concrete with mineral admixtures can be further improved by enhancing the pozzolanic or cementitious properties of the admixtures by processing.

This chapter mainly deals with the materials aspects of PFA obtained after the combustion of pulverized coal in the conventional boilers in thermal power stations and is divided into 13 sections. Section 1.2 reviews the

TABLE 1.2

Energy Consumption in Blended Cement Production

Sl No	Cement Type[a]	Comparative Energy Consumption (%) Existing (New) Plant[b]	
		Thermal[c]	Electrical[c]
i	PC	100 (100)	100 (100)
ii	PPC[d]	84 (84)	89 (89)
iii	PSC[d]	47 (47)	86 (79)

[a] PC, Portland cement; PPC, Portland pozzolan cement; PSC, Portland slag cement.
[b] The energy consumption in the manufacture of PC is considered as 100%.
[c] The ideal energy consumption for a new large capacity plant: thermal: 680–720 kcal/kg and electrical: 72–74 kWh/ton.
[d] The addition of 15% PFA in PPC and 50% blast furnace slag in PSC, as a replacement of cement is considered.

classification and Section 1.3 describes the physical characteristics of PFA: the particle shape, specific gravity, size and fineness, color, and the effect of unburned carbon (UBC). Section 1.4 discusses the chemical and mineralogical composition and Section 1.5 outlines the characteristics of PFA produced in the modern fluidized bed combustion process. Section 1.6 deals with the characteristics of PFA produced after co-combustion of bituminous coal and petcoke. Section 1.7 deals with the leaching characteristics and Section 1.8 discusses the radioactivity, toxicity and occupational health aspects of the PFA. Section 1.9 depicts the processing and quality improvement aspects—i.e. the collection, physical treatment, ultrafine PFA, and chemical activation—for PFA, which broadly satisfies the requirements of the national standards. Section 1.10 describes the processing of unusable PFA that does not satisfy the requirement of the national standards, from the point of view of using it in cement and concrete. Section 1.11 discusses the quality control of PFA at the user site. Section 1.12 discusses the different ways in which PFA is added to cement and concrete and the provisions of relevant national standards. Finally, Section 1.13 summarizes this chapter.

1.2 Classification

ASTM C618 [5] classifies PFA based on the source of mineral coal (Table 1.3). It defines two classes of PFA suitable for use in concrete—Class F and Class C. While the two classes have identical physical characteristics, they are distinguished by their chemical compositions. The Class F PFA, which normally results from the burning of anthracite or bituminous coal, is the more readily available of the two. The sum of silica (SiO_2), alumina (Al_2O_3), and iron oxide (Fe_2O_3) in Class F must constitute at least 70% of the total mass. It also has low (typically less than 10%) calcium oxide (CaO) content. Even though its crystalline mineral constituents are not reactive, Class F PFA has pozzolanic properties. The Class C PFA normally results from the burning of lignite

TABLE 1.3

Classification of PFA and other Pozzolans as per ASTM C618

Class	Description
N	Raw or calcined natural pozzolan, for example, some diatomaceous earths; opaline cherts and shales; tuff and volcanic ashes or pumices; some clays or shales requiring calcination
F	Flay ash with pozzolanic properties normally produced from anthracite or bituminous coal
C	FA with pozzolanic and cementitious properties normally produced from lignite or subbituminous coal. May have lime content more than 10%

TABLE 1.4

Some National Standards on PFA for Use in Cement and Concrete

Sl No	Standards Organization	Standard Number	Standard Title
i	Bureau of Indian Standards	IS 3812	Specification for FA for use as pozzolan and admixture
		IS 6491	Methods of sampling fly ash
		IS 10153	Guidelines for utilization and disposal of FA
ii	British Standards Institution	BS 3892-1	Specification for PFA for use with PC
		BS EN 450	FA for concrete—definitions, requirements, and quality control
iii	American Society for Testing and Materials	ASTM C311	Test methods for sampling and testing FA or natural pozzolan for use in Portland-cement concrete
		ASTM C618	Specification for coal FA and raw or calcined natural pozzolan for use in concrete
iv	German Institute for Standardization	DIN EN 450	FA for concrete; definitions, requirements, and quality control
v	Japanese Standards Association	JIS A 6201	FA for use in concrete
vi	European Committee for Standardization	EN 450-1	FA for concrete—Part 1: Definition, specifications and, conformity criteria
		EN 450-2	FA for concrete—Part 2: Conformity evaluation

or subbituminous coal, and the sum of silica (SiO_2), alumina (Al_2O_3), and iron oxide (Fe_2O_3) must constitute at least 50% of the total mass. The Class C PFA has a high calcium oxide content (between 10% and 30%) and almost all of its mineral constituents are reactive, giving it both pozzolanic as well as cementitious properties. Table 1.4 lists and Table 1.5 compares some national standards on PFA for use in cement and concrete.

1.3 Physical Characteristics

The physical characteristics of PFA—shape, specific gravity, size and fineness, and UBC content—affect its performance in the concrete in terms of volume, rheology, and water demand at a given slump, porosity, and reactivity. The following paragraphs briefly review these characteristics of the PFA produced after the combustion in conventional boilers and discuss how they may affect the performance of cement or concrete. The characteristics of the PFA produced in the modern fluidized bed combustion process as well as that produced after the co-combustion of bituminous coal and the petcoke are discussed separately.

TABLE 1.5

Comparison of Some National Standards on PFA for Use in Concrete[a]

Sl No	Particulars	ASTM C618 Type F	BS 3892 Part 1	IS 3812
i	Particle density (kg/m^3, min)	Not specified	2000	Not specified
ii	Blaine fineness (m^2/kg)	Not specified	Not specified	320
iii	Retention on 45 μm (325 mesh) sieve (%, max)	34.0	12.0	34.0
iv	Loss on ignition (%, max)	6.0	7.0	5.0
v	Water requirement (% of PC, max)	105	95	Not specified
vi	Moisture content (%, max)	3.00	0.50	2.00
vii	Soundness (autoclave, max)	0.8%	10 mm	0.8%
viii	Strength activity index (%)[b]	75	80	80
ix	$SiO_2 + Al_2O_3 + Fe_2O_3$ (%, min)	70.0	Not specified	70.0
x	SiO_2 (%, min)	Not specified	Not specified	35.0
xi	Reactive silica (%, min)	Not specified	Not specified	20.0
xii	CaO (%, max)	Not specified[c]	10.0	Not specified
xiii	MgO (%, max)	Not specified	Not specified	5.0
xiv	SO_3 (%, max)	5.0	2.0	3.0
xv	Alkalies as Na_2O (%, max)[d]	1.5	Not specified	1.5
xvi	Total chlorides (%, max)	Not specified	0.10	0.05

[a] The individual standards may be referred for more details.

[b] The 28 day compressive strength (N/mm^2) of blended cement mortar is expressed as the percent of that of the control PC mortar. The ASTM standard for the purpose: ASTM C311: "Standard test methods for sampling and testing fly ash or natural pozzolans for use in Portland-cement concrete."

[c] Not specified but generally below 10% when FA is produced from burning of anthracite or bituminous coal.

[d] The equivalent alkali content, expressed as Na_2O, is obtained as: $Na_2O + 0.658 K_2O$.

1.3.1 Particle Shape

The ash particles are formed due to the condensation of liquefied and incombustible inorganic matter left after coal burning in the firing systems obtained in the thermal power plants. The shape of PFA particles depends upon the condition in which the coal combustion and subsequent condensation take place. The two main types of burning processes found in the coal based thermal power plants are (a) pulverized coal firing system (peak temperature 1750–2000 K) and (b) fluidized bed combustion (peak temperature 1100–1200 K). The pulverized coal firing system is the one mostly used in the large thermal power plants. While the carbon burns in the high-temperature surroundings, the inorganic mineral matter gets sintered and liquefied at high temperatures. The melt flows down the walls of the furnace and about 25% gets collected as BA. The rest gets entrained in the up-flowing hot gas in the form of fine particles and gets trapped in the economizer, air-preheater,

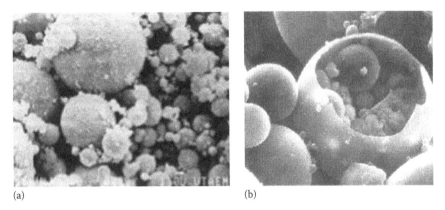

(a)   (b)

PLATE 1.1
Scanning electron micrographs of PFA: (a) cenospheres and (b) plerospheres.

mechanical separator, and, finally, battery of ESP. When the hot melt comes in contact with the hot gas, it gets divided into fine droplets. In the droplet, the silicate melt covers the UBC particles. The surface tension of the melt plays an important role in spheroidization of the PFA particles. The incorporation of carbon particles inside droplets favors the formation of gases such as CO, CO_2, SO_2, and H_2O and gives rise to cenospheres (ash particles hollow on the inside) and plerospheres (hollow ash particles with smaller particles inside) that are so common in PFA. Plate 1.1 shows typical micrographs of cenospheres and plerospheres. It has been suggested that the smaller particles fill in the larger ones, when the initially hollow particles are cracked or punctured during handling and not during the melting process [6]. The BA and the PFA distinctly differ in particle size, shape, and mineralogy.

The shape and surface characteristics of PFA particles affect the water requirement of concrete at the desired slump. The spherical particles reduce inter-particle friction (ball bearing effect) in the concrete mix, improve its flow properties and reduce water requirement. This phenomenon is commonly observed, when PFA replaces cement in concrete.

Cabrera et al. [7] reported the results of a comprehensive study carried out on the properties of 18 varieties of the PFA produced from bituminous coal in British thermal power stations. The results specifically deal with the variability of chemical and mineralogical composition of PFA, both within and between the sources (power stations), and physical properties such as particle size distribution, specific surface area, and particle shape. As the conclusion of a wide and in-depth study, the authors reported that the chemical and mineralogical properties of PFA fail to produce a parameter which explains or predicts the strength performance of these ashes in concrete. The authors proposed a new parameter, "shape factor," based on the specific surface area, which satisfactorily explains the variation in the workability and strength of concrete. The ash samples were divided in zones based on the shape factor. It was found that the more spherical the shape of the particles, the greater is their ability to reduce the water content of concrete mix at a given slump and improve the strength properties of concrete as a result. There is a need to carry out more work in this area and develop standard procedure and specification for the shape factor. The procedure followed by Cabrera et al. provides a guideline for this.

1.3.2 Particle Specific Gravity

The specific gravity of hydraulic cements is determined according to ASTM C188 [8]. This test method can also be used to determine the specific gravity of PFA particles. If it contains water-soluble compounds, the use of nonaqueous solvent instead of water is recommended. The specific gravity of PC and some mineral admixtures, commonly added to cement and concrete, is given in Table 1.6.

When cement is replaced by a mineral admixture of lower density, on a mass-to-mass basis, the volume of the mixture increases. If the strength and durability characteristics are kept reasonably constant, then such an addition may actually result in lowering the quantity of cementitious (in terms of mass) per unit volume of concrete. This aspect is important from the point of view of optimum use of cementitious materials in concrete. The following calculation gives a broad idea on the extent of volume increase due to such additions:

Consider the data on specific gravity, as given in Table 1.6.

The difference in the volume of PFA and PC may be calculated (considering average values) as follows:

$$\text{Volume difference per unit mass (cm}^3/\text{g)}: \left(\frac{1}{2.4}\right) - \left(\frac{1}{3.1}\right) = 0.094$$

Assume 10% replacement of PC by PFA in a concrete mix:

Percent increase in the mix volume:

$$\frac{(90/3.1 + 10/2.4) - (100/3.1)}{(100/3.1)} \times 100 = 2.92 \approx 3.0$$

TABLE 1.6

Specific Gravity of PC and Some Common Mineral Admixtures

Sl No	Mineral Admixture	Specific Gravity
i	PC	3.0–3.20
ii	PFA or FA	2.0–2.7
iii	Granulated blast furnace slag	2.9–3.0
iv	Silica fume	2.16–2.2
v	Rice husk ash	2.06–2.15
vi	Metakaolin	2.5–2.6

Thus, 10% replacement of PC by PFA on a mass-to-mass basis will result in approximately 3% increase in the volume of the mix.

The particle density is different from the bulk density. The bulk density needs to be considered when the powder is packed and transported. It determines the size of packaging, packing and handling equipment. The bulk density of PFA may vary between 0.8 and 1.2 g/cm^3, depending upon the level of packing.

1.3.3 Particle Size and Fineness

The particle size and fineness of PFA is considered as one of the most important properties and is mostly measured in two ways by the national codes:

a. *Specific surface area by Blaine apparatus*: In this method, the time taken by air to pass through a bed of FA is correlated to its specific surface area and is given in m^2/kg. The standard test method given at ASTM C204 [9] may be used for this purpose, except that the representative sample of PFA may be substituted for the hydraulic cement in the determination. However, ASTM does not specify any requirement of the surface area for the PFA for use in concrete. The Indian Standard IS 3812 Part 1 [10] specifies a minimum Blaine area of 320 m^2/kg of PFA for use in concrete.

b. *Residue on 45 μm sieve by wet-sieve analysis*: According to ASTM430 [11], this method measures the percentage of particles in PFA bigger than 45 μm (No 325) sieve, except that the representative sample of PFA may be substituted for the hydraulic cement in the determination. This method is prescribed by the national standards of many countries.

The difference between particle size and the particle size distribution should be clearly understood. Whereas the particle size may refer to a size of a single or an average of many particles lying in the narrow range, the size distribution refers to a range of the size of particles in a powder sample, often expressed as the mass (or mass fraction) of particles having a particular average particle size. Particles with identical specific surface areas may actually exhibit different size distributions [12]. The particle size distribution is an important parameter determining the cementitious activity in case of another mineral admixture, namely, blast furnace slag or the so-called ground granulated blast furnace slag (BFS or GGBS), which will be discussed later.

The PFA is generally collected in dry form from the hoppers installed under the dust collection equipment (mostly ESP) in the coal based thermal power generation units. The bag filters are also used in some places. Lee et al. [13] reported that the properties of PFA collected from ESP hoppers vary with the distance of collection point from the boiler. The authors found that more distant hoppers get PFA of larger specific surface area, smaller mean particle size as well as the residue on 45 µm sieve and higher glass content. Therefore in the ESP system, the hopper system itself affects the classification of PFA. The reactivity of PFA with $Ca(OH)_2$ also increased, going from the first hopper to the third one, due to the increase in the content of finer particles in the distant hoppers. In the investigation, authors also noted that the pozzolanic reactivity of PFA was more affected by the fineness than the glass content.

It is also seen that comminution (grinding or size reduction) operation improves the pozzolanic activity of PFA, if carried out in a controlled manner. The manufacture of blended cement using PFA, in the cement plants, frequently involves the inter-grinding of PFA with clinker. The results of the investigations carried out by Monk [14] revealed that the blended cement, manufactured by the intergrinding of PFA with clinker and gypsum as well as those manufactured by separate blending of PFA with ground clinker and gypsum, showed comparable properties with regard to water demand, slump, and compressive strength. The photomicrographs on a scanning electron microscope showed that the agglomerated spherical particles in coarse PFA get separated by intergrinding, which is thought to be largely responsible for the improvement in the water reducing property of PFA. Thus, the intergrinding process improves the properties of coarse PFA. The author concluded that the PFA, coarser than that currently recommended for use in concrete, can be used in blended cement manufactured by intergrinding, provided it satisfies the chemical requirements. It may be noted that the grinding process in the cement plant is required to be optimized in order to obtain maximum benefit in terms of the water requirement of the ground product, namely, the blended cement. It was earlier stated that the benefits of higher fineness resulting out of the grinding process—better workability and strength—can be reaped only as long as the spherical shape of the PFA particles is retained in the process.

The particles in raw PFA range mostly from 1 to 100 µm. The particles under 10 µm are the ones that contribute to the early (7 and 28 day) strength. The particles between 10 and 45 µm react slowly and contribute toward late strength (up to 1 year). The particles above 45 µm may be considered as inert and largely act as fine sand (filler).

1.3.4 Color

The PFA from bituminous coal is darker in color (gray) and that from lignite or subbituminous coal is lighter in color (buff to tan). The gray color can be attributed to the presence of UBC, more common in PFA from bituminous coal. In ash without any carbon or low carbon content, the color is likely due to the presence of iron compounds. When iron +3 compounds (i.e., compounds in which iron exhibits a valency of +3) are present, the ash is likely to have a brown color. If the iron is present as +2, the color would be bluish gray to gray. The gray would change to brown, when the ash is heated in the presence of air.

1.3.5 Unburned Carbon

The UBC particles in PFA are usually the largest contributor to the loss on ignition. The carbon particles do not take part in the chemical reactions during the cement hydration but influence its water demand for standard consistency. Carbon particles have strong affinity toward organic chemical admixtures, such as air-entraining agents (AEA). The AEA gets adsorbed on the carbon, which may adversely affect the air-void system within the hardened concrete. The degree of adsorption is dependent on the surface area, type of carbon (very coarse particles or soot) and the polarity of the carbon. The Foam Index (FI) is one of the most important quality control tests to determine the effect of UBC in PFA, in terms of the requirement of AEA. The capacity of carbon particles to adsorb AEA depends on the surface area. The carbon particles in ASTM Class C PFA (generally obtained by combustion of subbituminous coal) generally have more surface area than that of the carbon particles in Class F PFA (generally obtained by combustion of bituminous coal); hence adsorb more AEA. The excess carbon particles can be removed by beneficiation processes, such as sieving, air classification. However, the physical removal of carbon does not directly address the adsorptive characteristics of the carbon. Some PFA carbons have proven to be extremely adsorptive even at loss-on-ignition (LOI) levels that meet ASTM specifications. Some carbon removal techniques may alter chemical or physical characteristics of the mineral matter (while removing carbon). In order to determine if the carbon removal process degrades the final product, beneficiated PFA must be evaluated for the product compatibility.

Experience shows that PFA with less than 3%–4% carbon does not seriously affect the performance of organic chemical admixtures. The experiments carried out by Ha et al. [15], on the mortar with cement and PFA revealed that the increase in UBC content, beyond 8%, accelerated the corrosion of steel reinforcement. The experiments were conducted under aggressive, alternate wetting and drying conditions with 3% NaCl, for a period of 1 year. The corrosion rates were comparable to OPC for up to 6% UBC level.

1.4 Chemical and Mineralogical Composition

The chemical and mineralogical composition of PFA produced after the combustion of pulverized coal in the conventional boiler system is discussed here. The typical chemical composition of PFA, obtained in different regions, is given in Table 1.7 [3]. It is conventional to express the chemical composition in terms of the oxides, as seen in Table 1.7. However, it should be noted that PFA is a heterogeneous mixture of complex aluminosilicate glasses and some crystalline constituents. The aluminosilicate glass is amorphous and its structure gets modified due to the inclusion of alkaline and metal oxides like Na_2O, K_2O,

TABLE 1.7

Typical Chemical Composition of Low-Calcium PFA Obtained in Different Regions

Sl No	Country	Chemical Composition (Major Constituents)[a,b]								
		SiO$_2$	Al$_2$O$_3$	Fe$_2$O$_3$	CaO	MgO	Na$_2$O	K$_2$O	SO$_3$/S	LOI
i	France	48.1	24.68	6.5	1.41	1.82	0.56	4.06	—	11.7
ii	United Kingdom	50.0	28.1	11.7	1.62	1.54	0.28	0.62	—	2.27
iii	Germany	51.2	29.6	6.8	3.4	1.2	0.6	3.1	0.5	3.3
iv	United States	52.2	19.01	15.7	4.48	0.89	0.82	2.05	1.34	0.92
v	India	58.0	26.4	4.81	2.23	0.69	0.4	0.12	0.28	5.39
vi	Japan	59.6	31.2	2.3	1.4	0.5	0.7	0.5	0.2	1.0
vii	China	52.3	31.7	4.7	2.13	0.56	0.25	0.47	0.51	4.5

[a] The LOI often includes carbon content as well.
[b] The total of all percentages does not add to 100, as minor constituents have not been
mentioned.

MgO, CaO, and FeO. In a typical PFA of ASTM Type F, the glass content may
lie in the range of 75%–80% but in exceptional cases it may be as low as 50% or
as a high as 90%–95%. The reactivity of PFA is related to the amorphous phase,
that is, glass. Das and Yudhbir [16] found that a good correlation exists between
the glass content and the ratio of potassium to aluminum oxides (K$_2$O/Al$_2$O$_3$)
given by the chemical composition of PFA. The composition of glass in low
calcium PFA (ASTM Class F) is different from that of the high-calcium PFA
(ASTM Class C). The glass in the PFA of ASTM Class F has a highly polymerized network. The glass matrix depolymerizes when CaO content increases
above Al$_2$O$_3$ (alumina) content [6,17]. The major crystalline constituents of PFA
are quartz, mullite, magnetite, and hematite. The study of the mineralogy of
PFA helps in understanding its reactions with the calcium hydroxide liberated during the hydration of cement. The typical mineralogical composition of
some PFA obtained from different types of coal is given in Table 1.8 [18].

TABLE 1.8

Typical Mineralogical Compositions of Some PFA Varieties Obtained
from Different Types of Coal

Sl No	Type of Coal	Mineralogical Compositions of PFA					Loss on Ignition
		Glass	Crystalline Components				
			Quartz	Mullite	Magnetite	Hematite	
i	Bituminous	72.1	4.0	12.6	6.2	1.2	3.5
ii	Subbituminous	83.9	4.1	10.2	—	1.4	0.3
iii	Lignite	94.5	4.6	—	—	—	0.8

The total of all percent constituents does not add to 100, as some of the minor constituents have
not been indicated.

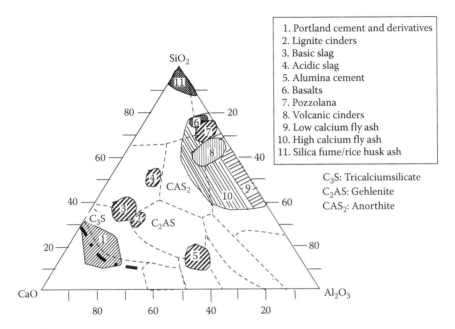

FIGURE 1.1
Ternary diagram of CaO-SiO$_2$-Al$_2$O$_3$ system depicting position of various resource minerals, cement minerals, and types of cement.

The CaO, SiO$_2$, and Al$_2$O$_3$ are the common oxide constituents of most of the cementitious or pozzolanic materials. The ternary diagram of CaO-SiO$_2$-Al$_2$O$_3$ system, given in Figure 1.1 [19], illustrates the common features and the differences in the chemical composition of various resource materials, cementitious/pozzolanic materials, and types of cement. As shown in Table 1.7, the oxide composition of PFA exhibits substantial regional variation. Another feature, typical to some regions (example: India), is that the PFA coming from the same source (thermal power station) may also show variation in composition, depending upon the consistency in the quality of coal, burning, and other conditions. The PFA coming from the same source as well that coming from different regions shows variation in chemical composition, in regions where the composition of coal differs far more widely and there is less standardization of the plant machinery.

The Indian PFA is characterized by relatively higher contents of SiO$_2$ and Al$_2$O$_3$ and lower contents of Fe$_2$O$_3$. The American PFA is mostly rich in Fe$_2$O$_3$, which acts as a flux during coal combustion. The higher Al$_2$O$_3$ in Indian PFA also implies higher fusion temperature; therefore, unless the ash is heated to higher temperatures, the glass content is likely to be lower. This perhaps could be an explanation for the comparatively lower glass content and lower activity of some of the Indian PFA.

As mentioned earlier, in the PFA Class F, the glass exhibits polymerization of silicate units. The entrapped gases inside the PFA particles may also be involved in the polymerization reaction, as shown in Reaction 1.1 [20]:

$$2(Si-O) + CO_2 = -(Si-O-Si)- + \left(CO_3^{2-}\right) \qquad (1.1)$$

The rate* at which the cementitious products are formed as a result of interaction between mineral admixtures (PFA, BFS, others) and the ordinary PC hydration products, depends upon the composition of PC, the mineralogical composition and particle characteristics of the admixture (contributing to its reactivity), the reaction temperature and the ionic concentration of reactants in the system.

In PC, the close relationship between the chemical and the mineralogical composition generally holds. However, it is not so in case of the mineral admixtures. It is only when the siliceous and the aluminous materials present in these admixtures, in noncrystalline and small particle form, hydrate at a slow rate in an alkaline medium to furnish silica and alumina for the reaction with lime, the formation of cementitious products takes place. This is contrary to the hydration of PC, where the principal silicates (C_3S and C_2S) and aluminate (C_3A) present essentially in the crystalline form react with water to provide the desired silica and alumina for the formation of cementitious compounds. Hence, while evaluating the suitability of mineral admixtures for blended cements, their mineralogical composition and particle size (and the size distribution) will have to be seen together with the chemical composition [21]. The glass forms a major component of mineralogical composition of both PFA and BFS. The reactivity of PFA in forming cementitious compounds is influenced to varying degrees by different parameters like glass content, basicity (capacity to release hydroxyl or OH⁻ ions during reaction), particle surface area, temperature, and size distribution [22].

1.5 Characteristics of PFA Produced in Fluidized Bed Combustion Process

The power plants, employing the recently developed circulating fluidized bed (CFB) combustion process, have come up in some parts of the world. They give higher power generation efficiency and reduced emission of polluting gases, like NO_x and SO_2. The CFB combustion has proven to be one of the most promising technologies for burning a wide range of coals and other fuels and handling wide variations in fuel quality, while still achieving strict

---

* Rate of reaction: In chemical reaction engineering terminology, the "rate" is often expressed as the moles of reactant (denoted by the concentration, moles per unit volume) disappearing, or the product forming, per unit time in a chemical reaction ($\partial c/\partial t$).

air emission requirements with high combustion efficiency. The low-grade fuels that have large ash content and very high sulfur do not normally find acceptance in conventional pulverized coal firing units (boilers). These fuels are burned efficiently in CFB systems. However, if the sulfur content is large, then the necessary sorbent, like limestone, has to be added to capture SO_2 in solid form. A desulfurization rate of up to 90% is achievable. In comparison to the conventional pulverized coal fired system, the CFB system has a relatively low combustion temperature (peak temperature 1100–1200 K). The physical and chemical characteristics of the PFA produced in the system are different. The extensive studies on these aspects are reported in the literature [23–27]. The two types of CFB processes that are in commercial use are atmospheric circulating fluidized bed (ACFB) and pressurized circulating fluidized bed (PCFB) combustion process.

Fukudome [27] reported that the ash particles from CFB have irregular shape or are less spherical in comparison to the PFA produced in conventional pulverized coal fired systems. The results of SEM and particle size analysis have shown that spherical or rounded PFA particles vary in size from 1.0 to 150 μm; those of irregular and angular shape are usually larger. The irregularity of the shape does not give the desired results in terms of lowering water requirement or the water-to-cement ratio, when PFA from CFB is added to cement/concrete, as generally seen in the case of spherical PFA particles obtained in a pulverized coal firing system in conventional boilers.

The lack of lime (CaO) in the PCFB ash is distinctly different from ACFB ash, which generally contains large amount of lime. The ACFB and PCFB PFA cannot be classified as ASTM Class F or C in many cases, because of low FAS (ferric oxide + alumina + silica content) and high SO_3 content.

1.6 Characteristics of PFA Produced after Co-Combustion of Bituminous Coal and Petcoke

The petroleum coke or the so-called petcoke is a byproduct from the oil refining industry. It has higher calorific value but a lower volatile content and normally higher levels of sulfur (5%–8% by mass) and nitrogen than traditional bituminous coal. While oil refineries consider it a waste product, to the cement manufacturer it represents an economical fuel alternative. The lower costs of petcoke can substantially reduce the production cost. Besides, the high calorific value makes petcoke ideal for firing in a cement kiln, notwithstanding the requirements of higher combustion temperature and longer combustion period. When co-fired with coal, petcoke can also provide flame stability and lower the operating cost of combustion units. The traditional bituminous coal is partially replaced by petcoke; the examples of up to 70% replacement have been reported [28,29]. However, the physical characteristics of PFA produced after the co-combustion of bituminous coal and petcoke are different. Its effectiveness and fitness for use as a mineral admixture in concrete need to be demonstrated.

The co-firing of coal with petcoke adds more UBC and coarseness to PFA [28]. The burning of petcoke results in a significantly lower quantity of ash, typically less than 0.5%, as compared with the 5%–20% normally associated with the burning of bituminous coal [30]. Consequently, the burning of petcoke in combination with coal is unlikely to have a significant impact on the composition of the inorganic material in the resulting PFA.

However, a significant amount of UBC remains in the petcoke PFA. In addition to higher levels of total carbon, petcoke PFA typically has elevated concentrations of nickel and vanadium. The carbon content of PFA is generally considered to interfere with the functioning of air-entraining admixtures (AEA) and prevent the formation of an adequate air-void system. The AEA consists of the solution of ionic or nonionic surfactants that gather at the water–air interface to stabilize the air bubbles. The organic molecules of AEA are adsorbed on the surface of the carbon particles present in the PFA and, thus, are not available to stabilize the air bubbles. An increase in carbon content would therefore be expected to reduce the effectiveness of the AEA. In practice, however, not all carbon is the same and characteristics such as surface area and surface chemistry must also be considered.

Yu et al. [31] investigated the adsorptive and surface properties of PFA from coal and pet coke with regard to the use of AEA. A range of PFA samples, from both coal and petcoke/coal co-firing, with a range of LOI values, were investigated for surface adsorptivity, as determined by FI test and surface area, by means of nitrogen absorption. It was observed that the inclusion of petcoke at 10%–15% replacement levels results in an increased LOI from 0.6% to 17.5%. Despite the considerably higher LOI for the petcoke blend, the surfactant adsorptivity remained unchanged at zero, which was below the Class F PFA with a lower LOI. The low adsorptivity of petcoke PFA is due to the very low surface area and coarse nature of the carbon particles, which is an intrinsic property of the material. Carbon from petcoke is dense and molecularly highly ordered, which limits microporosity and internal surface area. The negative effect of elevated carbon levels is therefore offset by the material's low adsorptivity.

The influence of petcoke on the strength development has been investigated by Lamers et al. [32]. The PFA without petcoke was shown to have a 28 and 91 day strength activity index of 89% and 90%, respectively. The petcoke PFA showed a slightly lower strength activity index of 85% at 28 days, with a somewhat higher value of 94% after 91 days. The use of petcoke PFA therefore appears to have no detrimental effects on either strength development or air entrainment compared bituminous coal PFA.

Another concern with respect to the use of petcoke PFA stems from the elevated levels of heavy metals, in particular vanadium. Vanadium can exist in either a bound or fused state or as a soluble free compound (vanadium pentoxide, V_2O_5) capable of reacting with the environment. Jia et al. [33] showed that vanadium, taken from a sample of PFA derived from burning 100% petcoke, exists primarily as $Ca_2V_2O_7 \cdot H_2O$, which is practically insoluble. Furthermore, vanadium can be incorporated into the C-S-H phase during the hydration of cement [34]. The inclusion of vanadium into the solid crystal structure thus removes it from possible reactions with the environment. The authors state that leaching experiments on cement reveal a low mobility of the metals incorporated into the clinker phases.

Scott et al. [30] examined the properties of PFA produced from the co-combustion of petcoke with bituminous coal. The inclusion of petcoke resulted in elevated total V_2O_5 concentrations but the concentration of soluble vanadium still remained below 100 ppm, as required by the Canadian Federal Government Transportation of Dangerous Goods Regulations (TDGR). The relatively insoluble form of vanadium pentoxide derived from the co-combustion of petcoke and bituminous coal, coupled with its incorporation into solid crystal structure of hydrated cement results in very limited possibility for leaching of the material into the environment. The PFA produced from co-combustion can therefore be used in concrete without any ecological concerns. This aspect needs to be demonstrated, particularly when the PFA generated from the co-firing does not satisfy the requirement of ASTM C618.

1.7 Leaching Characteristics

The leaching characteristics of PFA are important in both bound (concrete) as well as unbound (fill material) applications. A considerable amount of work has been carried out to determine the leaching characteristics of PFA [35–39]. The PFA is widely reported as having very low solubility. That is because most of the elements are bound within the largely insoluble "glassy" aluminosilicate matrix. Typically, 2% of PFA is found soluble and the elute mainly consists of calcium sulfate (gypsum), which is a naturally occurring compound found in many soils, with lesser contributions from sodium, potassium, and chloride ions. Most of the metals and metalloids present in the ash are either retained in the glass beads or firmly adhered to them, resulting in a very low leaching potential. A comprehensive study on the PFA obtained in United Kingdom, reported by Sear et al. [35], shows that the quantity of trace elements is less than 1% of the total. While the trace elements composition may indicate potential for environmental effects, the available leachable elements are minimal. The trace elements found in PFA are aluminum, arsenic, molybdenum, boron, nickel, phosphorous, cadmium, lead, cobalt, antimony, chromium, copper, tin, titanium, mercury, vanadium, and magnesium. Only a small fraction of these trace inorganic compounds, available on the surface in unbound form, are leachable in water. The leaching tests carried out by the authors, according to the German Standard (DIN 38414-S4) [40] and the United Kingdom's National Rivers Authority (NRA) extraction test, revealed that the trace elements in the leachates were mostly below detectable limits.

When used in structural fill applications, PFA has very low permeability ($\approx 10^{-7}$ m/s), which means there is very little passage of water through it and very little potential leachate. The trace elements are mostly held in aluminosilicate matrix and are not available to leach. The deposit of PFA is alkaline (pH 9–12), which further aids retention of metals. It has almost no biodegradable material and produces no gas product from degradation. When fully mixed with water, as an ash-slurry system, the pickup of trace elements has been found substantially less than the values obtained in standard leaching test. The dioxins are present in PFA but in very small concentrations, typically less than 20×10^{-9} g/kg. These levels are similar or less than the background levels found in soils.

The study on leaching characteristics of ASTM Class C PFA, obtained from the flue gas desulfurization (FGD) unit of Can Thermal Power Plant (Turkey), reported by Baba et al. needs special mention [41]. Besides the high content of sulfur (up to 26.09%) and ash (up to 44.6%), the Can coal also has trace concentration of potentially hazardous air pollutants (HAP) identified under the U.S. Clean Air Act and in particular exhibits high concentrations of arsenic, uranium and vanadium, higher than the average values for world coal. The authors observed that using the fluidized bed combustor (FBC) technology in power plants, the heavy metal contents in ash increases. The hazardous elements may be surface adsorbed on the glassy spherical FA particles. The elements that are surface adsorbed can be quite mobile. The leaching of these trace elements from wastes is controlled by the trace element concentration in coal and their modes of occurrence in coal. The results showed that water temperature, pH and limestone (added in FGD unit) were the most important factors affecting the leaching properties of the PFA obtained from the fluidized bed combustor. In addition, the result indicated that arsenic and selenium is more leachable than the other heavy metals in ash.

There is a need to establish the effects of large fill projects or ash lagoons, where the PFA cannot be fully isolated from the environment. Under such conditions, it may be required to monitor leachates using an approved method, periodically, to ensure no further testing should be required unless there is an inconsistency of material or a major change in fuel type. In case PFA is to be used within an area deemed to be environmentally sensitive, then conductivity and pH monitoring may also be carried out at the rate agreed between the parties concerned. The Energy and Resources Institute (TERI), New Delhi, India [42], has successfully grown Jatropha plants over the power station ash lagoons. It may be mentioned that Jatropha seeds are used for biodiesel manufacture. TERI has initiated large-scale plantations of Jatropha, that offer a unique advantage-the use of mycorrhiza as a natural inoculant to initiate early flowering and fruition of this much-sought-after plant. The versatility of this plant, it is hoped, shall help recover the wastelands.

1.8 Radioactivity, Toxicity, and Occupational Health

The radiation from PFA results from the concentration of natural minerals within coal. The carbon fraction is burned during the combustion and the radioactive minerals are left in the ash. The natural radioactivity of coal and of the ash results mainly from the radio nuclides from the decay series of uranium (U) and thorium (Th) as well as potassium 40 (K40). The K40 decays into calcium 40 or argon 40, both of which are stable nuclides that will not decay further. The PFA contains small quantity of potassium compounds, the K40 content of potassium is only 0.012% and from the radiation viewpoint K40 has little significance. In 2000, the UK Quality Ash Association commissioned the University of Nottingham to examine the leachates and radon emissions from PFA embankments constructed between 1967 and 2000. In addition a natural sand embankment was tested in a similar manner [43]. The radon and radium measurements were taken directly from the embankments by drilling into the embankments and sampling the air using a "sniffer" device. The principal investigators, Arnold, Dawson and Muller found that the radon was below the most onerous intervention level. The highest radon levels detected were in the PFA pore space. The radon levels in air adjacent to the PFA were extremely low. The age of the PFA and the PFA source did not appear to have a noticeable influence on the levels of contamination or of radon emissions.

The toxicity and occupational health hazards of PFA were studied by Born [44] and Meij et al. [45]. The purpose of the study was to review the in vitro (in laboratory) and in vivo (in living organism) data on coal FA and relate the findings to the role of crystalline silica, considering its classification as a human carcinogen. The coal mine dust was chosen as a reference, since it contains up to 10% of crystalline silica (α-quartz) and is well studied both in vivo and in vitro. The experimental studies, at both levels, showed lower toxicity, inflammatory potential, and fibrogenicity of PFA as compared to silica and coal mine dust. Although the studies suggested genotoxic effects of PFA, the data are limited and do not clarify the role of silica. The epidemiological studies on PFA exposed working populations have found no evidence for effects commonly seen in coal workers (pneumoconiosis, emphysema) with the exception of airway obstruction at high exposure. The authors concluded that the hazard of PFA is not to be assessed by merely adding the hazards of individual components. A closer investigation of "matrix" effects on silica's toxicity in general seems an obligatory step in future risk assessment on PFA and other particles that incorporate silica as a component.

The KEMA, Nederland BV, (KEMA is a leading energy consultant and a specialist in testing and certification in the Netherlands) carried out a study with regard to the health effects of PFA generated by the thermal power plants in the Netherlands [46]. Although the study was restricted to the thermal power stations operating in the Netherlands and conformed to the standards and the Threshold Limited Values (TLV) set by the Dutch health authorities, the broad conclusions of the study are relevant. The study revealed that the exposure of the personnel working in the power plant storage area to the respirable particulate matter was below the TLV, even in the event of accident. The people living in the vicinity of a coal-fired power station, with an open PFA storage facility, may be exposed to airborne PFA. In absolute terms, the concentrations involved are low and negligible in relation to normal background levels. Hence, the airborne dispersal of PFA does not exceed the recommended limits for fine particulate materials. The PFA originating from the co-firing of the examined secondary fuels (up to 10% replacement by dry mass, maximum) differ very little from PFA originating during the firing of bituminous coal only. No increased health risk is involved, as long as the requirement laid down for nuisance dust in the occupational environment are met. The PFA does not have any of the ill effects, for example, silicosis, normally associated with quartz. Hence, the TLV for quartz are not appropriate for the quartz found in PFA. The exposure to chromium-VI experienced by people living in the vicinity of power stations as a result of the airborne dispersal of PFA and other particulate material is low, in absolute terms, corresponding to the level considered by the Dutch government to pose a negligible risk. The level of exposure to chromium-VI experienced by power station personnel was also well below the TLV. The radiation doses associated with the exposure to PFA do not come anywhere near the relevant limits. It may therefore be concluded that radioactivity present in PFA does not represent a public health hazard. The PFA also contains very little dioxin, that is, less than 1 pg I-TEQ per gram (I-TEQ stands for International Toxicity Equivalents, a unit of toxicity introduced by the World Health Organization). The dioxin 2,3,7,8-TCDD, which is highly toxic to humans and animals, was not found in any of the PFA samples analyzed. Hence there is no reason to regard PFA as a "harmful" dust, as opposed to a "nuisance" dust. No increased health risk is involved as long as the requirements laid down for nuisance dust in the occupational environment are met. The assessment using KEMA Dust Assessment Methodology (KEMA-DAM®) indicated that exposure to all macroelements and trace elements was well below the relevant level.

The two aspects discussed in the preceding sections—the leaching characteristics and the radioactivity, toxicity, and occupational health effects of PFA—although cannot be directly correlated to its use as mineral admixtures in cement and concrete, are relevant from the point of view of PFA handling and storage at the user site. They also give an indication of the pollution and safety aspects that need to be considered during its utilization.

1.9 Processing for Quality Improvement and Assurance

The processing of PFA is carried out to improve and ensure consistency in its quality. It has significant advantages in terms of performance, for example, increased strength, greater pozzolanic reactivity, and lower water demand. It reduces the fineness variability significantly. The improved consistency normalizes the strength and other characteristics of PFA coming from different sources. The processing can be divided in two areas: first is the collection of PFA at the thermal power plant and second is its treatment, which includes the unit operations like size reduction, classification, dust free handling, quality control, storage, and dispatch. The discussion in this section is limited to the processing of PFA, which broadly satisfies the requirement of ASTM C618 [5]. The processing of "unusable" or "poor quality" PFA, generally not satisfying the requirement of the national standards, to make it suitable for use in concrete, is discussed in the next section.

1.9.1 Collection

The PFA is produced by coal-based thermal power plants. According to one report [47], fossil fuels (i.e., oil, natural gas, or coal) are used to generate 66% of electricity worldwide. The power plants in many countries still continue to dispose of PFA in the wet slurry form, in the so-called ash ponds. The most important aspect from the point of view of its use in concrete is that the PFA should be collected in the dry form. In recent times the dry PFA collection has gained momentum and the power plants are increasingly adopting separate collection of the PFA and the BA, keeping in view the difference in their characteristics and the applications. As an example, in India, a major FA producing country, the thermal power plants currently generate about 125×10^6 ton/annum PFA. The PFA utilization was just about 3% in 1994, which rose to about 15% by 2001, due to the sustained efforts by the Fly Ash Mission of the Government of India [48].

In the pulverized coal firing system, the mineral coal, in the "as received" condition, is pulverized in the grinding unit before burning in the furnace. As explained earlier, the ash formed is of two types—BA and the FA or the PFA, the latter so called because the fine particles of ash—get carried away (fly) with the flue gas and get collected at several locations along the flue gas path up to the chimney. The coarse fraction is collected in the economizer, air preheater, while the fine fraction (nearly 75%–80%) is collected in the electrostatic precipitator (ESP) system. Figure 1.2 shows a typical layout of the ash collection points and the approximate percentage of ash collected at each location [49]. The ESP has several fields and each field has a number of collection hoppers. A typical 210 MW generating unit, for example, will have an ESP with six fields and eight hoppers in each field, making total 48 collection hoppers. The coarsest PFA is collected in the Field-1 (nearest to the

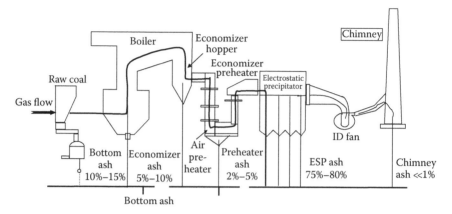

FIGURE 1.2
Ash accumulation and distribution in dry bottom furnace of thermal power generation unit.

gas entry) and the fineness of the collected PFA increases in the subsequent fields. The removal of PFA from the ESP hoppers can be done in two ways: (i) in the direction of gas flow or (ii) across the gas flow. In the first case, the material will be a mixture of the coarse and the fine fractions, as collected in all the fields. Whereas in the second, all hoppers in a plane perpendicular to the gas flow are interconnected, and the variation in the fineness of the PFA collected will be less and limited to that plane only. In the second case, it is possible to affect a preliminary separation between the "coarse" from the first two ESP fields and the "fines" in the subsequent fields.

The dry ash handling system at the thermal power stations typically consists of two stages. In the first stage, the ash from the economizer, air preheater, and the ESP is transported to a number of intermediate silos, located within a distance of 100–200 m from the last field hoppers of the ESP. In the second stage, the PFA from the intermediate silos is generally transported 800–2000 m away to large size bulk silo(s), from where it is finally disposed of, either to the user site or the ash processing unit. In some power stations, the dry PFA is given for utilization from the intermediate silos. The readers are advised to go through Refs. [49–53], to obtain more information on the working of ESP, PFA collection and handling systems in the thermal power stations.

1.9.2 Physical Treatment

The quality of PFA, as received from the thermal power plant, varies on account of the variations in the quality of fuel, that is, coal, plant design and engineering, operating parameters, and dust collection process. The utilization of PFA as an admixture in cement and concrete requires that its quality is consistently maintained as per the standards. The various operations carried out during the treatment do just that, besides improving the pozzolanic activity of PFA.

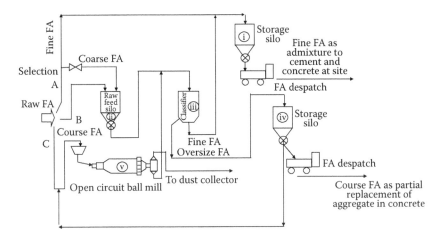

FIGURE 1.3
Alternatives for FA processing.

It is observed that the particle size (or the particle surface area) and the size distribution play an important role in deciding the pozzolanic activity of PFA. There is some UBC (normally up to 4%) present, depending upon the efficiency of burning in the thermal power plant. The carbon particles influence the water demand of cement and the requirement of the AEA. The under-10 μm fraction is the key parameter in processing. Thus the removal of the UBC particles and obtaining a proper particle size and the size distribution are the principal tasks of PFA processing.

Figure 1.3 shows a flow sheet, which has been developed to illustrate the possible alternatives for the treatment of PFA. The flow sheet shows three alternatives.

Alternative—A: The PFA discharged from the electrostatic precipitator is directly taken to the storage silo for further dispatch. This is possible if the segregation between the coarse and the fines is done at the collection stage in the thermal power plant, collecting the PFA across the gas flow and when it does not contain coal particles at undesirable levels.

Alternative—B: The raw PFA is first taken to an intermediate silo before feeding to a high efficiency air classifier. The fines (<45 μm) from the classifier are taken to the main storage silo for dispatch. The coarse oversize is taken to another storage silo, for dispatch to the construction site or RMC plant, where it could be used as a partial replacement of the fine aggregate in mortar and concrete.

Alternative—C: The coarse PFA is taken to a grinding unit. The discharge of the grinding unit is fed to the classifier. The fine PFA from the classifier is taken to the storage silo from where it is further dispatched to the user site. The coarse PFA from the classifier may be used as a fine aggregate in concrete or fed back to the grinding circuit for size reduction.

Barry [53] reported beneficiation trials on raw PFA complying with the requirements of CSA (Canadian Standards Association) Standard A23.5-M1982 on plant scale, using a high efficiency classifier to separate finer size fractions. The beneficiation was found to generally improve the quality of PFA, increasing both the glass content and the proportion of spherical particles. The result was improved pozzolanic activity, reduced water demand, and enhanced ability to control alkali-aggregate reaction. With finer particle size, higher proportions of cement could be substituted by PFA, without any loss of strength. Thus, considerable savings in cement could be achieved using beneficiated PFA.

1.9.3 Ultrafine PFA

As the name indicates, the ultrafine PFA is finer in comparison to that normally used as mineral admixture in cement and concrete. According to the quality control plan accepted by the Florida Department of Transportation (FDOT), United States [54], the sampling and testing of ultrafine FA (UFFA) shall follow the requirements of ASTM C311 and meet the quality requirements of ASTM C618 (see Table 1.4) as a Class F FA with certain modifications. The UFFA should have at least 90% particles below $8.5\,\mu m$ ($1\,\mu m = 10^{-6}$ m) size and at least 50% below $3.25\,\mu m$ size and the amount of material retained after wet sieving on $45\,\mu m$ sieve shall be less than 6%. Besides that, certain requirements have been specified by FDOT on the pozzolanic activity index, moisture content, and the LOI.

Chen et al. [55] characterized the UFFA obtained from combustion of three varieties of U.S. coal, using transmission electron microscopy (TEM). It was observed that the UFFA particles possess quite different morphology, composition, and microstructure in comparison to the coarser, micrometer size particles of PFA. Both crystalline and amorphous phases are observed. Discrete crystalline phases, down to $10\,\mu m$ size, were found rich in Fe, Ti, and Al. In some cases, alkaline-earth element aggregates in the form of phosphates, silicates, and sulfates were also seen. All the samples tested confirmed carbonaceous particles in the form of soot aggregates, with a particle size range of 20–$50\,\mu m$.

The properties of fresh and hardened concrete were studied, partially replacing cement with UFFA, mostly obtained from ESP and air classifiers. In general, UFFA is reported to exhibit higher pozzolanic activity [55], lower water demand in proportion with the fineness [56,57], larger slump and reduction in the slump loss [57,58], lower drying shrinkage and restrained shrinkage cracking [58–60], reduced requirement of total binder content in concrete to achieve target strength [56], reduced porosity thereby indicating better durability [55,60], and higher compressive strength [56,58,60]. Rathbone et al. [61] reported a novel development in the production of UFFA. The experiments were carried out at the

Centre for Applied Energy Research (CAER), University of Kentucky, United Kingdom. The UFFA was produced from the pond ash using hydraulic classification technology. Substantial reduction in the concrete permeability and the expected improvement in the concrete durability were observed. The authors recommended that the pond ash can be used to produce high-performance concrete at a competitive cost.

1.9.4 Chemical Activation

The activation of PFA by wet milling and leaching with sulfuric acid has been reported by Blanco et al. [62]. The application of activated PFA was studied to replace silica fume in producing high-strength concrete. The activated PFA was reactive and gave higher compressive strength, as a result of decrease in pore size in the hardened concrete. Poon et al. [63] reported the activation of low-grade FA (reject FA or r-PFA), which remains unused due to its high carbon content and large particle size. The authors found that addition of a small quantity of sodium sulfate or potassium sulfate (Na_2SO_4 or K_2SO_4) together with calcium hydroxide ($Ca(OH)_2$) significantly accelerated the hydration reaction and the compressive strength, as a consequence. While activating PFA chemically, the long term strength and durability aspect of concrete will have to be considered.

1.10 Processing of Unusable PFA

Notwithstanding the fact that the partial replacement of cement with PFA as a mineral admixture in concrete improves durability, its greatest impact is toward reduction in the emission of carbon dioxide (as a result of reduced production and consumption of PC) to the atmosphere on the one hand and the effective utilization of an industrial waste (generated as a result of power production using pulverized coal) available in large quantity, on the other. This aspect contributes toward the sustainable development of both the cement and the construction industry. However, PFA of a good quality only, satisfying the requirement of the national standards, is mostly used in cement and concrete presently, leaving behind huge quantity of nonstandard or unusable PFA, occupying substantial landfill space. The disposal cost has also increased over the years due to more stringent environmental controls and regulations. There is a need to develop technoeconomically feasible processes, to improve the quality of the unusable PFA so that it satisfies the requirement of the national standards for utilization in concrete. The processes that have been developed and tried on a large scale are briefly discussed below.

1.10.1 Principal Barriers in PFA Utilization

The PFA is rendered unusable on account of the following reasons:

a. *High UBC in PFA*: In order to comply with the NO_x emission limits set by the U.S. Clean Air Act Amendment, 1990 and similar regulations in many other countries, the utility burners in the thermal power stations have been converted to low NO_x burners in many places. The content of UBC increases as a result of the smaller dimensions of these burners and consequent reduction in fuel residence time. In modern power plants with circulating fluidized bed combustion (CFBC) boilers, sometimes bituminous coal is burned with high-sulfur petroleum coke (petcoke). It is observed that the addition of petcoke makes PFA particles coarse and increases UBC content (LOI) [28]. UBC in ash increases the water demand and the amount of air-entraining admixture. The ASTM C618 specifies an upper limit on carbon content of 6% for use in concrete. Frequent variations in PFA carbon content may also render it unusable, even when it satisfies the requirements of the standard.

b. *Activated carbon and mercury contamination in PFA*: The mineral coal sometimes contains high amount of mercury. With a view to reducing the CO_2 emission, the mineral coal is sometimes co-fired with 20%–25% secondary fuels. Although the mercury content of secondary fuels is negligible as compared to that of coal, the addition of secondary fuels to coal has a major impact on the mercury speciation in flue gas as well as the fuel ash [64]. Mercury increases toxicity. When such coal is burned in the power plants, activated carbon is added to flue gas to adsorb vaporized mercury and to prevent its emission in the atmosphere. The activated carbon, along with the adsorbed mercury, gets collected in the PFA. The activated carbon generally has very different characteristics in comparison to UBC in that it is much higher in porosity and internal surface area and is much more reactive. Even very small amounts of activated carbon renders an otherwise good quality ash, unacceptable to concrete manufacturers, due to its negative impact on air entrainment properties, namely, unwanted adsorption of air-entraining admixture on activated carbon.

c. *High ammonia content in PFA*: The power plants use ammonia to control NO_x or SO_3 emissions during the coal combustion. The excess ammonia precipitates on the PFA particles collected in the electrostatic precipitator. The release of ammonia during production and placement of concrete may be hazardous to the workers.

The unusable or nonstandard PFA must be beneficiated or processed before it is put to use, that is, as an admixture in concrete. The correct choice of whether or how to beneficiate a particular ash goes beyond the technological considerations alone. It must address the issues related to the available market and value of different products as well as the cost and performance of the technology.

1.10.2 Processing of PFA with High Unburned Carbon

Figure 1.4 illustrates common approaches for the separation of high UBC from PFA. There are two basic approaches: physical separation and combustion. As seen in Figure 1.4, the "physical separation" methods include sieving, air classification, and triboelectric separation and the "combustion" methods include furnace reinjection and carbon burnout. These methods can be used separately or in combination. It may be noted that the facilities for processing of PFA with UBC are required to be located near the thermal power station.

1.10.2.1 Physical Separation

The methods for the physical separation of ash produce more than one product stream. They generally yield low- and high-carbon products. Both these constituents can be collected and used commercially. The high-carbon fraction can be recycled back to the burner as fuel and the low-carbon fraction can be used as cement admixture. The product streams contain some carryover of undesirable constituents. The process with one feed stream and multiple output streams is often difficult to control by simple methods. In order to attain high purity and consistent product quality, multiple separation process steps are often necessary.

The physical removal of carbon does not directly address the adsorptive characteristics of the carbon. Some PFA carbons have proven to be extremely adsorptive even at LOI levels that meet ASTM specifications. Besides, the efficacy of physical separation processes in removing activated carbon is not fully established. Some carbon removal techniques may alter chemical or physical characteristics of the mineral matter while removing carbon. In order to determine if the quality of the PFA remaining after carbon removal degrades the final product, beneficiated PFA must be evaluated for product compatibility.

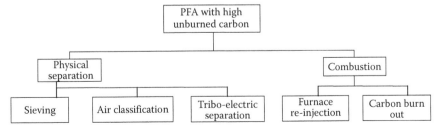

FIGURE 1.4
Approaches to remove carbon from PFA with high UBC.

The seasonality of both the power and the construction sectors is also important. The power (electricity) consumption in winter is more; hence, more FA is produced, when the cement and concrete consumption is usually low. Accordingly, storage space for the processed ash in the thermal power plant is needed for optimal operation and logistics all year round. Improved beneficiation and utilization schemes for PFA can transform it from a waste, unusable material, with associated disposal costs, to a valuable product.

1.10.2.1.1 Sieving

The simplest method to remove UBC from PFA is sieving. The efficiency of sieving is high when the average particle size (diameter) of the mineral matter is different from that of carbon. The ultimate separation performance depends on the initial size distribution of the raw (feed) ash. The technique is highly effective where the carbon is obtained in the larger particle size fraction and less effective with smaller particle size fraction. Some form of vibration of the screens assists sieving. The type of vibratory motion (circular or lateral), amplitude, frequency, and duration of vibration affect the performance. Laboratory testing to determine optimum screen configurations and vibration parameters is desirable prior to final equipment selection or installation. The sieving should be such that it minimizes carbon fracture during the process. It is important to include a stack of sieves upstream of the final sieve, in order to avoid blinding of the fine mesh by large particles. In general, LOI (carbon) reduction of 35%–45% is reasonable with yields in the 70% range, assuming that large particle carbon exists. Screen sizing is adjustable to allow selection of the large particle fraction but a smaller screen size negatively affects throughput rate. The industrial sieves are normally agitated at low frequency to help the particles distribute evenly over the surface and to help the small particles go through. Vibrating the mesh at ultrasonic frequencies, alternatively or complimentary to the low-frequency oscillation, can improve the rate of flow dramatically, preventing the product from blocking the holes in the mesh and helping to separate the small particles from the large. Thus, the efficiency of sieving can be increased by giving ultrasonic excitation to the sieves. Typically, a device called sonotrode that fits the sieve is excited by the ultrasonic processor. The sonotrode transmits the ultrasonic oscillations via the sieve frame to the screening surface. Refs [65–69] provide examples of a commercial application of the sieving technique. The efficacy of available commercial processes in removing mercury and activated carbon is not fully established. However, as the sieving process does not involve any chemical treatment, it should be possible to separate fine activated carbon particles (along with adsorbed mercury) from the high-carbon stream using an appropriate sieve size.

1.10.2.1.2 Air Classification

In air classification, separation is accomplished by engineering the aerodynamic drag and density characteristics of differing particles. When very small particles are entrained in air, aerodynamic drag dominates and even particles with widely different densities behave similarly. The PFA stored in large quantity, such as in hoppers and silos, often exhibits bulk characteristics that are different from that expected based on particle shape and chemical composition. The behavior of fine particles in PFA is strongly influenced by interparticle forces. The fine particles tend to agglomerate. Thus, the aerodynamic performance of bulk ash may often differ from the theoretical performance predicted on the basis of single particle analysis. The high efficiency air separator, cyclone, and fluidized bed separate small particles based on aerodynamic forces.

There are two air classification processes under development. The companies involved are EXPORTech Company, Inc., Pittsburgh, PA and Pittsburgh Mineral and Environmental Technology, Inc. (PMET), New Brighton, PA. The EXPORTech process involves using aerodynamic and magnetic force to separate carbon particles in PFA. In the PMET patented process, high-carbon ash is introduced into a jet mill, where it is subjected to size reduction by high-energy particle-to-particle impact (autogenous grinding), which separates free and adhered ash and carbon particles. The ground material is then carried upward in a controlled velocity air stream in which the high-density, low-carbon ash falls to the bottom of the unit and the low-density, high-carbon ash is carried out of the system through a rotating classifier wheel. The jet mill is supplied by Hosokawa Micron Powder Systems (HMPS). It is claimed that the process is suitable for processing a wide variety of PFA including that which is generated by the low NO_x burners and contaminated with ammonia, regardless of its characteristics and initial carbon content. The process can obtain a product with LOI content less than 3% from the initial LOI content of 12%. The technology is also competitive in terms of the cost, as claimed. The information on the commercial application of system was not available at the time of writing this book.

1.10.2.1.3 Triboelectric Separation

The method is based on triboelectrostatic charging of particles. In principle, charging occurs when two dissimilar materials are brought into direct contact with each other and are then physically separated. The static electrical charge (electrons) transfers from the material with the lower thermodynamic work function (carbon and steel) to the material with the higher thermodynamic work function (mineral ash). The polarity and the strength of the opposite charges differ according to the materials, surface roughness, temperature, strain, and other properties. The triboelectrostatic charging occurs in PFA containing UBC and the result of such charging process is carbon particles with significant positive charge and mineral (PFA) particles with significant negative charge. When these charged particles are exposed to a high-intensity electrostatic field, carbon and mineral (PFA) particles are collected at oppositely charged electrodes, where they are recovered. The performance of these methods appears to be sensitive to varying PFA resistivity, which changes significantly with the coal characteristics. The performance also appears to be sensitive to charge dissipation influences, such as relative humidity and high-carbon content. In general the separation system consists of a feeder or a charging mechanism, a high-voltage separation zone and a materials extraction system. At present, two commercial processes are available.

1.10.2.1.3.1 Triboelectrostatic Process with Mechanical Belt Transport   The process offered by Separation Technologies, Inc. (STI), Needham, MA, is a dry, triboelectrostatic separation process utilizing a mechanical belt transport [70–75]. The technology is fully operational and commercially available. The first STI system at Brayton Point Station in the New England Power system was commissioned in 1995 and has been successfully operational since that time. The company claims that the yield or recovery rates of 2% LOI material from widely varying feed ash can exceed 80%, depending upon the initial LOI and the level of liberation of the ash/carbon particles. The system does not produce any emissions. The processed ash is marketed under the brand name ProAsh. The STI process operates at ambient temperature and does not involve any chemical or gas treatment. Therefore when mercury is present in the raw PFA, it does not undergo any chemical change and the initial distribution of mercury in carbon particles remains almost unaltered. Thus, mercury concentration is expected to be higher in smaller carbon articles with larger surface area and a significant fraction would reside with the high-carbon reject product and make it moderately attractive for mercury recovery [65]. The company offers an add-on process to its separation technology that utilizes a proprietary chemical treatment to reduce ammonia concentrations from as high as 1000–20 mg/kg. The system is designed to optimize the capital and operating costs.

1.10.2.1.3.2 Triboelectrostatic Process with Pneumatic Transport   The process offered by Tribo Flow Separations (TFS), Lexington, KY is a dry, triboelectrostatic process using warm, low relative humidity air for transport [76–78]. The technology has been developed by the Center for Applied Energy Research (CAER) of the University of Kentucky, beginning in 1992. The system consists of a feeder mechanism, high-voltage separation zone and a vacuum based materials extraction system. The system is operational on a full scale, commercial basis since 2002. According to the tests conducted by the Company, when PFA containing mercury is fed to the system, mercury gets concentrated in the high-carbon product stream. The TFS system charging line air temperature can be controlled up to 700 K, which will eliminate most common types of surface ammonia in PFA, as claimed by the company.

1.10.2.2 Combustion

As shown in Figure 1.4, there are two methods to remove UBC by a combustion process. The first method is to reintroduce the high-carbon PFA in the boiler furnace in the power plant and the second is to burn the excess carbon in the FBC. In both the cases, the heat value of carbon is recovered. It removes carbon without removing the mineral constituents. Unlike the physical separation processes discussed in Section 1.10.2.1, it has a single product stream. It is possible to control the fraction of UBC in the product PFA to any desired level. As the process is carried out at high temperature, it is possible to remove surface ammonia, mercury, or burn activated carbon. With regard to mercury, it may be removed along with PFA to get encapsulated in concrete or recovered from the gas stream from the FBC as a value added product. Removal of activated carbon renders PFA without any problems related to the adsorption of the AEA, as indicated by the FI. In order to make the combustion process self-sustaining, the fraction of UBC in PFA should be more than 7%–8%. The presence of activated carbon in PFA may make combustion easier to sustain due to its high reactivity.

1.10.2.2.1 Carbon Removal with Furnace Reinjection

In most thermal power stations, the PFA is collected in dry form in the ESP. The ash is collected in hoppers at the bottom of ESP. Typically, an ESP line may consist of six to eight modular units (called stages), arranged along the gas flow, each with a positively charged electrode plate to collect the PFA and a dust collection hopper at the bottom. The dust collection system in a power plant may comprise of six to eight such parallel ESP lines. The heavier and coarser particles are collected in the initial stages and the lighter and finer toward the end. The PFA collected in the hoppers is pneumatically conveyed and stored in silos. Table 1.9 gives a broad idea about the fineness, carbon content (LOI), and the mass of dust collected under different ESP stages in a typical power plant.

As shown in Table 1.9., the fineness and the carbon content of PFA increase with the distance along the gas flow (stages). Second, nearly

TABLE 1.9

PFA Collection in Power Plant ESP

Sl No	Particulars	ESP Stage					
		I	II	III	IV	V	VI
i	Fineness (d_{50}, micron)[a]	32	25	20	15	10	5
ii	Loss on ignition (LOI, %)	4	5	9	13	14	15
iii	Dust collection (cumulative %)[b]	50	70	80	85	90	100

[a] d_{50} indicates the size of those particles that are collected with 50% efficiency in the ESP.

[b] Cumulative percent of the total mass of dust collected.

80% mass of the dust gets collected in the first three stages. It should be noted that the attractiveness of any ash beneficiation approach will depend on the actual cost of processing the ash at the power plant, the disposal cost (when ash is not used), and the market value of the product streams. The management of ash collected in ESP is important from that point of view. Consider the distribution of ash as given in Table 1.9. The low carbon ash (LOI < 6%, Stage I and II) can be stored for direct use in concrete, high-carbon ash (Stages IV–VI) may be recirculated for reinjection or re-burning in the boiler and that of intermediate quality (6%–9% LOI, Stage III) may be subjected to physical separation to bring down the carbon content below the acceptable limit. Thus only 20%–25% ash, with high carbon content, gets recirculated to the boiler for re-burning, contributing toward its thermal efficiency.

1.10.2.2.2 Carbon Removal with Fluidized Bed Combustor or Carbon Burn Out

The patented, so-called Carbon Burn Out (CBO™) process developed by Progress Materials, Inc., ST. Petersburg, FL, is fully operating and commercially available [79,80]. The UBC in high-carbon PFA is used as a fuel in the FBC. The heat is recovered from the exhaust gas and exhaust ash streams, using a heat exchanger. This recovered heat is fed back to the main utility boiler in the power plant for beneficial use.

1.11 Quality Control

The ready mix concrete plants require that the PFA, which they purchase, should satisfy certain minimum quality requirements. The PFA used in concrete should be consistent and uniform in quality. It should be monitored by a quality assurance/quality control (QA/QC) program that complies with the recommended procedures in the national standard. One such procedure is given in ASTM C311 [81]. These procedures establish standards for methods of sampling and frequency of performing tests for fineness, LOI, specific gravity, and pozzolanic activity such that the consistency of a PFA source can be certified. Many state transportation agencies in United States, through their own program of sampling and testing, have been able to prequalify sources of PFA within their own state (or from nearby states) for acceptance in ready mix concrete. The prequalification of PFA from different sources provides user with a certain level of confidence, in the event PFA from different sources are to be used in the same project.

As a typical example, the quality control requirements of the European Standard BS EN 450 for PFA used in concrete [82] may be considered.

The standard stipulates the following main requirements of PFA in the production of structural concrete conforming to BS EN 206:

- The fineness of the FA shall be less than 40% retained on the 45 µm sieve.
- The fineness shall not vary by more than ±10% of the mean value declared by the manufacturer.
- The ratio of mortar prism strength for the water-to-cement ratio of 0.50 shall be greater than 75% at 28 days and 85% at 90 days.
- The LOI is limited to 5.0%.

The test procedure and the frequency of testing for each property are specified. The BS EN 450 permits a wider range of fineness and performance tests at fixed water-to-cement ratio, rather than equal workability.

In Japan, the PFA obtained from the coal fired power station undergoes three-stage quality control to verify if it satisfies the requirement of the Japanese Standard [83]. These three stages are (a) recovery quality control: to determine the acceptability of PFA collected under the hopper of the electrostatic precipitator, (b) manufacturing quality control: to test PFA quality before and after the classification and after the storage silo before transferring to the homogenizing silo, and (c) delivery quality control: to test PFA quality before delivering to the land or marine transport vehicles. The parameters to be tested are specified.

The aim of these tests is to identify the variability during the production and respond quickly to the changes critical to the performance of PFA in concrete. The PFA is difficult for sampling. It is important that the samples are representative and consistent. The ash handling system should be provided with suitable sample drawing points. The PFA quality can be affected by major changes in the quality of coal or the process conditions in the thermal power plant, in which case the ash processing unit should be kept informed so that they make suitable arrangements.

1.12 Addition of PFA to Cement and Concrete

The type and quantity of PFA added to cement and concrete, as a partial replacement of cement, is governed by the provisions of the national standards. The PFA is added directly to ready mix concrete at the batching unit, interground with cement clinker, or blended with PC at the cement manufacturing unit to produce blended cement. The new ASTM Standard Specification on blended cement, C 595-06 [84], was released in August 2006 with some important changes to the nomenclature. The specification was simplified with the intent of making it easier to use. The new Portland-pozzolan cement, Type IP(X) can include between 0% and 40% pozzolans, replacing old Types IP and I(PM). The letter "X" stands for the nominal percentage of the supplementary cementitious material (SCM) included in the blended cement. The new standard facilitates communication of the total amount of SCM in concrete mixtures, which may also contain other added SCM (like MK) and may be subject to the restrictions on the total amount of SCM in the concrete. The ASTM C618 [5] defines two classes of FA for use in concrete: (a) Class F, usually derived from the burning of anthracite or bituminous coal and (b) Class C, usually derived from the burning of lignite or subbituminous coal. The standard also delineates requirements for the physical, chemical, and mechanical properties for these two classes of FA. The Class F FA is pozzolanic, with little or no cementing value, whereas Class C FA has cementitious as well as pozzolanic properties. The extent of replacement of cement with PFA is governed by the performance requirement of concrete. The design of any concrete mix, including FA concrete mix, is based on proportioning the mix at varying water-cementitious ratios to meet or exceed requirements for compressive strength (at various ages), entrained air content and slump or workability needs. The mix design procedures stipulated in ACI 211.1 [85] provide detailed, step-by-step directions regarding trial mix proportioning of the water, cement (or cement plus FA), and aggregate materials. PFA has a lower specific gravity than PC, which must be taken into consideration in the mix proportioning process.

BS EN206 [86], the European Standard, gives the specification for the use of PFA in concrete. This European Standard applies to concrete for structures cast in situ, precast structures and structural precast products for buildings and civil engineering structures. It specifies requirements for the constituent materials of concrete, properties of fresh and hardened concrete and verification of these properties, limitations for concrete composition, specification of concrete, delivery of fresh concrete, production control procedures and conformity criteria, and evaluation of conformity. The standard recognizes two types of additions:

- Type I additions—these are suitable for use in concrete but do not count toward the cement content of a mix.
- Type II additions—these are cementitious or pozzolanic and count toward the cement content.

The standard gives two basic concepts for utilizing Type II additions: the "k-value concept" and the "equivalent performance concept." Series of rules have been stipulated for these concepts. It should be noted that "other concepts" for additions are also permitted if their suitability is established, by either the European Technical Approval or the National Standard. The PFA is required to conform to BS EN450 or BS 3892: Part 1 [87] and the maximum amount of PFA to be taken into account is limited to 33% of the cement content.

The Japan Industrial Standard on Portland Fly-ash Cement [88] allows 5%–30% addition of PFA to cement. The Japan Industrial Standard on PFA for use in concrete [89] divides PFA in four classes. The earlier standard was revised in 1999 to widen the utilizable amount of PFA as a mineral admixture. The important aspects of the current standard are (a) PFA with high LOI ranging from 5% to 8% is specified as Class-III, (b) PFA with low Blaine fineness ranging from 1500 to 2500 cm^2/g is specified as Class-IV, and (c) high-quality PFA with LOI less than 3% and Blaine fineness more than 5000 cm^2/g is specified as Class-I. The Class-II PFA, with Blaine fineness of 2500 cm^2/g or higher and LOI less than 5%, is used in Japan for most applications. The PFA is added in different proportions in concrete, based on the applications.

The Indian Standard Specification for Portland pozzolan cement [90] allows addition of 15%–35% PFA to cement as replacement. The Indian Standard Specification on PFA [10] covers the extraction and the physical and chemical requirements of FA for use as a pozzolan for part replacement of cement, for use with lime, for use as an admixture, and for the manufacture of Portland pozzolan cement conforming to IS: 1489. The addition of PFA to concrete is governed by the provisions of the Indian Standard Code of Practice for plain and reinforced concrete [91]. The code recommends the use of PFA Grade 1 of IS 3812, with the upper limit of addition as fixed by the provisions of IS 1489: Part I.

The review of the national standards reveals that most of them allow addition of PFA to cement up to the extent of 35%–40%, to manufacture blended cement. Although some national standards allow flexibility in PFA addition, based on the performance requirement of concrete and even high volume of PFA (>50%) can be added to the concrete, the addition of PFA at the batching unit necessitates additional quality control and storage facilities at the concrete plant. With a view to overcome this problem CANMET, Canada, in the 1990s, undertook a major project to develop blended cement incorporating a high volume of ASTM Class F PFA (HVFA). The results of the investigation were reported by Bouzoubaa et al. [92]. The blended cement was made by grinding approximately 55% low-calcium PFA together with ASTM Type III PC clinker and a small amount of gypsum. The mechanical properties and the durability characteristics of concrete made with HVFA cement as well as that in which HVFA was added as a separately batched material at the mixer were comparable or superior to that of concrete made with commercially available ASTM Type III cement; the only exception was inferior de-icing salt scaling resistance of the former. In fact one variety of coarse PFA, which failed the requirements of ASTM C618, was also successfully used to produce HVFA cement. The large-scale manufacture of HVFA cement has not been reported so far. However this area is open for further trials.

In another study Zhang et al. [93] reported development of high performance cement blending three components: high-activity mineral admixture, cement, and low-activity mineral admixture or inert filler, with gap-graded particle size distribution and arranged in the fine, middle, and coarse fractions, respectively. The PFA and BFS were used as high-activity mineral admixtures and steel slag as inert filler in the study. The large-scale manufacturing and field experience on using such cement in concrete are yet to be reported. However the development opens a new possibility of manufacturing blended cement with low clinker content.

Sonebi [94] reported the development of less expensive, medium strength self-compacting concrete (SCC), using PFA. The SCC mix generally has a high content of fine fillers, including cement to produce high compressive strength concrete, which increases the cost and narrows its field of application to special concrete only. However, the application of PFA in SCC can help in widening its scope of application to general concrete construction.

1.13 Summary

This chapter mainly deals with the material aspects of PFA or the so-called FA, used as a mineral admixture in cement and concrete, a product of the pulverized coal firing system, through conventional boilers, mostly used in the thermal power plants. Notwithstanding the greater utilization of PFA (and BA) in recent times in cement and concrete, in bricks, and for land filling, a large quantity of ash still lies unutilized. As per several estimates, the cement industry contributes about 5% of the global generation of carbon dioxide, which is a green house gas. At the global level, the cement industry will be required to reduce the CO_2 generation by 30%–40% by 2020 and about 50% by 2050, over the 1990 measure. The answer to the problem of green house gas emission, on account of cement manufacturing, lies in reducing the output of clinker (raw cement before grinding, in the form of coarse particles obtained from the manufacturing process) and overcoming the loss in clinker production by the use of PFA and other supplementary cementitious or pozzolanic materials in cement and concrete.

The ASTM C618 classifies PFA, based on the source of mineral coal. It defines two classes of FA suitable for use in concrete—Class F and Class C. While the two classes have identical physical characteristics, they are distinguished by their chemical compositions. In the more readily available Type F PFA, the sum of $SiO_2 + Al_2O_3 + Fe_2O_3$ must constitute at least 70% of the total mass of ash. The Type F is pozzolanic and has low CaO content (<10%), whereas the Type C has both pozzolanic and cementitious properties, with high CaO content (10%–30%).

The shape of PFA particles depends upon the condition in which the coal combustion and subsequent condensation take place. The BA and the FA distinctly differ in particle size, shape, and mineralogy. The shape and surface characteristics of PFA particles affect the water requirement of concrete, at the desired slump. The spherical particles reduce inter-particle friction (ball bearing effect) in the concrete mix, improve its flow properties, and reduce water requirement. This phenomenon is commonly observed, when PFA replaces cement in concrete. The specific gravity of PFA lies in the range of 2.0–2.7 and bulk density in the range of $0.8–1.2 g/cm^3$. When cement is replaced by a mineral admixture of lower density, on a mass-to-mass basis, the volume of mixture increases. If the strength and durability characteristics are kept reasonably constant, then such an addition may actually result in lowering the quantity of cementitious (in terms of mass) per unit volume of concrete. This aspect is important from the point of view of optimum use of cementitious materials in concrete.

The PFA is generally collected in dry form from the hoppers installed under the dust collection equipment (mostly ESP) in the coal based thermal power generation units and the ESP hopper system affects its classification. The pozzolanic activity of PFA is more affected by the fineness than the glass content. The comminution (grinding or size reduction) operation improves the pozzolanic activity of PFA, if carried out in a controlled manner. The manufacture of blended cement in the cement plants frequently involves the inter-grinding of PFA with clinker. The intergrinding process improves the properties of coarse PFA. The benefits of higher fineness resulting out of the grinding process—better workability and strength—can be reaped, only as long as the spherical shape of the PFA particles is retained in the process. The particles in raw PFA range mostly from 1 to 100 μm. The particles under 10 μm are the ones which contribute to the early (7 and 28 day) strength. The particles between 10 and 45 μm react slowly and contribute toward late strength (up to 1 year). The particles above 45 μm may be considered as inert and largely act as fine sand (filler). The UBC particles in PFA are usually the largest contributor to the loss on ignition. When PFA is added to cement, carbon particles influence its water demand for standard consistency. The organic chemical admixtures such as AEA get adsorbed on the carbon, which may adversely affect the air-void system within the hardened concrete. Experience shows that less than 3%–4% carbon in PFA does not seriously affect the performance of cement.

It is conventional to express the chemical composition of PFA in terms of the oxides. It is a heterogeneous mixture of complex aluminosilicate glasses and some crystalline constituents. In a typical PFA of ASTM Type F, the glass content may lie in the range of 75%–80% but in exceptional cases it may be as low as 50% or as high as 90%–95%. The reactivity of PFA is related to its amorphous phase, that is, the glass content. While evaluating the suitability of mineral admixtures for blended cement, their mineralogical composition and particle size (and the size distribution) will have to be seen together with the chemical composition. The glass forms a major component of mineralogical composition of both PFA and BFS. The reactivity of PFA in forming cementitious compounds is influenced to varying degrees by different parameters like glass content, basicity (capacity to release hydroxyl or OH⁻ ions during reaction), particle surface area, temperature, and size distribution.

Power plants employing the recently developed CFB combustion process have come up in some parts of the world. It is reported that the ash particles from CFB have an irregular shape or are less spherical in comparison to the PFA produced in conventional pulverized coal fired systems. The irregularity of the shape does not give the desired results in terms of lowering water requirement or the water-to-cement ratio in concrete. The CFB FA cannot be classified as ASTM Class F or C in many cases, because of low FAS (ferric oxide + alumina + silica content) and high SO_3 content.

The petroleum coke or the so-called petcoke is a byproduct from the oil refining industry. It has higher calorific value but a lower volatile content and normally higher levels of sulfur and nitrogen than traditional bituminous coal. While oil refineries consider it a waste product, to the cement manufacturer it represents an economical fuel alternative in terms of lower cost and higher calorific value, in comparison with the traditional bituminous coal. The partial replacement of bituminous coal with petcoke, up to 70%, has been reported. The physical and chemical characteristics of PFA generated after co-firing of petcoke with bituminous coal, differ from that generated from burning of bituminous coal, in terms of UBC and elevated levels of heavy metals, in particular vanadium. Therefore, the utility of such PFA as an admixture to cement and concrete needs to be demonstrated in terms of strength, durability, and ecological concerns, on a case-to-case basis, particularly when the PFA generated from the co-firing does not satisfy the requirement of ASTM C618. The use of PFA in concrete has not been found to pose any risk in terms of leachability of harmful chemicals, radioactivity, toxicity, or occupational health.

The processing of PFA is carried out to improve and ensure consistency in its quality. It has significant advantages in terms of performance, for example, increased strength, greater pozzolanic reactivity, and lower water demand. It reduces the fineness variability significantly. The improved consistency normalizes the strength and other characteristics of PFA coming from different sources. The ash formed in thermal power stations is of two types, BA and the FA, the latter so-called because the fine particles of ash get carried away with the flue gas and get collected at several locations, in the dust collection equipment, along the flue gas path up to the chimney. The removal of the UBC particles and obtaining a proper particle size and the size distribution are the principal tasks of PFA processing.

The ultrafine fly ash (UFFA) is reported to exhibit higher pozzolanic activity, lower water demand in proportion with the fineness, larger slump and reduction in the slump loss, lower drying shrinkage and restrained shrinkage cracking, reduced requirement of total binder content in concrete to achieve target strength, reduced porosity thereby indicating better durability, and higher compressive strength.

The quality of low-grade FA (or reject FA), that remains unused due to its high carbon content and large particle size, can be improved using chemical activation method. While activating PFA chemically, the long term strength and durability aspect of concrete will have to be considered.

The PFA is rendered unusable on account of high UBC, activated carbon and mercury contamination, and high ammonia content. The unusable or nonstandard PFA must be beneficiated or processed before it is put to use, that is, as an admixture in concrete. The choice of beneficiation process should be based on, besides the technological considerations, cost and performance issues as well. There are two common approaches for the separation of high UBC from PFA: physical separation and combustion. The physical separation methods include sieving, air classification, and triboelectric separation and the combustion methods include furnace reinjection and carbon burnout. Situated near a thermal power station, these methods can be used separately or in combination.

2

Blast Furnace Slag

2.1 Introduction

The American Society for Testing and Materials (ASTM) defines blast furnace slag (BFS) as "the non-metallic product consisting essentially of calcium silicates and other bases, developed in a molten condition simultaneously with pig iron in a blast furnace" [1]. Normally 1 ton of pig iron generates 250–350 kg of slag. In the blast furnace, the process is optimized to maximize the production of pig iron, that is, with effective removal of slag. The smooth flow of molten slag out of the furnace is ensured by adding a basic agent such as lime to reduce its viscosity. When the blast furnace is tapped to release molten iron, it flows from the furnace, with molten slag floating on its upper surface. These two materials are separated using a weir, the molten iron being channeled to a holding vessel and the molten slag to a point where it is to be treated further. The slag consists of, principally, silicates and aluminosilicates of lime and magnesia. The BFS produced from the blast furnace in the integrated steel plant is also called ferrous slag; the nonferrous varieties are the byproducts of copper, nickel, and lead melting. The BFS from different plants differ in chemical, mineralogical, and physical constitution. The final form of the BFS depends upon the method of cooling and can be produced in the following forms:

Air-cooled: The molten slag solidified under ambient conditions in pits near the furnace. It is crystalline in nature. It has very limited cementitious properties and is used as an aggregate for all types of construction.

Foamed or expanded: The molten slag is treated with a limited quantity of water to increase vesicularity and decrease unit weight. The processes and cooling rates vary widely. The product may be highly crystalline or extremely glassy. It is used primarily as a lightweight aggregate.

Pelletized: The molten slag is cooled and solidified with water and air quenched in a spinning drum to form pellets. The cooling process is controlled to obtain crystalline or amorphous (glassy) product, which could be used as an aggregate or ground for cementitious applications, respectively.

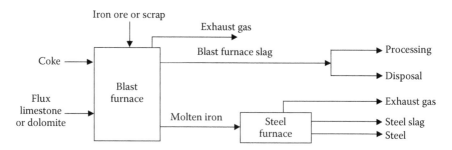

FIGURE 2.1
Blast furnace slag production.

Granulated: The granulated slag is produced when the molten slag, flowing out from the blast furnace, is rapidly cooled and solidified by quenching in water to a glassy (amorphous) state. The quenching process results in the formation of sand-sized fragments, usually with some friable clinker-like material. The product has cementitious properties. Figure 2.1 shows the schematic diagram of BFS production.

The shape and size of granulated (or pelletized) slag depend upon the chemical composition, temperature, and the time of quenching and method of production. The granulated slag is latently hydraulic or cementitious material. When ground to very fine cement-sized particles, ground granulated blast furnace slag (GGBS) can be suitably used as a mineral admixture, as a partial replacement of cement.

This chapter focuses mainly on the materials aspects of the GGBS as a mineral admixture for cement and concrete. Section 2.2 summarizes the various processes for the granulation of BFS as obtained in the modern integrated steel plants and describes the physical characteristics of granulated slag. Section 2.3 covers the production of GGBS, which includes moisture reduction and grinding and particle size and size distribution of GGBS. Section 2.4 discusses the chemical and mineralogical composition of cementitious BFS. Section 2.5 discusses the quality control aspects of GGBS as related to some national standards. Section 2.6 describes the different ways in which BFS or GGBS is added to cement and concrete and the provisions of the relevant national standards. Section 2.7 mentions the quantitative determination of BFS in cement. Finally, Section 2.8 summarizes this chapter.

2.2 Granulation of BFS

The hydraulic activity of BFS refers to its capacity to react with calcium hydroxide in the hydrated cement paste and form strength-giving compounds, mainly the calcium silicate hydrates (C-S-H), which depends largely on its glass content. In order to convert the inorganic mineral phases present in the molten state at 1600–1800 K, into amorphous glass, molten slag must be rapidly cooled (quenched) to a temperature below 1150 K. The process of quenching and converting the molten slag into granules of 3–5 mm size is known as granulation or pelletization. The modern blast furnaces produce 250–350 kg slag per ton of pig iron, depending upon the iron content of the ore.

2.2.1 Granulation Process

When the BFS is cooled slowly, say in air, most of it gets crystallized. The crystallized BFS does not possess any notable hydraulic properties. Eventually, through crystallization, the mineral phases reach lowest energy levels, with a heat of crystallization of 200 kJ/kg. The process of quenching (rapid cooling) converts these phases into glass, which is at a relatively higher energy level. The granulated BFS is called latent hydraulic material because, when activated by lime (or sulfate), it forms solid hydration products to attain a lower energy level. The hydration products of finely ground granulated BFS, that is, GGBS, are principally the same as those obtained in the hydration of Portland cement.

The granulation of slag on an industrial scale started in 1953; since then the process has continuously improved in terms of glass content, particle size, and moisture content of the granulated product. Figure 2.2 illustrates different alternatives for the granulation of molten slag. It illustrates four schemes of slag granulation. Scheme I is the "pelletizer" process developed in Canada [2]. It has been used successfully in many parts of the world. The process involves first quenching the molten slag with water sprays and then passing it over a rotating drum. The fins on the drum break the flow and throw slag in air for sufficient time to cause pellet formation due to surface tension. The sphericity of granules is improved by this process and the flowability is better. Schemes II and III are indeed the variants of the so-called classical process, which is used more commonly. It essentially consists of quenching the molten slag in water continuously flowing in a rotary granulator or a channel. Scheme IV is the modern dry granulation process. The water + air or water quenching of slag have certain drawbacks, like (a) requirement of a large quantity of water to granulate the molten slag at a high temperature, (b) pollution of water due to the alkaline compounds in slag, (c) emission of sulfide from the slag into air due to water quenching, (d) requirement of drying of slag after water quenching, and (e) wastage of thermal energy of high-temperature slag without recovery. The dry process reportedly removes these drawbacks. The process involves atomizing molten slag and then rapid cooling of the particles to produce glassy slag. The molten slag coming out from the blast furnace enters the heat-recovery vessel through a launder, covered to reduce heat loss. The launder delivers the molten slag directly into the rotary-cup air-blast atomizer located in

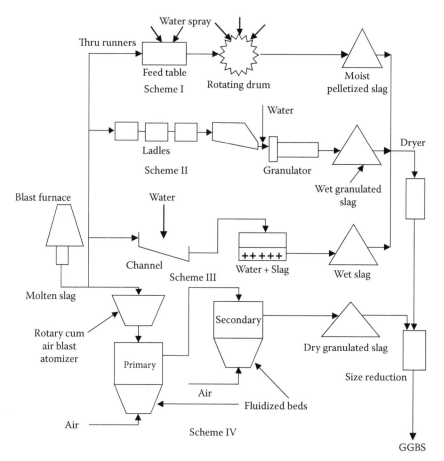

FIGURE 2.2
Schemes for slag granulation.

the center of the cylindrical granulator chamber. The chamber may be up to 20 m diameter. On atomization, the slag particles are projected radially outward and slightly upward in a spray and impinge on the vessel wall. The particles cool as they travel through air and then are cooled further in the fluidized bed coolers. The particles, flowing down the vessel wall, fall directly into the primary fluidized bed. The slag particles then overflow into the secondary fluidized bed, where the temperature of granulated slag is brought down to the discharge temperature. The fluidized beds provide the rapid cooling necessary for the formation of a glassy slag product. The fluidized bed is a convenient method of containing the slag particles as it prevents the agglomeration of hot particles. Slag particles with a mean diameter of about 2 mm can be produced by this process, which makes it easy to handle [3]. The arrangement for moisture reduction of BFS is generally made in the grinding plant itself.

As mentioned earlier, the reactivity of slag depends upon the glass content. The finer grains of granulated slag are rich in glass. However, to obtain the finer grains, pressure of the water jets for quenching needs to be increased; that requires a higher capacity pump with higher power consumption. The fine granules also pose associated problems in handling and in the grinding process. Thus, the grain size of BFS is optimized keeping these aspects in mind. The results of these optimization experiments carried out in the Tata Iron and Steel Company, Jamshedpur, India, are reported by Piplal et al. [4]. Regourd [5] reports that the rate of cooling (quenching) affects the hydraulic activity of slag. The quenching is responsible for the creation of chemically active sites on the grain surface of slag. These active sites are responsible for the initial hydration of slag.

2.2.2 Physical Characteristics of Granulated BFS

The granulated slag, resembling coarse sand, has a particle size mostly passing 4.75 mm or ASTM E11 No 4 sieve [6,7]. The maximum size may go up to 10 mm. The size distribution and the composition of glass and crystalline phases depend upon the type of granulation process. In order to ensure the reactivity, the British Standard specification for GGBS, for use with Portland cement [8], requires glass content to be not less than 67%. High molten slag temperature and the optimum quenching conditions would result in glass content higher than 95%. The bulk density is in the range 1.0–1.3 ton/m³. The pelletized slag (Scheme I, Figure 2.2) contains a larger proportion of spherical particles in comparison to the granulated slag (Schemes II and III, Figure 2.2), which improves its flowability. The moisture content of pelletized slag is generally lower, due to lower water consumption. It is claimed that the pelletized slag presents many advantages over granulated slag, such as higher glass content, lower drying cost, and less gas emission during pelletization [9]. In comparison to clinker, granulated slag is difficult to grind, that is, slag grinding consumes more energy.

Lang [10] reported that some properties of granulated slag change after prolonged storage. The content of combined water increased but the glass content remained unchanged. The setting time of cement produced using the granulated slag stored outdoors became longer with increasing storage time.

2.3 Ground Granulated Blast Furnace Slag

The processing of BFS starting from slag granulation, moisture reduction and finally grinding (comminution), leads to the production of GGBS. It is fine, off-white, and lighter in color than both Portland cement and granulated slag (Plate 2.1). The moisture reduction is generally integrated with

PLATE 2.1
Ground granulated blast furnace slag (GGBS).

grinding process. GGBS can be added to cement or concrete mix as a partial replacement for cement, as stipulated by the national standards. The following sections briefly discuss various aspects related to the moisture reduction and grinding of BFS leading to the production of GGBS.

2.3.1 Moisture Reduction and Grinding

The various grinding systems for the production of GGBS, in the modern cement plants, are schematically illustrated in Figure 2.3 [11]. The choice of the appropriate system depends on considerations like the moisture content of feed BFS, whether the plant is new or old (retrofitting/modernization), availability of space, requirements of the product characteristics, and, finally, the availability of the capital. Table 2.1 [12] compares the alternative systems of GGBS production, given in Figure 2.3, in terms of the requirements of energy, space, capital and maintenance cost, and product quality. It may be mentioned that the grading of the systems is notional and is based on the general experience. All the systems are currently obtained in the industry. A production unit may find a particular system more useful than the others, based on the local and plant-specific considerations.

As shown in Figure 2.3, the size reduction of BFS granules can be carried out in the ball mill, roller press, or vertical roller mill (VRM) or by a suitable combination of these equipment. In comparison to the cement clinker, the BFS is moist, fine grained, more abrasive, and harder to grind. The moisture content should be less than 4%, if grinding is done in a ball mill. In the roller press, moisture content up to 1%–2% is desirable for satisfactory performance. The VRM can dry and grind BFS up to 10% moisture content. It is desirable to dry BFS in a separate unit, for higher moisture contents. Wiegmann et al. [13] reported that for the same fineness of the finished material, the throughput of grinding plant is higher, when grinding moist

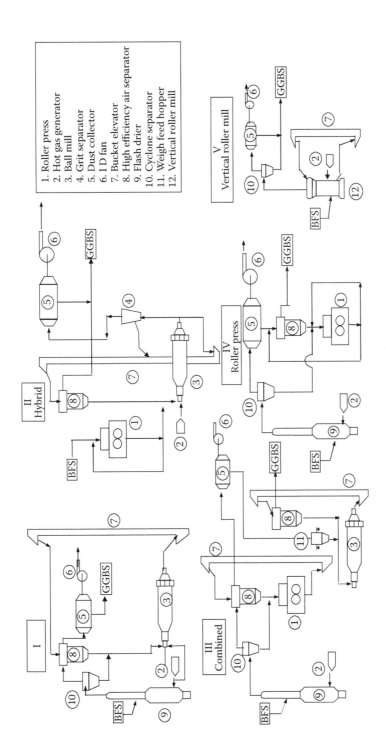

FIGURE 2.3
Alternatives for GGBS production.

TABLE 2.1

Comparison of Modern Production Systems for GGBS (Best System = 1.0)

Sl No	Grinding System	Energy			Relative Cost[a]		Product Quality
		Drying[b]	Grinding	Space	Capital	Maintenance	
i	Ball mill in closed circuit with high efficiency air separator	1.5	2.0	2.0	1.0	1.0	1.0
ii	Roller press and ball mill in hybrid mode	1.4	1.3	3.0	1.5	1.3	1.0
iii	Roller press and ball mill in combined mode	1.5	1.2	3.0	1.7	1.3	1.0
iv	Roller press in finish mode	1.5	1.0	2.0	1.0	1.0	1.0
v	VRM in finish mode	1.0	1.1	1.0	1.3	1.3	1.0

[a] Cost indices are based in India.
[b] Moisture content of BFS at feed = 10%.

granulated BFS, in comparison to that of pre-dried granulated BFS. The moisture acts as a grinding aid. However, at the same time, it also reduced the compressive strength at 2, 7, and 28 days, as a result of prehydration of BFS, the extent of reduction depending upon the feed moisture content. This can be counteracted by increasing the fineness of ground BFS or by increasing the clinker content in the blend. If the fineness is increased to compensate for prehydration, there is a drop in the throughput of mill. Increasing the clinker content shall put a limit on maximizing the use of BFS. The authors found drying and grinding in VRM as the most energy efficient from the point of view of drying energy requirement, as it required only 65% of the drying energy in comparison to that required for a flash drier. The combined grinding method (Option III in Figure 2.3) is the most expensive from the view point of capital investment. In the manufacture of Portland slag cement, separate grinding of the constituents (cement + additives and BFS) and subsequent blending is found to give energy savings and better product quality in most cases.

The activity of BFS is proportional to the glass content. The activity of inferior quality BFS (lower glass) can be improved by increasing its fineness but at a higher expense of the grinding energy, which increases exponentially with fineness, at higher fineness. On the other hand, BFS with higher glass can be ground relatively coarser (to save energy); that will give the advantage of lower water demand with good activity [14].

2.3.2 Particle Size and Size Distribution

The dried and ground BFS is known in practice as GGBS. It has a specific gravity in the range of 2.9–3.0, lower in comparison with the Portland cement (Chapter 1). The fineness is one physical property of GGBS with which the strength characteristics of GGBS are conventionally correlated. It is commonly expressed as Blaine surface area (cm^2/g). The measurement can be carried out quickly and therefore normally used to monitor the grinding operation. However, specific surface area is not a definitive characteristic of fineness. The ground materials having identical surface area can have significantly different particle size distribution. That the particle size distribution affects the strength properties of cement, irrespective of its type, has been observed by several workers [15–18]. Whereas the impact of particle size distribution on Portland cement or Portland slag cement is observed directly through strength and other properties, on GGBS this is seen indirectly, that is, when it is blended with cement or concrete; for example, it is commonly observed that the cement with narrow particle size distribution exhibits improved strength properties in comparison to that having wider distribution.

The particle size distribution is conventionally expressed by the Rosin-Ramler-Sperling-Bennett (RRSB) function [19], as given in Equation 2.1:

$$Q(x) = 1 - \exp\left[-\left(\frac{x}{x'}\right)^n\right] \tag{2.1}$$

or

$$-\ln[1 - Q(x)] = \left(\frac{x}{x'}\right)^n \tag{2.2}$$

Each distribution has a unique value for the position parameter (x') and the slope (n), in logarithmic plot of cumulative percent mass distribution [$Q(x)$] and the particle size (x) (Equation 2.2). The position parameter (x') is the particle size at a cumulative mass distribution of 63.2%. A typical RRSB plot is shown in Figure 2.4. The position parameter characterizes the fineness of the particle population in terms of a characteristic diameter. The smaller the position parameter, greater is the fineness of the particle population. On the contrary, the slope indicates the width of the distribution. The larger the slope, narrower is the distribution. The values of the position parameter and the slope of the curve are calculated by mathematical fitting of the measured values of size distribution in the RRSB function. The commonly used methods for that purpose are linear regression, weighted linear regression, and nonlinear approximation [20,21].

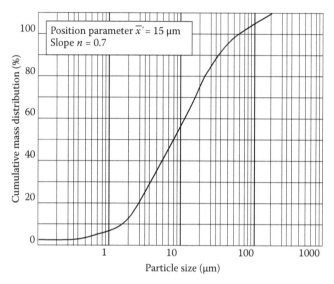

FIGURE 2.4
Typical RRSB plot.

The foregoing discussion makes it clear that the fineness expressed in terms of unit surface area (cm^2/g, Blaine) is not a unique property, as the cements that have identical unit surface areas can have different particle size distributions, represented by RRSB parameters. The fineness of cement can be unambiguously characterized by defining the surface area (cm^2/g, Blaine) and one of the two parameters related to size distribution, the slope (n) or the position parameter (x'), in the RRSB diagram. It will be in consonance with the observed findings, if the same criteria are applied in the case of GGBS also. The German Standard, DIN 66145: "Graphical representation of particle size distributions, RRSB-grid," contains the mathematical equation of the distribution function and its application.

2.4 Chemical and Mineralogical Composition of Cementitious BFS

The reactivity of BFS, among the mineral admixtures, is most sensitive to the glass content. The glass in granulated slag is its active part; it is responsible for the formation of chemical phases—C-S-H and calcium aluminate hydrate—after reacting chemically with the alkalies released during cement hydration. The slag granulation process used today can produce granulated slag with a degree of vitrification (glass formation) greater than 95% by mass, depending upon the method of cooling and the temperature at which cooling is initiated. The tests on slag reactivity have shown that the total glass content of 100% does not give the highest reactivity. Some microscopic crystal nuclei in the glass structure are required to promote glass dissolution before the chemical reaction with the alkali [22–25]. The different processes of slag granulation are discussed in Section 2.2.1. The pelletizer process (Scheme I, Figure 2.2) is characterized by a lower speed of cooling, compared to the classical granulating process (Schemes II and III), which permits formation of a certain proportion of the crystalline phase in the glassy phase. Several studies have proved that pelletized slag may have higher hydraulic activity as compared to granulated slag produced from the same origin. The explanation of this superiority can be attributed to the microcrystals that increase the heterogeneity of the glass structure. It may be noted that in cement containing BFS, the mechanical strength depends not only on the nature of slag but also on the clinker, which is associated with it [26].

The BFS belongs to the quaternary system CaO-SiO_2-Al_2O_3-MgO. The mineralogical compounds formed after the crystallization of BFS are given in Table 2.2. In the granulation process, molten slag is rapidly quenched from a temperature of 1600–1800 K to below 1150 K, and the glass formation takes place. In the BFS glass, different combinations of the minerals (Table 2.2) are found. The melilite glass composition is important from the point of view of slag cement hydration. The slag cement, containing slag with melilite glass

TABLE 2.2

Mineralogical Composition of BFS

Sl No	Mineralogical Compound[c,d]	Chemical Formula[a]
i	Pseudo-wollastonite	CS
ii	Dicalcium silicate	C_2S
iii	Rankinite	C_3S_2
iv	Gehlenite[b]	C_2AS
v	Anorthite	CAS_2
vi	Monticellite	CMS
vii	Akermanite[b]	C_2MS_2
viii	Merwinite	C_3MS_2
ix	Diopside	CMS_2
x	Spinel	MA
xi	Forsterite	M_2S

[a] Notations: $C = CaO$, $S = SiO_2$, $A = Al_2O_3$, and $M = MgO$.

[b] Gehlenite and Akermanite form a solid solution Melilite having a general composition $C_2A_xM_{1-x}S_{2-x}$ in which x varies between 0 and 1.

[c] In addition to the aforementioned minerals, all BFS contain calcium sulfide and minor constituents such as alkalies or iron and manganese oxides.

[d] In a BFS, different combinations of these minerals occur.

structure having x between 0.5 and 0.8, is likely to have high compressive strength [27,28]. Thus, the hydraulic activity of slag depends not only on the content of glass but also on its structure.

The network theory of Zachariassen [29] provided a suitable basis to describe the role of different oxides in the structure of glass in PFA and BFS, ultimately leading to a better understanding of the difference in their reactivity. The oxide of silicon (SiO_2), having a tetrahedral structure, is the smallest building unit in glass. It is capable of forming polyhedral groups (polyhedra). The anion O^{2-} forms bridge between two polyhedra. The introduction of large cations into these groups causes the breakage of oxygen bridges. The cations are located in the cavities of the glass network and their number is statistically controlled. Applying the Zachariassen theory to PFA and BFS glass, the cations could be classified as follows:

a. Network former: Si with coordination number 4
b. Network modifier: Ca with coordination number 6
c. Intermediates that may be part of the network, that is, network formers (coordination number: 4) or modify the network, that is, network modifiers (coordination number: 6), but cannot form glass on their own: Al, Fe, Mn, Mg. The aluminum and the magnesium are the two important elements in the glass network. These elements are amphoteric in nature; this means that they exhibit both sixfold as well as fourfold coordination. In the former type, they play the role of network modifiers and that of network formers in the latter

In comparing PFA and BFS glass structures, it should be noted that the ratio of network formers (SiO_2 + Al_2O_3 + Fe_2O_3) to network modifiers (Na_2O + K_2O + CaO + MgO) in PFA is 4–9, while that in BFS it is approximately 1 [30]. In sixfold coordinated cations (network modifiers), the binding energy is less. Thus, increasing the content of network modifiers, in comparison with the network formers, makes the glass less stable and increases its hydraulic activity, as it happens in BFS. It has also been seen that higher the content of network modifiers, smaller is the polymerization level of network forming SiO_2 tetrahedra. That again means that the glass has lower stability and better chemical reactivity. However, it should be noted that high lime (network modifier) content can adversely affect the glass formation during quenching (granulation) process. Gribko et al. [28] found that the two specific glass modifying agents, aluminum and magnesium, in sixfold coordination increased the hydraulic activity of BFS. Satarin [31] reported the study of 15 Russian BFSs with comparable Al_2O_3 and MgO contents, which showed the highest chemical reactivity when the ratio of Al and Mg in sixfold to that in fourfold coordination was 0.35. It should be noted that BFS does not contain any free MgO, even at a content as high as 20%. During the process of hydration, MgO in BFS glass is incorporated into the hydration products and no $Mg(OH)_2$ or the expansion phenomena are observed in the concrete on that account. Although the opinions on the influence of glass structure on the reactivity of BFS or the strength slag cement may differ, there seems to be a general agreement on one aspect, that the more disordered the structure of glass the higher is its hydraulic activity. The reactivity of BFS in forming cementitious compounds is influenced in varying degrees by different parameters like glass content, basicity (capacity to release hydroxyl or OH⁻ ions during reaction), particle surface area, temperature, and size distribution [32].

2.5 Quality Control

Some of the national standards stipulating the criteria for quality control of GGBS are as follows:

ASTM C-989-09: "Standard specification for slag cement for use in concrete and mortars" [1]. The American standard specifies three grades of GGBS by strength: Grade 120, Grade 100, and Grade 80. The Grade 120 provides the highest strength and is targeted by the majority of producers in the United States. The grade primarily depends upon the quality of the slag and the fineness to which it is ground. It is determined by the strength of mortar, when GGBS is mixed with equal mass of Portland cement. The three Grades, 80, 100, and 120, are classified according to their slag activity index, which is the average compressive strength of the slag-reference cement cubes (*SP*) divided by the average compressive strength of the reference cement cubes (*P*) and multiplied by 100:

$$\text{Slag activity index (\%)} = \frac{SP}{P} 100$$

The standard mentions the mix proportions for each type of cube. It also stipulates the requirement of residue left on a No. 325 sieve (45 μm) to be 20% and that of air content in the slag mortar not to be greater than 12%. The chemical requirements are stipulated as follows: the sulfur and ion sulfate contents are not to exceed 2.5% and 4.0%, respectively. The standard stipulates the quality monitoring program with minimum sampling and testing frequencies.

ASTM C595-06: "Standard specification for blended cements" [33]. The new standard Portland BFS cement, Type IS(X), can include between 0% and 95% GGBS, replacing the old Types IS, I(SM), and S.

BS 6699: 1992: "Specification for ground granulated BFS for use with Portland cement" [8]. The British Standard specifies the conformity criteria or manufacturer's autocontrol for GGBS, which include the following three aspects: (a) requirement in terms of characteristic values of fineness (not less than $275\,m^2/kg$, Blaine), glass content (not less than 67% m/m), GGBS and 30% ordinary Portland cement mixture's compressive strength, initial setting time and soundness, chemical composition (insoluble residue and magnesia), and the moisture content; (b) acceptable percentage of defects; and (c) probability of accepting GGBS not conforming to the requirement.

2.6 Addition of BFS and GGBS to Cement and Concrete

GGBS can be added as a partial replacement for cement in the following two ways:

a. In the cement manufacturing unit
   i. Separate grinding: BFS and clinker (along with gypsum) are ground separately and later mixed in the desired proportion in a mechanical blending unit
   ii. Intergrinding: BFS, clinker, and gypsum are mixed in the desired proportion and later ground together to the desired fineness, to make slag cement
b. At the concrete making site, GGBS of the desired fineness is added to concrete, to replace cement partially

ASTM C595-06 [33] defines the blended cement using GGBS, conforming to the requirement of ASTM C989 [1]. The new standard Portland BFS cement, Type IS(X), can include between 0% and 95% GGBS, replacing the old Types IS, I(SM), and S. The ASTM C989 [1] defines three varieties of GGBS: Grades 80, 100, and 120. The use of Grade 80 GGBS yields concrete with compressive strength lower than the same concrete made without GGBS, at all ages. Hence it may be used in special applications, such as mass concreting, where the heat of hydration (and the resultant temperature and temperature gradient in concrete mass) is important. The use of Grade 100 and 120 yields concrete with compressive strength equivalent or higher than the same concrete made without GGBS. While substitution of GGBS up to 70% of the Portland cement in a mix has been reported, there appears to be an optimum substitution that produces maximum strength. This is typically around 50% of the total cementitious material but depends on the grade of GGBS used. The research shows that the scaling resistance of concrete decreases with GGBS substitution rates greater than 25% [34]. The ACI 318 Code [35] covers the materials, design, and construction aspects of structural concrete used in buildings and where applicable, in nonbuilding structures. The Code also covers the strength evaluation of existing concrete structures.

In United Kingdom, the BS 6699 [8] specifies the GGBS for use with Portland cement. The addition of BFS to cement is covered under the provisions of BS EN 197–1 [36]. The standard allows the addition of BFS up to 95% in CEMIII/C variety of cement, although that variety of cement is not generally manufactured in United Kingdom. The use of GGBS in concrete is covered under the BS EN 206-1 [37] and its complimentary Standard BS 8500 [38].

The Japanese industrial standard [39] on Portland BFS cement limits the slag content up to 70%. The Indian standard IS 455 [40] on Portland slag cement also allows the addition of slag in the range of 25%–65%. The manufacturers of GGBS in India mostly follow the British Standard BS 6699 [8] and the addition of GGBS to concrete is covered under the provisions of IS 456 [41], which puts an upper limit of the cement replacement up to that specified in IS 455 [40].

Although most world standards limit the addition of BFS in slag cement up to 70%, Tomisawa and Fuji [42] reported that the higher contents of GGBS could be advantageous for certain applications. They found that, when GGBS content is more than 70%, the heat of hydration reduces significantly. In addition, when fineness of GGBS is increased, higher compressive strength can also be achieved. The blended cement with high fineness and high (above 70%) content of GGBS results in more compact pore structure, in comparison to Portland cement, due to formation of finer hydration products. The authors have not reported the durability studies.

As seen earlier, in the cement plants, BFS is either interground with Portland cement or ground separately (GGBS) and later blended with Portland cement. The grinding operation is mostly carried out in the ball mill or the VRM. Plate 2.2 shows a typical scanning electron micrograph of Portland cement and that of slag cement particles [43]. As clearly visible, the GGBS particles have an angular shape characterized by sharp edges

(a)   (b)

PLATE 2.2
Scanning electron micrograph of (a) Portland cement and (b) slag cement particles. (Courtesy of Prof. S. Tamulevičius for Skripkiunas, G. et al., *Mater. Sci. (Medziagotyra)*, 11(2), 150, 2005.)

and angles. However, it is seen that the effect of GGBS particle shape on the concrete slump is minimal [21].

The compressive strength and water demand of slag cement are related to fineness (size or surface area) and particle size distribution of the slag component [20,21]. The compressive strength increases as the Blaine fineness and the slope (n) increases and the position parameter (x') deceases, in the RRSB diagram (Section 2.3.2). When slag cement is manufactured by intergrinding, the particle size distribution of the "composite ground material" depends on its grindability and nature; the more easily ground additives, like clinker and gypsum, favor a wider particle size distribution. That is the reason why intergrinding of clinker and BFS produces slag cement with a relatively wider particle size distribution, in comparison to separate grinding and subsequent mixing of clinker and slag, for identical fineness. In intergrinding, clinker and gypsum get more finely ground in relation to the hard-ground slag [44]. In general, the composite cement with wider particle size distributions has lower mixing water demand [45]. In practice, it will be desirable to evaluate both the fineness as well as size distribution, besides other parameters, while comparing the competitive varieties of GGBS or slag cement.

It is generally observed that cement with a steep particle size distribution has higher water demand, at standard consistency. That happens because narrower (steep) particle size distribution results in lower packing density of the cement and GGBS particles [18]. In practice, it means such cement will require a relatively higher water-to-cement ratio to achieve a certain workability, in comparison with that required for cement exhibiting flat RRSB slope.

Zhang and Han [46] reported that the addition of 35% ultrafine slag (8460 cm²/g, Blaine) significantly reduced the viscosity and the yield stress of cement paste. When coarsely ground GGBS (surface area < 3000 cm²/g, Blaine) replaces cement in concrete, strength is reduced in comparison with concrete containing OPC only, for a certain total binder content. However, this reduction in strength can be compensated by increasing the total binder content. It is desirable, as it still reduces the total content of Portland cement in concrete [47]. The addition of finely ground GGBS (5000–6000 cm²/g, Blaine) can increase the strength of concrete substantially. Tan and Pu found that incorporating the combination of finely ground PFA and GGBS increases the compressive strength of concrete significantly at all ages, quantitatively similar to that of silica fume concrete [48].

It is reported that finely ground slag cement exhibits even better durability characteristics, such as sulfate resistance [49] and carbonation [50], in comparison with the coarsely ground slag cement. The work carried out by Bougara et al. [51] showed that the difference in the performance of GGBS having similar glass content, fineness, and size distribution could be attributed to their respective chemical compositions.

The structural engineer, with an understanding of these parameters, should be able to prepare a concrete mix design that maximizes the use of GGBS without sacrificing the vital parameters of workability, strength, and the durability. The computation of "efficiency factor" may help in doing this; see Refs. [52–54] for details. A simpler way is to carry out cube tests on concrete as per the relevant standard. Experience shows that it is possible to obtain concrete strength comparable with that containing only Portland cement by increasing the total cementitious (binder) content (by volume), commensurate with the level of replacement of cement with GGBS [55].

2.7 Quantitative Determination of Blast Furnace Slag in Cement

The engineer, frequently, is required to determine the content of BFS (or fly ash) in the slag cement (or fly ash cement) purchased from the supplier for the purpose of quality control or to prepare a mix design. The method reported by Cantharin [56] appears to be suitable for routine checking of cement containing slag. The method incorporates accurate determination of BFS in cement by density separation and chemical correction. The author reports the accuracy of determination of the order of ±0.5% in case of BFS and ±2% in case of PFA. In comparison to other methods, this method has the advantage that no assumptions regarding the content of reference elements and SO_3 are required. The readers are advised to refer to the original article to know the details.

2.8 Summary

The BFS is the nonmetallic product consisting essentially of calcium silicates and other bases, developed in a molten condition, simultaneously with pig iron in a blast furnace. Normally 1 ton of pig iron generates 250–350 kg of slag. The granulated slag is produced when the molten slag, flowing out from the blast furnace, is rapidly cooled and solidified by quenching in water to a glassy (amorphous) state. The quenching process results in the formation of sand size fragments, usually with some friable clinker-like material. The product has cementitious properties. When ground to very fine cement-sized particles, GGBS can be suitably used as a mineral admixture, as a partial replacement of cement.

The granulated slag, resembling coarse sand, has the particle size mostly passing 4.75 mm. The size distribution and the composition of glass and crystalline phases depend upon the type of granulation process. The ground granulated BFS is called latent hydraulic material. When activated by lime (or sulfate), it forms solid hydration products. The processing of BFS starting from slag granulation, moisture reduction, and finally grinding (comminution), leads to the production of GGBS. It is fine, off-white, and lighter in color than both Portland cement and granulated slag. The hydration products of GGBS are principally the same as those obtained in the hydration of Portland cement. The rate of cooling (quenching), during granulation process, affects the hydraulic activity of slag. In order to ensure the reactivity, the British Standard specification requires the glass content to be not less than 67%.

The strength characteristics of GGBS (when blended with cement) are related to its fineness, expressed in terms of Blaine surface area (cm^2/g). The particle size distribution also has been found to affect the strength. However, specific surface area is not a definitive characteristic of fineness. The ground materials having identical surface areas can have significantly different particle size distributions. The particle size distribution is conventionally expressed by the RRSB function. Each distribution has a unique value of the position parameter, x' (characteristic particle diameter) and the slope, n. The smaller the position parameter, greater is the fineness of the particle population, whereas the larger the slope, narrower is the distribution. Thus, the fineness of GGBS can be unambiguously characterized by defining the surface area (cm^2/g, Blaine) and one of the two parameters related to size distribution, the slope (n) or the position parameter (x'), in the RRSB diagram.

The slag granulation process used today can produce granulated slag with glass content greater than 95%, depending upon the method of cooling and the temperature at which cooling is initiated. The different processes of slag granulation are available commercially. In the BFS glass, different combinations of the minerals are found. The Melilite glass composition is important from the point of view of slag cement hydration. Thus, the hydraulic activity of slag depends not only on the content of glass but also on its structure. The more disordered the structure of glass the higher is its hydraulic activity. The reactivity of BFS in forming cementitious compounds is influenced in varying degrees by different parameters like glass content, basicity (capacity to release hydroxyl ions during reaction), particle surface area, temperature, and size distribution.

The GGBS can be added as a partial replacement of cement in two ways: (a) in the cement manufacturing unit, where GGBS and cement are blended either after separate grinding or through intergrinding, or (b) directly at the concrete making site. The national standards stipulate the criteria for quality control of GGBS. Most standards limit the addition of BFS in slag cement up to 70%. Irrespective of the content or the method of blending, it is seen that the particulate characteristics of GGBS—the particle size, shape, and size distribution—affect the quality of cement and of concrete, both in fresh and hardened state. The intergrinding of clinker and BFS produces slag cement with a relatively wider particle size distribution, in comparison to separate grinding and subsequent mixing of clinker and slag, for identical fineness. The GGBS particles have angular shape, characterized by sharp edges and angles; however, its effect on the concrete slump is minimal. In practice, it will be desirable to evaluate both the fineness as well as size distribution, besides other parameters, while comparing the competitive varieties of GGBS or slag cement. It is found that incorporating the combination of finely ground PFA and GGBS increases the compressive strength of concrete significantly at all ages, quantitatively similar to that of silica fume concrete.

3

Silica Fume

3.1 Introduction

Silica fume (SF), also called condensed silica fume (CSF), is a mineral admixture, mostly composed of submicron particles of amorphous silicon dioxide. The term "silica fume" is adopted by the American Concrete Institute [1]. The SF used in concrete is a byproduct of the smelting process in the production of silicon metal and ferrosilicon alloys, containing more than 75% silicon. The smoke from the smelting furnace contains silicon monoxide (SiO (g)). It gets oxidized in air during condensation and is collected in the dust collectors as SF (SiO_2 (s)). SF is pozzolanic in nature, which means it does not gain strength when mixed with water. When added to concrete, SF acts in two ways. As a filler it improves the physical structure, occupying the space between hydrated cement particles, and as a pozzolan, it reacts chemically with the calcium hydroxide released during the hydration of cement, forming strength-giving compounds to impart greater strength and durability to concrete. Bridge construction, marine structures, parking structures, water supply, and sewage facilities, all benefit from the use of SF. SF is always used with a chemical admixture in concrete. The special properties of SF improve the rheology of fresh concrete benefiting concrete pumping and the stability of concrete mix. Special varieties of concrete like high-strength concrete, lightweight concrete, and shotcrete are better when made with SF. The important characteristics that make SF an effective pozzolanic material are (a) fineness (15,000–25,000 m^2/kg, N_2 adsorption method), with an average particle diameter 100 times smaller than that of Portland cement, (b) spherical shaped particles that improve the rheology of concrete, and (c) glassy structure and high amorphous silica that enhances reactivity with cement. The quality of SF is specified by ASTM C1240 and AASHTO M 307. Important references to obtain additional information are Kuennen [2], Holland [3], Helland et al. [4], and Malhotra et al. [5].

Section 3.2 gives a brief on the types of SF commonly available. Section 3.3 deals with the physical characteristics that include particle size and size distribution (Section 3.3.1), specific gravity and bulk density (Section 3.3.2), and specific surface (Section 3.3.3). Chemical and mineralogical composition are discussed in the Section 3.4. Aspects related to toxicity and occupational health are mentioned in Section 3.5. Section 3.6 summarizes quality control and the subsequent Section 3.7 discusses different aspects related to the addition of SF to cement and concrete. Finally Section 3.8 summarizes this chapter.

3.2 Types of Silica Fume

SF, as obtained after condensation from the furnace, is grey in color, somewhat similar to cement or PFA (Plate 3.1). The image of SF particles taken under a scanning electron microscope is shown in Plate 3.2. As shown in Plate 3.2, SF particles appear to be agglomerated and round. SF is commercially available in the following four types. The end use determines in what form SF should be supplied.

a. *Undensified SF (bulk density, up to 350 kg/m³)*: It is an "as produced" powder and stored in a separate silo in the manufacturing unit. It occupies a large volume due to low bulk density. It is sticky and self agglomerating, hence inconvenient to handle and transport pneumatically. It also generates more dust. Undensified SF is often available in bulk or bags near the smelters where it is produced. It is mainly used in making pre-bagged products like grouts or repair mortars.

b. *Densified SF (480–720 kg/m³)*: The densification of SF is done in the storage silos installed at the smelting plant. Air is blown through the screen provided at the bottom of the silo. This air gently rises

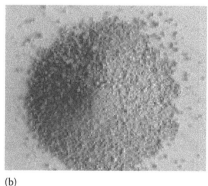

(a)   (b)

PLATE 3.1
(a) Condensed SF. (b) Densified SF.

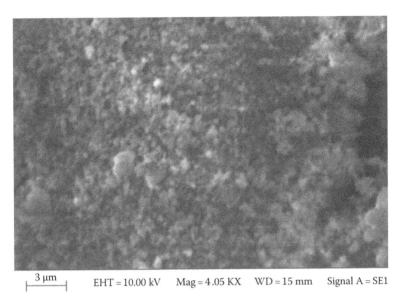

3 μm | EHT = 10.00 kV | Mag = 4.05 KX | WD = 15 mm | Signal A = SE1

PLATE 3.2
SEM image of condensed SF.

through the bed of SF causing the individual silica particles to rub against each other. As the particles rub, the naturally occurring van der Waals forces on their surface cause adherence to each other. The longer the air is allowed to flow through the fume bed, the greater the degree of agglomeration and correspondingly the density. Once the desired density is reached, the airflow to the silo is stopped. The newly densified SF can be shipped in bulk in pneumatic trucks, in jumbo bags or in small (usually 25 kg) paper bags. The value of the densification process is most evident in shipments over long distances and especially for off-shore customers because high bulk density reduces the cost of transport. The particles of densified SF look like small beads, flow like water, and produce very little dust (Plate 3.1).

c. *Pelletized SF (about 1000 kg/m³)*: The pellets of 10–25 mm diameter are formed by adding water to SF on disc pelletizer. The pelletized form is not suitable for direct use in concrete, as the hard pellets do not break and disperse in a concrete mixer. It can be interground with Portland cement clinker to produce blended SF cement. The densified SF should not be confused with pelletized SF.

d. *SF slurry (bulk density 1300–1400 kg/m³)*: Transporting SF as a water-based slurry is sometimes more economical. The slurry contains 50%–52% SF (mass basis) dispersed in water and transported in bulk tankers. The storage tanks may require agitation to prevent slurry from stiffening and protection from freezing. It is easy to use at the batching unit, once the required dispensing equipment is available. The dispensing equipment used for the purpose is larger and more complex in comparison to that used for chemical admixtures, due to the nature and the quantity of slurry required. SF slurry is no longer available in the U.S. market [3].

3.3 Physical Characteristics

SF is essentially amorphous silicon dioxide and is dark gray in color. It takes part in the pozzolanic reactions occurring during hydration of cement, due to its amorphous nature. It also acts as a filler in concrete. Besides its amorphous nature, the reactivity of SF in concrete depends upon the size of particles, size distribution and how well the particles get dispersed in concrete after mixing. The sand present in concrete is also silicon dioxide, chemically similar to SF but crystalline in nature, hence does not react with hydrated cement. The physical characteristics of SF— particle size and size distribution, surface area, and specific gravity and bulk density—are discussed below.

3.3.1 Particle Size and Size Distribution

The undensified SF particles (as produced) are spherical in shape and on average 0.02–0.25 μm in diameter, nearly 100 times smaller than the average cement particle. It is amorphous (85%–98%) and glassy. However, it should be noted that SF is primarily obtained in the form of fused or linked clusters of spheres; isolated spheres are extremely uncommon. The agglomeration of particles is observed under the scanning electron microscope (Plate 3.2). These strongly bonded particle clusters are formed at the source (and recognized as its characteristic), during oxidation and condensation processes of SF in the smelting furnace. As a result of agglomeration, the actual mean particle size of SF dispersed in water varies between 1 and 50 μm, depending upon the source, bulk density, and age [6]. A better dispersion of SF in water can be obtained using ultrasonic technique; however, that should be considered only as a limiting case and the actual dispersion during intermixing with concrete will be less. St. John [6] suggested, when SF is used to make high-strength concrete or mortar, its ultrasonic dispersivity should be measured, and it should be accepted only when a mean particle size of 1 μm or less is obtained. The author further recommends that, when water soluble alkali content of concrete exceeds 3 kg/m³, the ultrasonic dispersion of SF should be measured to ensure that only a small amount of agglomerates greater than 50 μm are present. Alternatively, consideration may be given to reduce the alkali content of concrete to a safer level, if excessive amount of large agglomerates are detected. Wolsiefer [7] reported about the development of a test method with ultrasound to determine dispersivity or dispersibility of SF.

The densified SF is a collection of coarse agglomerated particles, with grain size up to several mm. The agglomeration is supposed to be reversible, which requires the SF particles to mix with other constituents in the presence of chemical admixture, when added to concrete. However, it has been reported [6,8,9] that densified SF, as commonly supplied, is not dispersible into individual SF spheres. Whilst the densified SF from some sources can be dispersed by moderate ultrasonic treatment into small clusters or chains of spheres; others resist such treatment and mostly remain as large agglomerates. Ordinary concrete mixing is not efficient in breaking down the agglomerates by crushing and shear action of aggregate [10], although the number of such agglomerates is reduced [9]. Thus the assumption that the densification process is "reversible" does not fully represent the actual situation. Generally air densified SF with bulk density beyond 700–720 kg/m^3 is difficult to disperse [11]. The size of undispersed agglomerates remaining in concrete after mixing often exceed the size of Portland cement particles, thus limiting any potential benefits attributed to the fine-particle filler effect. It is thus necessary to ensure that mixing is adequate to break the agglomerates. The coarse aggregate is not present during mixing in certain cases such as that of dry mix shotcrete and roof tiles, which may result in inadequate breaking of agglomerates. In those cases, use of undensified SF may be more appropriate. The information on mixing of densified SF is given in *Silica Fume User's Manual* [3]. The undispersed SF agglomerates may take part in ASR under certain conditions [12–14].

When the well-dispersed, superfine SF particles act as filler, they fill the void space between the irregular-shaped cement particles, and thus bring in refinement of the pore structure in the hardened concrete. The improvement of particle packing due to SF is advantageously used in oil-well cementing [15].

Less than 1% of SF particles are crystalline. The needle-shaped crystalline particles are typically of nanometer (1 nm = 10^{-9} m) size, approximately 200 nm long and 20 nm wide. Amorphous silica particles are spherical in shape with a diameter in the nanometer scale. The undensified SF can therefore be classified as ultrafine for the practical purpose.

3.3.2 Specific Gravity and Bulk Density

The specific gravity of SF is about 2.2, which is lower than 3.0–3.20 of the Portland cement. Thus when SF is added to concrete, in proportion to the quantity of cement on mass basis, the volume is larger in comparison to the Portland cement. The increase in paste volume and relatively slow hydration reaction may contribute to increased shrinkage, especially in high-strength concrete [16]. This aspect should be taken into account while proportioning the concrete mix.

As seen earlier, the types of SF differ in bulk density. The bulk density of undensified SF (as produced) depends upon the process parameters where it is made. The bulk density of undensified SF is the minimum and that of the SF slurry is maximum. Thus, SF slurry shall occupy nearly four times less and densified SF nearly two times less volume in comparison with the undensified SF, for the identical mass. This aspect has an impact on the transportation cost.

3.3.3 Specific Surface

The specific surface of SF lies in the range of $15-30\,m^2/g$. The particles of SF are very fine; hence, it is difficult to measure the specific surface with air-permeability apparatus. The measurement is done with the more sophisticated "nitrogen adsorption Brunauer, Emmett, and Taylor (BET) method." There is no direct correlation between the specific surface area measured by different methods.

The addition of SF to concrete vastly increases the solid-to-liquid interfacial area. For example, in a mix with $320\,kg$ cement of $300\,m^2/kg$ surface area and 5% SF with $20,000\,m^2/kg$ surface area per m^3 concrete, the total surface area of cement particles is $96,000\,m^2$ and that of SF is $320,000\,m^2$, nearly three times more than that of cement. The increase in surface area results in the increase in water demand to achieve the desired workability, the ball bearing effect due to spherical particles notwithstanding. Adding more water shall adversely affect the durability and hence cannot be practiced. Thus, the advantages due to the addition of SF, that is, increased strength and durability, can be obtained only by keeping the water-to-cement ratio low, using the high-range water reducing agent (HRWR) or the superplasticizer. The dosage of air-entraining agent needed to maintain the required air content when using SF is slightly higher than that for conventional concrete because of high surface area and the presence of carbon. This dosage is increased with increasing amounts of SF content in concrete.

The increased solid-to-liquid interfacial area also affects the rheology. The plastic viscosity decreases with SF addition up to 5% and increases for further addition [17]. There is little or no bleeding. Therefore, construction workers must be trained to obtain the best results with SF concrete.

3.4 Chemical and Mineralogical Composition

SF contains 85%–98% silica in the amorphous form, and the crystalline matter may contain quartz, cristobalite (polymorph of quartz, meaning it has the same chemistry as SiO_2 but has a different crystal structure), and silicon carbide. Besides silica, the chemical analysis reveals oxides of iron, aluminum, calcium, magnesium, potassium, sodium, and sulfur in small quantity, depending on the source. In the light, gray color amorphous silica, crystalline matter, and impurities are seen as glassy particles. Transmission electron microscopy (TEM) is used to analyze the particle size, crystallinity, and composition of crystalline phases present in SF. TEM techniques of bright field (to count particles) and conical dark field (to determine the crystallinity of the particles) and the energy dispersive x-ray spectroscopy (EDS) have been successfully used to evaluate the composition of the crystalline particles. Table 3.1 shows the typical chemical composition of SF samples collected from two different sources and, for the sake of comparison, one micronized amorphous silica sample collected from a natural source.

As shown in Table 3.1, the SF samples received from the silicon metal manufacturing, ferrosilicon alloy manufacturing and the natural source have different chemical compositions. In general, the SF obtained from ferrosilicon furnace contains more iron and magnesium oxides than that from a furnace producing silicon metal. The amorphous silica obtained from the natural source is from a silica quarry of the Isparta region in Turkey, reported by Davraz and Gunduz [18]. SF is a stable material in all forms and does not react with water.

The amorphous silica present in the SF reacts with calcium hydroxide produced during cement hydration. The reaction may induce three changes: (a) an increase in the content of the principal strength giving

TABLE 3.1

Typical Chemical Composition of SF Obtained from Different Sources

Sl No	Constituent	Source and Composition (%)[a]		
		Silicon	Ferrosilicon	Silicon Quarry
i	SiO_2	97.5	96	92.48
ii	C	0.4	0.5	—
iii	Fe_2O_3	0.03	0.5	0.09
iv	Al_2O_3	0.29	0.20	2.6
v	CaO	0.20	0.20	0.31
vi	MgO	0.10	0.50	0.0
vii	K_2O[b]	0.20	0.50	0.04
viii	Na_2O	0.10	0.20	1.08
ix	Cl	0.01	0.01	—
x	SO_3	0.10	0.15	0.09
xi	Moisture	0.20	0.50	—
xii	Loss on ignition [$T = 975°C$]	0.70	0.70	1.85

[a] The total of all percentages does not add to 100, as some minor constituents have not been mentioned.

[b] The soluble alkali content, that is, Na_2O and K_2O, is sometimes expressed as Na_2O. The equivalent alkali content, expressed as Na_2O, is obtained as Na_2O + $0.658K_2O$.

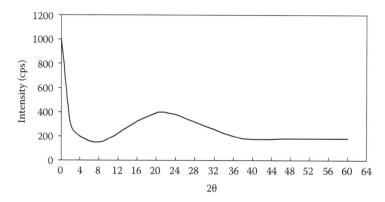

FIGURE 3.1
Typical x-ray diffractogram of SF.

compound in concrete, that is, C-S-H with lower CaO/SiO_2 ratio, (b) a reduction in pH value, and (c) a dilution effect on the C_3A [19]. These changes have a positive effect on the durability characteristics of concrete with SF.

The mineralogy of SF can be studied with x-ray diffraction (XRD) analysis. The resulting diffractogram, with the visible amorphous halo, characterizes the spectrum of amorphous SF. Figure 3.1 shows a typical x-ray diffractogram of SF.

3.5 Toxicity and Occupational Health

SF comes under the category of hazardous materials. The national organizations of health and safety of most of the countries [20] mandate the supply of information necessary to ensure the health and safety of personnel handling hazardous materials. Such information is usually contained in the hazards data sheet required to accompany shipment of such materials.

At the ambient temperature, SF is stored dry in closed containers. Dust generation should be avoided while handling, and adequate provisions should be made for ventilation according to the required standards [21] to keep the exposure in processing areas below the permissible exposure limit (PEL) or the threshold limit value (TLV).

SF is generally considered a nuisance dust of low toxicity; consequently, it is considered to pose minimal risk of pulmonary fibrosis (silicosis). However, prolonged exposure to SF dust concentrations above the recommended exposure limits should be avoided, unless protective equipment is used. It may contain trace amounts (<1%) of crystalline silica, which has been shown to cause silicosis and has been identified by IARC (International Agency for Research on Cancer, WHO) and NTP (U.S. National Toxicology Program) as a possible human carcinogen. SF may cause skin irritation in some cases;

hence, adequate skin protection (gloves) and similarly eye protection (goggles) to prevent possible irritation and damage are recommended. The inhalation of SF may cause coughing and sneezing. High exposure can cause a flu-like illness with headache, fever, chills, aches, chest tightness, and cough. Repeated exposure may cause lung damage (fibrosis).

Some published work on nano particles has shown that their respiratory effects include inflammatory responses. It is possible that co-exposure to ultrafine or nano amorphous silica particles as well as the crystalline silica particles might trigger silicotic effect, even at low concentrations of the latter [22]. The Safety in Mines Research Advisory Committee (SIMRAC), South Africa [23] carried out investigations to ascertain whether crystalline phases were present in SF and at what levels and to evaluate the role of the ultrafine (nano) nature of SF in its overall toxicity. The study concluded that the presence of even very small quantity of crystalline silica along with the amorphous silica should affect the overall interpretation of the association between exposure to SF and the potential for development of silicosis. The study recommends reassessment of the occupational exposure guidelines for SF at the national level.

3.6 Quality Control

The quality control of SF is carried out as per the procedure stipulated by different national standards [24–28]. The "silica fume" is so called because it is a condensed form of the fumes collected as fine powder in the dust collectors of silicon metal or ferro-silicon alloy industries. The ferro-silicon alloy industry produces alloys of different compositions and the composition of the condensed fume (SF) collected also changes accordingly. The dust collector is often attached to furnaces producing different types of alloy and the composition of alloy itself changes according to the market demand. Thus, the SF collected is a mixture of SF coming from different streams, whose composition keeps changing. Therefore, the user is required to check the composition carefully or buy SF from a source that controls these variations.

The parameters that decide the quality of SF are (a) silica, alkali, and carbon content or loss on ignition, (b) content of crystalline and amorphous silica (to be determined by x-ray diffractogram), (c) specific surface area (to be determined by nitrogen adsorption), and (d) pozzolanicity or pozzolanic activity [29]. The dispersibility or dispersivity (to be determined by ultrasound) is also an important parameter, as discussed earlier. Sieve analysis is required when woodchips are used during the production of silicon metal. Aitcin [29] recommends that chemical analysis, x-ray diffractogram, and bulk density measurements should be routinely used for quality control, when densified SF is used. When SF slurry is used, its solids content is required to be checked.

The method to determine the amount of SF in prepackaged cement is reported by Coleman et al. [30], using combined heat treatment and XRD. The authors claim it is reasonably accurate. The conventional air permeability or Blaine apparatus will not be appropriate to measure the fineness of SF blended cement, in view of the large specific surface. The nitrogen adsorption BET method may be appropriate.

The national standards [24–28] on SF specify silica content above 85%, silica being a reactive ingredient. The moisture content is required to be less than 3% and the loss on ignition is specified below 4%–6%. The limit on the loss on ignition is put primarily to control the content of unburned carbon or other materials. The minimum requirement of the specific surface is $15 \, m^2/g$ (BET); the European Standard [25] specifies a range of $15–35 \, m^2/g$. The oversize retained on $45 \, \mu m$ (No 325) sieve is specified below 10%. In addition, the ASTM Standard [24] and the Indian Standard [28] also put a limit of 5% on the variation from average on $45 \, \mu m$ sieve. The limits on the oversize and the variation therein aim to control the content of foreign materials in SF. The requirement of pozzolanic activity index (PAI) is specified in all standards, either for 7 day accelerated or 28 day standard curing. Whereas the ASTM Standard specifies a minimum PAI of 105% for 7 day accelerated curing, the European Standard puts the minimum requirement at 100% for 28 day standard curing. The Canadian Standard [26] puts a maximum limit of 0.2% on autoclave expansion. In case of SF slurry, the European Standard [25] allows a maximum deviation of 2% on the dry mass declared. Besides these, the ASTM Standard [24] also gives certain "optional" and "report only" requirements, such as the variation in the demand of air-entraining admixture (uniformity), the reactivity with cement alkalies, sulphate resistance expansion, specific gravity and bulk density, and total alkalies.

3.7 Addition of SF to Cement and Concrete

There are two ways to add SF in cement and concrete:

a. As a partial replacement of cement
b. As a constituent material in concrete in addition to cement

The SF blended cement is a high-performance cement. SF replaces Portland cement in blended cement. Since SF is so effective in increasing the performance of cement, it may also be added in small amounts to other types of blended cement, as a secondary additive to develop the overall quality. The blended cement contains 7%–10% SF, as a replacement of Portland cement. While manufacturing the blended cement, the quantity of SF to replace Portland cement can be estimated based on the total content of reactive silica, obtained from the chemical analysis of cement and SF. The characteristics of a coarsely ground cement may also be improved by blending with SF. Thus, the energy saving in finish grinding (cement grinding) operation can be achieved through SF blending [31].

The commercial use of SF blended cement (SF + Portland cement) has been reported from Canada. Thomas et al. [32] also reported the commercial application of factory-blended ternary cement consisting of Portland cement, SF, and slag in Canada. The Electric Power Research Institute (EPRI), United States, recommends [33] the use of ternary blends containing Portland cement, SF, and FA for the effective control of alkali silica reaction (ASR). These blends are also found to improve the durability of concrete— sulphate and chloride resistance—besides ASR [34]. However, SF blended cement is not commonly used, as it has not been found economical under local conditions.

The difference in the impact of SF addition vis-à-vis the other mineral admixtures, FA and BFS, to concrete should be clearly understood. The FA and BFS are mostly added to concrete as a replacement of cement, resulting in the overall reduction in the quantity of Portland cement used. Their addition improves the properties of fresh concrete and the strength and durability of concrete in the hardened state. It also economizes the mix. Thus, the addition of FA and BFS to cement and concrete leads not only to the fruitful utilization of these industrial wastes but also to the sustainable development of cement and construction industry. On the other hand, SF is added to concrete as a constituent material, in addition to cement. The aim is to improve its strength and durability. There is a substantial difference in the cost of SF and the Portland cement. The addition of SF results in the larger dose of superplasticizer as well as the air entraining agent to concrete. Thus, the addition of SF increases cost and therefore needs to be adequately justified on the considerations of strength and durability of concrete. In general, the addition of SF is practiced to obtain high-strength concrete. Therefore, SF is only selectively used in ASTM Class I and Class II high-performance concrete (50–100 MPa) and generally preferred for Class III (100–125 MPa) variety of concrete. There is an optimum SF replacement; extra SF only acts as an inert filler, as there is not enough calcium hydroxide from cement hydration to react with it pozzolanically [35]. Theoretically nearly 25%–30% SF is required to react with the entire calcium hydroxide liberated after cement hydration. The dosage of SF is estimated once the strength and application of concrete is decided. However high dosages are seldom used as large amount of superplasticizer will be required to achieve the desired slump. Normally 3%–10% (mass-to-mass of cementitious) SF is added to high-performance concrete. Aitcin [29] suggests a dosage of 8%–10% SF in the high-performance concrete. The transport and handling of water-based SF slurry is easy and more economical and hence should be preferred when the required dispensing equipment at the batching unit and a regular supply of SF slurry is available [36].

3.8 Summary

SF, a byproduct of the smelting process in the production of silicon metal and ferrosilicon alloys, is mostly composed of submicron particles of amorphous silicon dioxide. It is available in four forms commercially; the appropriate form is decided by the end use. The undensified SF (bulk density, up to $350\,kg/m^3$) is often available in bulk or bags near the smelters where it is produced. It is mainly used in making pre-bagged products like grouts or repair mortars. It is inconvenient to handle, generates dust, is self-agglomerating, and is difficult to transport. The particles of densified SF ($480–720\,kg/m^3$) look like small beads, flow like water, and produce very little dust. It can be shipped in bulk in pneumatic trucks, jumbo bags, or in small (usually 25 kg) paper bags and is suitable for long distances, especially for off-shore customers, as high bulk density reduces the cost of transportation. The pelletized ($10–25\,mm$) SF (about $1000\,kg/m^3$) is not suitable for direct use in concrete, as the hard pallets do not break and disperse in a concrete mixer. It can be interground with Portland cement clinker to produce blended SF cement.

The undensified SF particles (as produced) are amorphous and glassy, have spherical shape, and on average have $0.02–0.25\,\mu m$ diameter, nearly 100 times smaller than the average cement particle. As a result of agglomeration, the actual mean particle size of SF dispersed in water varies between 1 and $50\,\mu m$. When SF is used to make high-strength concrete or mortar, its ultrasonic dispersivity should be measured and it should be accepted only when a mean particle size of $1\,\mu m$ or less is obtained. The densified SF is sometimes difficult to disperse. When the well dispersed, superfine SF particles act as filler, they bring in refinement of the pore structure in the hardened concrete.

The lower specific gravity ($2.2 < 3.2$) of SF, in comparison to cement, increases paste volume and relatively slow hydration reaction may contribute to increased shrinkage, especially in high-strength concrete. Higher particle surface area of $15–30\,m^2/g$ vastly increases the solid-to-liquid interfacial area and results in higher water demand to achieve the desired workability and slightly higher requirement of air-entraining agent. Thus, the advantages due to the addition of SF, that is, increased strength and durability, can be obtained only by keeping the water-to-cement ratio low, using the HRWR or the superplasticizer.

The addition of SF to cement leads to (a) an increase in the content of the principal strength giving a compound in concrete, that is, C-S-H with lower CaO/SiO_2 ratio, (b) a reduction in pH value, and (c) a dilution effect on the C_3A. These changes have positive effect on the strength and durability characteristics of concrete.

The parameters that decide the quality of SF are (a) silica (>85%) and carbon content or loss on ignition (<4%–6%), (b) content of crystalline and amorphous silica, (c) specific surface area (>15 m^2/g), and (d) pozzolanicity or pozzolanic activity (ASTM 100%–105%).

The SF blended cement is not commonly used. The SF is generally added to concrete as a constituent material, in addition to cement. Due to large surface area, it results in the larger dose of superplasticizer as well as the air-entraining agent in concrete. Therefore, SF addition needs to be adequately justified on the considerations of strength and durability on the one hand and cost of concrete on the other. Normally 3%–10% (mass-to-mass of cementitious) SF is added to high-performance concrete.

4

Rice Husk Ash

4.1 Introduction

Rice husk ash (RHA), used as a pozzolanic admixture in cement and concrete, is obtained from the combustion of rice husk (RH) under certain conditions of the surrounding environment, temperature and residence time* in a combustor, and subsequent size reduction. It contains low unburned carbon, and silica is mostly in amorphous form. RHA manufactured in the modern fluidized or cyclonic bed reactors has high surface area of the order of 20–40 m²/g, comparable with that of silica fume (SF). It is a material with proven pozzolanic characteristics and is added to cement and concrete as a partial replacement of Portland cement. However, its application has not been widely commercialized as yet, mainly on account of the nonavailability of RHA of the desired pozzolanic characteristics on a large scale on the one hand and the lack of awareness about the potential for RHA as a mineral admixture on the other.

In the literature we find a mention of high ash (inorganic matter) content of RH and its utilization [1]. The RH produced in farms and rice mills is conventionally employed as a fuel or dumped as waste. Systematic studies on the controlled combustion of RH and the utilization of RHA in cement and concrete were first reported by Mehta and Pitt [2–5]. Subsequently many researchers found that the use of properly treated RHA improves the performance and the durability of concrete; Nehdi et al. gave a good list of references [6].

The addition of SF increases the cost of concrete and its supply is also limited (Chapter 3). RHA possesses the potential to replace SF in high-strength and high-performance concrete. RHA manufactured through controlled burning of RH shows performance comparable with that of SF, when added to concrete in binary (Portland cement [PC] + RHA) or ternary (PC + RHA + FA) blends, in terms of strength and reduced permeability toward the external deteriorating agents [7]. The major characteristics of RHA are its high water demand and coarseness in comparison to SF. In order to improve these characteristics, RHA needs to be ground finer

---

* Residence time expresses how fast something moves through a system. At steady state, it is the average time a substance spends within a specified region of space, such as a reactor.

into particle size range of 4–8 μm (1 μm = 10^{-6} m) and a superplasticizer is added to reduce water requirement.

RH is presently considered as an agricultural waste and used as fuel, as mentioned earlier, where its pozzolanic value lies unutilized. Thus, the incorporation of RHA in concrete as a mineral admixture adds value, both from the economical and ecological point of view.

Section 4.2 briefs on the relevance of RHA as a contributing factor toward the sustainability of construction industry in the rice-producing countries. Section 4.3 summarizes the properties of RH. Section 4.4 deals with the production of RHA. It incorporates the characteristics of RH combustion and the modern methods to produce pozzolanic RHA on the industrial scale. Section 4.5 discusses the physical and chemical characteristics of RHA. Section 4.6 summarizes various aspects related to the addition of RHA in cement and concrete. Section 4.7 gives a brief summary of this chapter.

4.2 Relevance of RHA for the Sustainability of Construction Industry

Rice, produced from paddy, is a cereal grain and the most important staple food for a large part of the world's human population, especially in tropical Latin America, the West Indies, and east, south, and southeast Asia. According to one estimate, the world paddy production is expected to touch 847–915×10^6 ton, by the year 2030, from the current (2008) level of 683×10^6 ton; out of which around 600–774×10^6 ton paddy and from that around 120–155×10^6 ton RH shall be produced in the Asian countries [8,9]. The abundant availability of RH in the rice-producing countries provides us a huge scope to recover its heat value to generate power and to use the RHA produced in cement and concrete on a large scale.

The production of RHA with cogeneration of power as well as its application in cement and concrete, both contribute toward the reduction of green house gas (GHG) emissions. It is found that the generation of power through the combustion of RH reduces carbon emissions, in comparison to coal, oil, and natural gas [10]. RHA is added to cement and concrete as a partial replacement of cement to the extent of 30%. Thus, it reduces the consumption of PC and to that extent contributes toward the reduction of CO_2 emission, which is a GHG, in the manufacture of PC. The reduction of GHG through such practices has been provided with incentives under the United Nations (UN) framework.

The Kyoto Protocol is a part of the United Nations Framework Convention on Climate Change (UNFCCC) and has set an agenda for reducing global GHG emissions. If CO2 emissions can be shown and verified to be reduced due to different practices, then Certified Emission Reductions (CERs) are issued under the Clean Development Mechanism (CDM) of UNFCCC. These CERs are tradable in the primary and secondary market and generate revenue for the CERs holding party. The readers are advised to go through the UNFCCC documents to obtain more information on the subject [11]. When RHA is used in cement and concrete manufacture as a cement substitute, there is potential to earn CERs [12]. There are other environmental benefits of substituting Portland cement with RHA. The need for quarrying and mining such primary raw materials and fuel as limestone, clay, and coal is reduced, and thus overall reduction in emissions of dust, CO_2, and acid gases is attained. As the cement and concrete industry uses RHA with amorphous silica, the health issues, mainly associated with the fine crystalline silica, are minimal. The large-scale application of RHA in the construction industry requires industrial and economic policy planning and efforts in the following areas:

a. Creation of general awareness about the benefits of using RH in power generation and RHA in cement and concrete. In India, the government took lead promoting the utilization of pulverized fuel ash (PFA) in cement and concrete, through the Fly Ash Mission. It is time that similar missions are taken up to create awareness about the less known mineral admixtures, such as RHA.

b. RH is produced by the farmers in their paddy fields. The RHA producing unit will require continuous supply and adequate storage of RH. Thus, a viable method of collection and transportation of RH from the paddy fields to the RHA producing unit will have to be put in practice.

c. Identification of a techno-economically feasible method to produce and process RHA along with the cogeneration of power to suit the local conditions.

d. Formulation of national Standards on the quality assessment and the use of RHA in cement and concrete.

In summary, the application of RHA in the construction industry shows tremendous potential for the rice-producing countries, both from the point of view of promoting the sustainable development of construction industry and as a valuable input for the economic growth of these countries. The techno-economic aspects related to the production and the application of RHA need to be properly addressed.

4.3 Rice Husk

The paddy grown in the field is associated with the production of essentially two byproducts: rice husk or hull (RH) and rice bran. RH is the outer shell covering the rice kernel. It is obtained when paddy is threshed to separate rice

PLATE 4.1
Rice husk.

and the husk. The unground RH is grayish brown in color, tough, woody, and abrasive (Plate 4.1). It contains moisture, about 20% ash (of which about 95% is silica), lignin, cellulose, pentosans, little protein, and very small amount of vitamins. It has a specific gravity of about 0.75 and the bulk density of about $100\,kg/m^3$, occupying eight times more space than the paddy of equal mass [13]. The amorphous silica in the husk is formed when the orthosilicic acid ingested from ground water gets polymerized in the rice plant [14]. The observation under scanning electron microscope (SEM) indicates that the amorphous silica occurs principally on the external surface of the husk and to a lesser concentration on the inner surface. The crystalline silica sometimes observed on the husk is probably due to contamination by sand [15].

The proportion of RH in the paddy and that of ash in the RH shows regional variation, mainly depending upon the type of paddy and geographical factors. RH is the raw material for the production of RHA. On average, 1 ton of paddy produces about 200 kg of husk, which upon combustion produces about 40 kg of ash [7]. Thus, paddy has potential to produce 4% its mass of ash. The ash of RH contains the highest proportion of silica, of all the plant residues.

4.4 Production

The controlled combustion of RH produces RHA with silica in amorphous form. The RHA thus produced is ground to required fineness (surface area) to obtain desired pozzolanic activity. The effectiveness of RHA can be improved by chemically treating RH before burning [16]. The following paragraphs discuss the characteristics of RH combustion and the modern methods to produce pozzolanic RHA on the industrial scale, with no chemical treatment, as reported in the literature. The chemical treatment of RH before burning is seldom done, when RHA is produced on large scale for its application in cement and concrete. Besides, it also adds to the cost. Hence it is not included in the scope of this book.

4.4.1 Characteristics of RH Combustion

RH is difficult to burn due to its typical silica-wood composite morphological structure, high mineral content, large bulk volume, and a pronounced tendency to cake and agglomerate during combustion [17]. The important physical and mineralogical characteristics of RHA—the content of amorphous silica, surface area, grindability, and carbon—content depend upon the temperature, environment, and the duration of combustion. Along with the other changes, the combustion in air is marked by the weight loss, due to the removal of organic matter [18,19]. It is exothermic* and the fuel calorific value ranges between 13.8 and 15×10^6 J/kg (3300–3600 kcal/kg). Table 4.1 summarizes the characteristic changes that occur during the combustion of RH. As shown in Table 4.1, the combustion process is divided into four stages marked by the temperature range, weight loss, and corresponding physicochemical and other changes. Isothermal heating at a minimum of 402°C is required for complete destruction of organic matter from RH and to liberate silica [20]. On combustion, the cellulose–lignin matrix burns away, leaving a porous silica skeleton, which on grinding gives fine particles having large surface area (Plates 4.2 and 4.3). Nair et al. observed that "magic angle spinning nuclear magnetic resonance spectroscopy (MAS NMR)" could be effectively used to accurately determine the temperature at which significant crystal formation occurs and to directly observe the surface silanol[†] (Si-OH) sites in the amorphous silica, which are thought to be responsible for the pozzolanic activity of RHA [21]. The laboratory experimentation carried out by the authors showed the most reactive RHA is produced after incineration at 500°C for 12 h and subsequent quick cooling of the sample. Asavapisit and Ruengrit [22] report the characteristics of RHA obtained after burning RH at different temperatures in the laboratory muffle furnace, in the range of 400°C–800°C and subsequent rapid cooling in air and grinding. They found that the carbon content of RHA after incineration at 650°C for 1 h was low and the 28 day pozzolanic activity index was maximum. At 800°C, although the silica content was found higher and the carbon content lower,

---

* In the chemistry terminology, combustion is a chemical reaction involving oxidation. An "exothermic" reaction is a chemical reaction that releases net energy in the form of heat, during the process from start-up to the completion.
† Silanol is a compound in which hydroxyl group is attached to a silicon atom (Si-OH). It has a very strong polar interactivity. It occurs on the surface of silica and also as a group of chemical compounds.

TABLE 4.1

Combustion of RH

Sl No	Particulars	Combustion Stage			
		I	II	III	IV
i	Temperature range (°C)	<100	100–350	350–700	>700
ii	Cumulative weight loss (%)[a]	4–6	51–65	Up to 100	—
iii	Physicochemical change	Liberation of physically held water	Lignin, cellulose decomposition and volatiles liberation	Carbon char burning and amorphous silica formation	Amorphous silica conversion to crystalline form
iv	Internal pore surface area	—	Increases	Increases up to 500°C, later decreases	Decreases
v	Grindability[b]	—	—	Increases up to 500°C, later decreases	Decreases

[a] The cumulative weight loss is expressed as the percentage of the weight of total organic matter present in RH. 1 ton of RH contains approximately 0.8 ton of the organic matter.

[b] The grindability is indicative of the amount of energy required to grind RHA, as produced, to the required fineness per unit mass. The increase in grindability indicates decrease in the energy requirement.

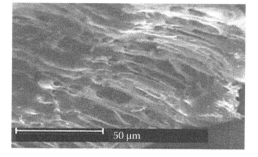

PLATE 4.2
Scanning electron micrograph of RH after combustion.

the pozzolanic activity index reduced substantially, inferring reduction in the content of amorphous silica. In order to obtain silica in the amorphous form, it is generally recommended that the combustion temperature remains in the range of 500°C–700°C, with optimum time duration appropriate for the process, to obtain RHA with low carbon content, high surface area and maximum grindability. The modern fluidized bed burners, producing RHA on a large scale and on a continuous basis, typically operate at 700°C–750°C with a residence time of few seconds [6,19,23,24]. The thorough mixing and improved heat transfer in the fluidized bed results in low carbon content and

PLATE 4.3
RHA after grinding.

very short residence time for burning. At higher temperature, the amorphous silica gets converted to the crystalline form—quartz, cristobalite, and tridymite—depending upon the temperature, burning duration, and the cooling conditions. The size of silica crystals grows with the time of burning [25].

4.4.2 Modern Methods to Produce Pozzolanic RHA

In order to use RHA as a mineral admixture in cement and concrete, it is required to be produced under controlled conditions of the surrounding environment, temperature, and residence time. Figure 4.1 shows different processes available for the purpose. Reviews on the existing thermal treatment processes are given by Rozainne et al. [26] and Nair et al. [27]. As shown in Figure 4.1, the combustion of RH is divided into two types: "controlled" and the "uncontrolled." The large-scale production of RH requires controlled combustion. The rotary kiln process has been discussed in the literature [28,29] and also patented [30]; however, no industrial application has been reported so far. The traditional method of RH combustion in the inclined (stepped or perforated) grate furnace is quite inefficient and converts only about half of the energy available in the husk. Moreover, the furnace requires large area [31]. It produces RHA of a poor quality with 10%–30% carbon

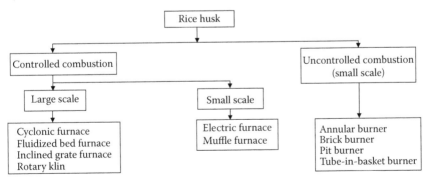

FIGURE 4.1
Production of RHA by thermal treatment.

content. In the worst cases, the ash contains char with up to 50% combustible matter and hard-burned crystalline ash and inert fused lumps formed in the localized hot spots in the combustion chamber [17]. The electric furnace and the muffle furnace, although coming under the controlled combustion category, require large quantity of electric power for combustion and gas heating, and hence are suitable for small-scale RHA manufacture. The process that fruitfully utilizes both the fuel and pozzolanic value of RH is the most suitable for the large-scale manufacture of RHA. The modern fluidized or cyclonic bed processes produce RHA with above 85% amorphous silica and very low carbon content in the range of 1%–4% [3,6]. The fluidized bed process is discussed in the following paragraphs. There is also a need to identify a process to manufacture RHA on a small scale near the paddy fields and suitable for adoption in the rural areas. The RHA produced in this manner can be appropriately used to fulfill the requirement of pozzolanic admixture for the local construction activity. Keeping that in view, one process to manufacture RHA on a small scale, the annular oven process, is also discussed.

4.4.2.1 Fluidized Bed Process for Large-Scale Production of Rice Husk Ash

The fluidized bed process is more efficient in comparison to other processes, as it ensures complete mixing and higher heat transfer rate during the combustion of RH. As a result, using the modern fluidized bed process, it is possible to produce RHA with consistently low ash content (<4%), at a very small residence time (<2 min) in the combustion chamber. Figure 4.2 depicts the process. In the combustion chamber, RH is held in the fluidized state in the silica sand medium, passing air from the bottom. As mentioned in Section 4.4.1,

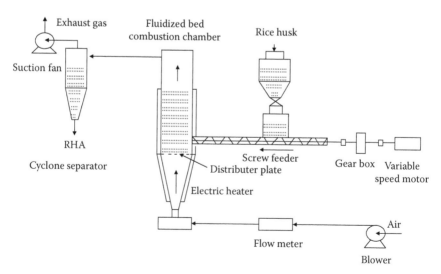

FIGURE 4.2
Fluidized bed process for large-scale production of RHA.

the combustion of RH is exothermic with high calorific value. The external heating is required only during the start-up; the process is later self-sustaining (autogenous). The process can be operated continuously for large-scale operation. RHA is easily removed from the fluidized bed by entrainment in the gas stream, from which it can be separated by a simple particle separating system such as a cyclone (see Figure 4.2). The control of critical process parameters, such as temperature and the residence time is achieved through the control of the ratio of air and RH feed in the fluidization chamber. The main advantage of the process is its flexibility to produce RHA of different grades (level of carbon content) depending upon the market demand. Rozainee et al. [26] reported the production of RHA with 100% amorphous silica at very low (<2%) carbon content using the fluidization process in their laboratory.

4.4.2.2 Annular Oven Process for Small-Scale Production of Rice Husk Ash

The four main available processes to produce RHA on a small scale are shown in Figure 4.1. All the four are affordable and simple options for the production of good quality RHA for building materials application. The first three processes—the annular, brick, and pit burner—were evaluated by Nair et al. [27], and the annular oven process was found to be the best option. The fourth process, the tube-in-basket burner or the so-called TiB burner [17], is a variant of annular burner provided with an arrangement to control the flow of secondary air into the inner (burner) tube. All the processes are batch operated and essentially come under the uncontrolled combustion category. The schematic of the annular oven process is given in Figure 4.3. The temperature inside the oven does not exceed 700°C

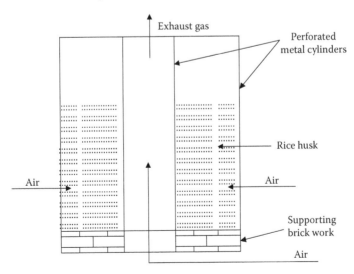

FIGURE 4.3
Annular oven process for small-scale production of RHA.

and it takes about 6–15 h for the complete combustion, depending upon the capacity, design, and operating parameters [17,27,32]. Allen reported that one burning takes about 15 h to produce 4 kg of RHA, using a somewhat similar burning process [32].

4.5 Physical and Chemical Characteristics

RHA produced after combustion under the controlled conditions of environment, temperature, and residence time possesses very high internal surface area (50–100 m^2/g) due to porous nature of the particles. The rate of cooling after combustion affects the characteristics. The rate at which RHA gets cooled after combustion depends upon the type of process used for its manufacture. It is seen that slow cooling increases particle size. However, the size of particles cannot be interpreted in terms of the degree of crystallinity or surface area, as the particles may have a high meso- and microporosity [20]. It is necessary to grind RHA to the desired fineness before adding to cement or concrete.

RHA produced under uncontrolled conditions of combustion generally has physical and chemical characteristics inferior in comparison to that produced under controlled conditions. Under the uncontrolled combustion, the desired burning temperature range of 500°C–700°C is not always maintained and sometimes it crosses 900°C, leading to the formation of crystalline quartz. When sufficient oxygen is not available during combustion and when the residence time for the RH in the combustion chamber is not adequate, organic material in the form of unburned or partially burned RH remains within the bulk RHA after the burning process. These variations have been reported in the literature [33,34]. Under such conditions, the pozzolanic activity of RHA may get affected.

RHA, produced under proper conditions, is soft and easily pulverized to the desired (<45 μm) size (Plate 4.3). Unlike SF or FA, the particles of RHA are angular and remain porous even after the size reduction (Plate 4.4). Therefore, the reactive surface area is not fully measurable by the conventional Blaine apparatus used for cement and also not by the particle size distribution measured by the laser apparatus. The internal surface area on account of its porosity is required to be measured and that is done by the more sophisticated "nitrogen adsorption BET method."

There is evidence to show that unground or improperly ground RHA exhibits low pozzolanic activity and high water requirement due to water absorption by the porous particles. The improvement in both the properties is observed after appropriate size reduction by grinding, even when the combustion process results in slight crystal formation and residual carbon

PLATE 4.4

30 μm — Scanning electron micrograph of ground RHA.

content [34–37]. However, the grinding of RHA to a high degree of fineness should be avoided, as it requires higher energy and the pozzolanic activity does not always show corresponding increase [38].

In the size reduction process, the internal surface area on account of the porosity of the cellular particles decreases due to the particle breakage but that due to the reduction in particle size increases, as the particles are ground finer [39,40]. The fine particles affect refinement in the pore structure of concrete, act as nucleation points for hydration products and restrict growth of the crystals (calcium hydroxide) generated in the hydration process [37]. All these aspects contribute toward improving the hydration process, reduce porosity, increase strength, and durability of concrete [41].

The water requirement of cement mortar with RHA replacement is higher in comparison to that without RHA replacement, due to the internal porosity, irregular shape, and higher fineness of the particles. The shape and fineness of the particles also play a critical role in deciding the rheological properties of fresh concrete, namely, the yield stress and the plastic viscosity. Any deviation from the spherical shape of particles implies corresponding increase in the plastic viscosity for the same phase volume [42].

The amorphous silica in the RHA reacts with the calcium hydroxide, that is, $Ca(OH)_2$, generated during the hydration of cement to form the strength giving calcium-silicate-hydrate or C-S-H. The particle size or specific surface area is important, as this reaction is surface-assisted. Further, the dissolution of amorphous silica in calcium hydroxide is strongly influenced by the surface structure, where the charge on the silica surface plays an important role depending on the binding or release of protons at the surface silanol (Si-OH) groups [43]. The combustion at 500°C is found to produce highly reactive RHA, as the concentration of surface silanol groups is found high at that temperature, especially after quick cooling [20]. Thus, it can be said that the pozzolanic activity of RHA is high when it contains maximum silica in the amorphous form with a high surface area (fine particle size) and the presence of many silanol groups (Si-OH) on the surface.

The results of the autogenous, early-age or drying shrinkage of the cement paste and concrete (for the sake of convenience the term "drying shrinkage" will be used in the later text), after the addition of RHA as cement replacement, are subject to different interpretations by the researchers. The drying shrinkage of concrete is defined as the macroscopic volume change occurring when there is no moisture exchange between the material and the exterior surrounding environment. It is the result of the volume reduction associated with the chemical reactions that occur during the hydration of cementitious materials in concrete. The study of results reveals that it depends upon the particle size, internal surface area (BET measurement), internal relative humidity of the sample, and the pozzolanic activity of RHA. Whether the shrinkage is comparable, higher or lower in comparison with the plain concrete, after the addition of RHA is decided by the complex interplay of these factors [7,39,44–48]. The following broad observations could be made based on the study of results as reported in the literature:

a. RHA with high pozzolanic activity, contributing high initial strength, is likely to render comparable or lower the drying shrinkage in relation to that of the plain concrete.

b. When the RHA samples from different sources are compared, the drying shrinkage is likely to vary inversely with the internal surface area, other conditions remaining identical. This positive effect is perhaps due the porous cellular structure, which stores water and acts as an internal reservoir, providing a source of curing water to the cementitious paste volume in its vicinity.

c. When the RHA samples from the same source are compared, the drying shrinkage is likely to vary inversely with the median particle size, other conditions remaining identical. The finer (smaller median article size) RHA is likely to exhibit higher drying shrinkage.

d. When the shrinkage behavior of concrete with RHA is compared under different surrounding conditions, the samples placed under lower humidity conditions are likely to exhibit higher absolute shrinkage.

The concentration of impurities and particularly the toxic elements in RHA are negligible or low [49]. Hence, the use of RHA in concrete can be considered as safe.

4.6 Addition of RHA to Cement and Concrete

The ground RHA can be added to cement or concrete as a partial replacement of Portland cement. Table 4.2 gives the important physical and chemical properties of ground RHA produced by different processes. The values given in Table 4.2 are not exact but intended to give an idea to the engineer regarding possible range of properties of RHA to render concrete with improved performance. The range of properties in Table 4.2 have been worked out based on the actual values reported in the literature [5–7,22,38–40,46,50]. The ASTM C618 [51] covers the use of fly ash and raw or calcined natural pozzolan as mineral admixtures in concrete. As shown in Table 4.2, RHA in most cases satisfies the ASTM C618 requirement for Class N pozzolan in terms of

TABLE 4.2

Physical and Chemical Properties of Ground RHA for Partial Replacement of Cement in Concrete

		Property Values	
			ASTM C618 Class N
Sl No	Particulars[a]	RHA	Pozzolan
i	*Physical properties*		
	Color	Gray	
	Specific gravity	2.05–2.16	—
	Bulk density (ton/m^3)	0.4–0.5	—
	Fineness, retained on 325 mesh (45 μm) sieve (max %)	1	34
	Fineness, median particle size (μm)	3.8–8.3	—
	Fineness, nitrogen adsorption (m^2/g)	20–40	—
	Pozzolanic activity index—cement (min %)		
	7 day	99–117	75
	28 days	92–97	75
ii	*Chemical properties*		
	Silicon dioxide (SiO$_2$)	87–96	—
	Oxides of silicon, aluminum, iron (SiO$_2$ + Al$_2$O$_3$ + Fe$_2$O$_3$, min %)	87–98	70
	Available alkalis as Na$_2$O (max %)[b]	0.75–3.5	1.5
	Sulfur trioxide (SO$_3$, max %)	0.01–0.32	4.0
	Moisture content (max %)	0.8–1.9	3.0
	Loss on ignition (max %)	2–12	10.0

[a] The particulars given in the bracket refer to the units or the minimum/maximum limits as specified for the ASTM C618 Class N pozzolan.

[b] The equivalent alkali content, expressed as Na$_2$O, is obtained as Na$_2$O + 0.658K$_2$O. Applicable only when specifically required by the purchaser for mineral admixture to be used in concrete containing reactive aggregate and cement to meet a limitation of the alkali content.

the retention on 45 μm sieve (<34%), moisture content (<3%), loss on ignition (<10%), sum of the mineral oxides SiO_2 + Al_2O_3 + Fe_2O_3 (>70%), sulfur trioxide (<4%), and the strength activity index (>75%). Hence, RHA can be classified as an ASTM C618 Class N pozzolan.

The level of cement replacement depends upon improvement in the strength and durability properties of the blend. The improvement in these properties has been observed, when RHA was used both with and without water reducing agent in concrete, replacing cement up to 30% on mass basis [5,7,40,52]. Although the laboratory experiments on blending of RHA with cement have been successfully carried out, there is no report of large-scale manufacturing of blended cement with RHA, in a cement plant [40].

As shown in Table 4.2, the RHA is ground finer than cement, with a median particle size <10 μm (1 μm = 10^{-6} m) and the BET surface area of 20–40 m²/g. Improved performance of blended system can be obtained, when fine RHA is blended with relatively coarser cement, through higher packing factor (filler effect) [53].

The silica content (mostly amorphous) could lie between 87% and 96% and the loss on ignition between 2% and 12%. The 7 day pozzolanic activity index is around 100% or even higher. The water requirement is high in general and the superplasticizer is added to obtain the desired slump. The influence of RHA on increasing the water requirement can be reduced by forming ternary mixtures, suitably adding FA along with RHA as a replacement of PC [54].

The RHA has lower specific gravity in comparison with cement; hence, the volume of blend shall increase depending upon the level of mass of cement replaced. This aspect should be kept in mind while preparing the mix design. When the choice is available, RHA with high pozzolanic activity, exhibiting high initial strength, may be used to minimize the drying shrinkage.

4.7 Summary

The RHA, used as pozzolanic admixture in cement and concrete, is obtained from the combustion of RH under certain conditions of the surrounding environment, temperature and residence time, and subsequent size reduction. RH is presently considered as an agricultural waste and used as fuel, where its pozzolanic value lies unutilized. The abundant availability of RH in the rice-producing countries provides us a huge scope to recover its heat value to generate power and to use the RHA produced in cement and concrete on a large scale. The production of RHA with cogeneration of power as well as its application in cement and concrete, both contribute toward the reduction of GHG emissions. The large-scale application of RHA in the construction industry requires industrial and economic policy planning.

RH is the outer shell covering the rice kernel. It is obtained when paddy is threshed to separate rice and the husk. It contains about 20% ash, of which about 95% is silica, besides other constituents. On average, 1 ton of paddy produces about 200 kg of husk, which upon combustion produces about 40 kg of ash.

The important physical and mineralogical characteristics of RHA—the content of amorphous silica, surface area, grindability, and carbon content—depend upon the temperature, environment, and duration of combustion. The surface silanol (Si-OH) sites in the amorphous silica are thought to be responsible for the pozzolanic activity of RHA. In order to obtain silica in the amorphous form, it is generally recommended that the combustion temperature remains in the range of 500°C–700°C, with optimum time duration appropriate for the process, to obtain RHA with low carbon content, high surface area, and maximum grindability. The modern fluidized or cyclonic bed processes produce RHA with above 85% amorphous silica and very low carbon content in the range of 1%–4%. There is also a need to identify a process to manufacture RHA on a small scale near the paddy fields and suitable for adoption in the rural areas.

RHA produced under uncontrolled conditions of combustion generally has physical and chemical characteristics inferior in comparison to that produced under controlled conditions. RHA, produced under proper conditions, is soft and easily pulverized to the desired size, generally finer than cement, with a median particle size lesser than 10 μm or lesser and the BET surface area of 20–40 m^2/g. The silica content (mostly amorphous) could lie between 87% and 96%. Unlike SF or FA, the particles of RHA are angular and remain porous even after the size reduction. There is evidence to show that unground or improperly ground RHA exhibits low pozzolanic activity and high water requirement due to water absorption by the porous particles. The fine particles of RHA affect refinement in the pore structure of concrete. The water requirement of concrete with RHA is higher. The use of RHA in concrete can be considered as safe from the toxicity point of view.

RHA can be classified as an ASTM C618 Class N pozzolan. The level of cement replacement depends upon improvement in the strength and durability properties of the blend. The improvement in these properties has been observed, when RHA was used both with and without water reducing agent in concrete, replacing cement up to 30% on mass basis. There is no report of large-scale manufacturing of blended cement with RHA in a cement plant. The water requirement of concrete is high in general and the superplasticizer is added to obtain the desired slump.

5

Metakaolin

5.1 Introduction

Metakaolin (MK) or the so-called high-reactivity metakaolin (HRM) is a mineral admixture, which is relatively new to the cement and concrete industry. It has the potential to improve the strength and durability of concrete. The term "high reactivity" is used to distinguish a whitish, purified, manufactured, thermally activated kaolinite (a constituent of kaolin clay) or MK from the less reactive varieties of calcined clay pozzolan containing impurities, which cannot be activated to a pozzolanic form at the temperature used to produce MK.

MK is a natural pozzolan, which is defined in ACI 116R [1] as "...either a raw or calcined natural material that has pozzolanic properties (for example, volcanic ash or pumicite, opaline chert and shales, tuffs and some diatomaceous earths)." It conforms to ASTM C618 [2] Class N pozzolan specifications. The chemical and mineralogical compositions of some natural pozzolans are given by Mehta [3]. MK is unique in that it is not the byproduct of an industrial process nor is it entirely natural; it is derived from a naturally occurring mineral and is manufactured specifically for cementing applications. Unlike a byproduct pozzolan (such as pulverized fuel ash [PFA] or silica fume [SF]), which may have variable composition, MK is produced under carefully controlled conditions to refine its color, remove inert impurities, and tailor particle size. Thus, a high degree of purity and pozzolanic reactivity can be obtained. It has great promise as a mineral admixture for cement and concrete.

The natural pozzolan in the form of calcined earth, blended with lime, has been in use as a building material for thousands of years [4]. A more recent example is the construction of a reservoir in Amazon basin in 1960, where approximately 300,000 ton of locally available clay was calcined and blended with the Portland cement (PC). These structures have not suffered from the alkali-silica reaction, despite the fact that highly reactive aggregate was used [5]. The performance of concrete depends mainly on the environmental conditions, the microstructure, and the chemistry of hydration. The last two factors are strongly affected by the concrete components.

The pozzolanic nature of MK has been reported in the literature [6,7]. When used as a partial replacement for PC, MK reacts with calcium hydroxide (CH), produced during hydration, to form strength giving compounds. HRM is an engineered, high-strength, pozzolanic material. It is an economical alternative to SF and can be utilized in high-performance concrete. Its use has grown rapidly since 1985 [8]. The comprehensive review of the studies on the use of MK as a partial pozzolanic replacement for cement in mortar and concrete has been presented by Sabir et al. [9] and Siddique and Klaus [10].

MK is generally produced by the calcination (also called thermal activation) of raw kaolin clay. The natural sources of kaolin are found in several countries. A material is considered to constitute kaolin when the amount of kaolinite in it (such as rock) is greater than 50% [11], besides the other mineral constituents. The calcination process, under the moderate temperature range of 600°C–800°C, transforms a crystallized, organized phase (kaolinite) into a disorganized transition phase (MK) through crystal lattice failure. The dehydroxylation and disorganization of the kaolinite generate anhydrous and amorphous aluminosilicate ($Al_2Si_2O_7$ or $Al_2O_3 \cdot 2SiO_2$), which is a reactive admixture with pozzolanic properties. The kaolinite can also be converted into amorphous and reactive phases by grinding kaolin under controlled conditions [12].

MK is an off-white powder and the oxide composition, as seen earlier, consists mainly of silicon dioxide (SiO_2) and aluminum oxide (Al_2O_3). It generally contains only a small quantity of free calcium oxide and alkalies and therefore it does not have a capacity to hydrate by itself. On reaction with CH (produced by cement hydration) at ambient temperature, it produces C-S-H gel and alumina-containing phases (C_4AH_x, like C_4AH_{13}) in various concentrations, including gehlenite hydrate (C_2ASH_8) and hydrogarnet (C_3AH_6) [14,15].

The pozzolanic admixtures such as FA do not show beneficial effects until later in the hydration process. MK is relatively fine and possesses some latent hydraulic reactivity. It overcomes the dilution effect to a large extent, contributing to both heat and strength generation at early ages. There are three elementary factors influencing the contribution that MK makes to concrete strength when it partially replaces cement in concrete. These are the filler effect, the acceleration of cement hydration, and the pozzolanic reaction of MK with CH [16].

Besides the application aspects, the research work on MK continues in two other areas, the kaolin structure and the kaolin-to-MK conversion, including the analytical techniques required for the investigation [17].

MK finds usage in many forms of concrete:

- High performance, high strength, and lightweight concrete
- Precast concrete for architectural, civil, industrial, and structural purposes
- Fiber cement and ferrocement products
- Glass fiber reinforced concrete
- Mortars, stuccos, repair material, pool plasters

Advantages of using MK are as follows:

- Increased compressive and flexural strengths
- Reduced permeability
- Increased resistance to chemical attack
- Increased durability
- Reduced effects of alkali-silica reactivity (ASR)
- Reduced shrinkage due to particle packing, making concrete denser
- Enhanced workability and finishing of concrete
- Reduced potential for efflorescence
- Improved finish, color and appearance

Section 5.2 discusses different processes to manufacture MK: the thermal activation of kaolin (Section 5.2.1), mechanical activation of kaolin (Section 5.2.2), and the calcination of waste paper sludge (Section 5.2.3). Section 5.3 summarizes the physical and chemical characteristics of MK. A brief on the quality control aspects of MK is given in Section 5.4. Section 5.5 discusses different aspects related to the addition of MK to cement and concrete. Section 5.6 summarizes this chapter.

5.2 Production

Figure 5.1 illustrates different methods to produce metakaolinite or metakaolin (meta prefix in the term is used to denote change): the thermal and mechanical activation of kaolin and the calcination of waste paper sludge in the incinerator. The thermal activation processes are commonly used on the industrial scale and, hence, will be discussed in detail in the following paragraphs. In thermal activation, the raw material input for the production of MK is kaolin clay. It is a white, clay mineral (Plate 5.1). The primary constituent of kaolin is kaolinite (40%–70%), that is, hydrated aluminum disilicate $(Al_2Si_2O_5(OH)_4$ or $Al_2O_3 \cdot 2SiO_2 \cdot 2H_2O)$. The other constituents include quartz and cristobalite (SiO_2), K-alunite $(KAl_3(SO_4)_2(OH)_6)$, illite or muscovite-like micas (layered alumino-silicate incorporating magnesium, potassium, and iron) and rutile (TiO_2) [18]. It is a "phyllosilicate" or layered silicate material [19] and one of the highly priced industrial mineral clay varieties. This electrically neutral, crystalline layer structure, which is a common characteristic of clay minerals, leads to fine particle size and plate-like morphology and allows the particles to move readily over one another, giving rise to physical properties such as softness, soapy feel, and easy cleavage [20].

Kaolin clay has been traditionally used in ceramics for centuries. It is also extensively used in the paper industry as a filler, opacifier (a chemical agent

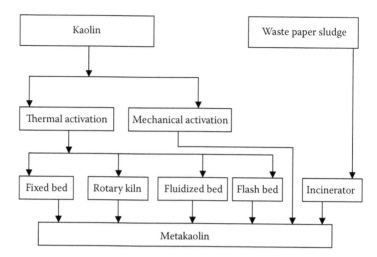

FIGURE 5.1
Production of MK.

PLATE 5.1
Kaolin rock.

added to a material to make it opaque), and as an important input to high-end coatings. The other smaller applications of kaolin are in the refractory, rubber, paint, plastic, chemical, and pharmaceutical industries. Its application in the construction industry is a relatively recent development.

5.2.1 Thermal Activation of Kaolin

MK is largely manufactured by thermal activation of kaolin, which refers to calcination or dehydroxylation of kaolinite constituent of the kaolin clay, carried out in a processor under controlled conditions. The conversion takes place according to the chemical reaction

$$Al_2O_3 \cdot 2SiO_2 2H_2O \ \rightarrow \ Al_2O_3 \cdot 2SiO_2 + 2H_2O \uparrow \qquad (5.1)$$

TABLE 5.1

Thermal Treatment of Kaolin: Physicochemical Changes

Sl No	Temperature (°C)	Physicochemical Changes
i	<100	Dehydration, low temperature release of absorbed water in pores or on the surface, etc.
ii	100–400	Weight loss taking place as a result of the loss of adsorbed water, the burning of organic matter, and the reorganization of octahedral layers in kaolinite structure
iii	400–650	Dehydroxylation and activation. Kaolinite gets converted to reactive metakaolinite or MK $Al_2O_3 \cdot 2SiO_2 \, 2H_2O \rightarrow Al_2O_3 \cdot 2SiO_2 + 2H_2O \uparrow$
iv	>500	Decomposition of K-alunite to Al_2O_3 begins and goes up to about 900°C
v	>650	Recrystallization: Conversion of MK to inert ceramic materials like silica (SiO_2), spinel ($2Al_2O_3 \cdot 3SiO_2$), and mullite ($2[3Al_2O_3 \cdot 2SiO_2]$) begins and goes up to about 1300°C

The broad physicochemical changes that occur during the thermal treatment of kaolin are given in Table 5.1. Under ambient conditions, kaolin is stable. As shown in Table 5.1, when heated to temperatures of 400°C–650°C, kaolinite gets calcined and loses most of the mass (about 14%) in the chemically bound hydroxyl ions. The heat treatment, thermal activation or calcination breaks down the structure of kaolinite such that the alumina and silica layers become puckered and lose their long-range order. The result of dehydroxylation and disorder is MK, a highly reactive transition phase. MK is an amorphous pozzolan with some latent hydraulic properties (Plate 5.2), well suited for use as an admixture in cement and concrete [21].

The key to producing high-quality MK is to achieve a near-complete dehydroxylation, without overheating. The thermal treatment beyond a defined

PLATE 5.2
Metakaolin (calcined kaolin).

point results in sintering and the formation of mullite, which is nonreactive. In other words, optimally altering kaolinite to MK state requires that it is thoroughly roasted but never burnt. The temperature of thermal activation in the processor should correspond to the range between the endothermal dehydration of the clay and the exothermal recrystallization of high temperature phases. This range can be determined using the differential thermal analysis (DTA) technique [22]. The temperature in the processor is typically maintained in the range of 650°C–900°C [23,24]. The actual temperature depends upon the type of processor, as the process parameters are different. Badogiannis et al. [17] report the removal of SO_3 from the poor quality kaolin containing higher proportions of k-alunite by treating the kaolin samples at higher temperatures up to 850°C.

The development of pozzolanic properties in calcined kaolin mainly depends upon the nature and abundance of kaolinite in the raw material, on the calcination conditions and on the fineness of the final product. The impurities act as diluents [22,25–27]. The mineralogical characteristics of raw kaolin—grain size, nature of impurities, and especially the crystallization state—also affect. The issue of the effect of the crystallization state of raw kaolin on the properties of MK—on the C-S-H formation and the development of compressive strength—in the lime reactivity test is not yet clearly resolved [25,28]. Extensive investigations are required on the subject. It is also reported that less crystalline (more disordered) kaolin gets easily dehydrated, that is, it requires less time for dehydroxylation at a given temperature [29].

As shown in Figure 5.1, the thermal activation of kaolin incorporates four processes: fixed bed, rotary kiln, fluidized bed, and flash bed calcination. The fixed bed requires the longest time for calcination, up to 6h, and the rotary kiln requires up to 3h [23]. The fluidized bed process reduces the time for calcination to few minutes. The kaolin gets harder after heat treatment. MK, before it is used as an admixture, requires size reduction (grinding) to obtain the desired fineness. The energy required for the process is more, when the size reduction is carried out after the heat treatment, in comparison to that required for the raw kaolin. The calcination time is reduced to a few seconds, using flash bed calcination process on finely ground raw kaolin. It consists of very rapid heating, calcining, and cooling of powdered material suspended in a gas [24]. The calcined product may not necessarily be ground further. Thus, flash calcination can simplify industrial production installations and decrease the energy cost of grinding. In the heat treatment processor, as the residence (calcination) time reduces, the processing temperature increases to achieve higher calcination rate. While longer residence time and lower temperature is generally quite effective at dehydroxylation of kaolin, it is seen that higher calcining rate affects the pozzolanic reactivity of MK. Salvador [24] found that flash-calcined kaolin had higher initial reactivity and gave better compressive strength in comparison to that treated in a fixed bed.

The findings reported by Shvarzman et al. [30] are interesting. The authors found that MK containing less than 20% of amorphous phase can be considered as an inert material from the standpoint of pozzolanic activity. Whilst the chemical activity, defined by the authors as the ability to react with portlandite or $Ca(OH)_2$ in the presence of an excess of water, was found to be a linear function of the amorphous phase content in the range of 50%–100%, similar trend was not observed in case of the pozzolanic activity or the strength activity index. The strength activity index of the samples was determined according to ASTM C311 and European Standard EN-450. In contrast to the chemical activity, the increase of amorphous phase content over 55% did not lead to additional growth in strength activity index. The phenomenon could probably be related to the $Ca(OH)_2$/admixture molar ratio, morphology of the untreated clay, water/binder ratio, and the action of admixture as microfiller or some other factors, which were outside the purview of the study. The authors suggested that even with the partial dehydroxylation of kaolinite accompanied with about 55% amorphization, the material may be considered as a very active pozzolanic admixture. The findings direct toward the possibility of reducing energy consumption during the large-scale manufacture of MK by heat treatment. The work underlines the need for more research in the area.

The large-scale manufacture of MK across the United States, under the supervision of Ash Grove Cement Company, can be mentioned as an example [31]. MK production by heat treatment of kaolin does not involve releasing large amount of CO_2 into the atmosphere in comparison with that released during cement production. It is estimated that 1 ton of MK produced releases about 175 kg of CO_2 in the atmosphere, through different processes such as extraction of raw materials, kiln fuel burning, etc. [13].

5.2.2 Mechanical Activation of Kaolin

The kaolinite can also be converted into amorphous kaolin or MK by mechanochemical treatment or mechanical activation. The large-scale application of this process has not been reported so far. It comprises of a grinding operation under such condition that changes occur in the crystal structure of the materials ground, along with the size reduction. The treatment results in the generation of active surface sites, besides the surface area and the change takes place in the physicochemical behavior of the materials [12]. The mechanical activation effect is observed, when the size reduction operation is carried out in an equipment using impact and friction forces among particles and between particles and the media, as obtained in vibratory, oscillatory, or planetary mills. The energy transferred through the grinding operation is applied to bend and/or break the crystals [32,33]. The time required for complete amorphization in the mill depends upon the composition of natural kaolin. The pozzolanic activity of the product is a function of the mineralogical composition of kaolin and the grinding time. It has been found that the quartz particles in kaolin help the grinding process and contribute towards its mechanical activation [34]. In the past, the mechanical activation effect has been used to improve the pozzolanic or cementitious activity of mineral admixtures [35]. As the process does not involve any combustion or burning, it is expected to reduce the greenhouse gas generation in comparison to that in thermal activation process.

Vizcayno et al. observed that the mortars with MK, obtained by mechanical activation, were somewhat more porous than those with MK obtained through heat treatment of kaolin [36]. Some more work is required in this area.

5.2.3 Calcination of Waste Paper Sludge

The kaolin is used in paper manufacture as a filler and opacifier. When waste paper is recycled, nearly 20%–35% of it is lost as sludge. The sludge typically contains kaolin, calcite, muscovite, talk, quartz, and cellulosic matter. Highly reactive MK can be produced, calcining the waste sludge [37].

The waste paper sludge, when calcined at 600°C–750°C [37,38] to remove the organic matter, produces a reactive pozzolanic material containing MK and calcite. The decomposition of calcite is not desirable. At higher temperatures (800°C and above), calcite gets decomposed to CaO (lime) and CO_2. The lime formed after decomposition, further reacts with alumina and silica (in MK) to form inert products such as gehlenite and anorthite and less MK is available for the pozzolanic reaction. The reactive surface also decreases at higher temperature [39].

The rate of heat evolution during hydration of cement with MK manufactured from paper sludge is similar to that obtained with the MK manufactured from natural kaolin [40]. Mozaffari et al. [41] report the utilization of ash produced after the incineration of waste paper sludge in an industrial-scale combustor (200 ton/day), at 850°C–1200°C and subsequent rapid cooling. The ash contained free lime and cementitious compounds such as α-dicalcium silicate (α-C_2S) and bredigite ($Ca_7Mg(SiO_4)_4$). The ash was utilized to activate ground granulated blast furnace slag (GGBS) and the concrete prepared using that mixture (no cement) gave 28-day compressive strength of 15–25 MPa. The influence of variations in the composition of sludge on the pozzolanic activity and the behavior of cement based systems containing calcined sludge is under investigation [37,42,43].

The calcined wastepaper sludge as a source of MK offers a route to utilize the waste material and also an alternative to the production of MK from natural kaolinite resources, that is, kaolin clay and the associated environmental burden [40].

5.3 Physical and Chemical Characteristics

The range of physical and chemical characteristics of MK for the partial replacement of cement in concrete are given in Table 5.2 [13,15,44–62]. The characteristic values are given vis-à-vis the requirement of ASTM C618 Class

TABLE 5.2

Physical and Chemical Characteristics of MK for Partial Replacement of Cement in Concrete

Sl No	Particulars[a]	Characteristic Values[c,d]	
		Metakaolin (Dry, Calcined)	ASTM C618 Class N Pozzolan
i	*Physical characteristics*		
	Specific gravity	2.5–2.6	—
	Bulk density (ton/m³)	0.3–0.4	—
	Color	Off-white	
	TAPPI (GE) brightness	79–86	
	Fineness, retained on 325 mesh (45 μm) sieve (max %)	<0.1–0.2	34 (1, wet sieving)
	Fineness, average particle size (μm)	1.0–9.5	—
	Specific surface, nitrogen adsorption (m²/g)	9.5–18	—
	Pozzolanic activity index—cement (min %)		
	7 day	—	75 (85)
	28 days	—	75
	Increase of drying shrinkage of mortar bars at 28 days (max %)	—	(0.03)
ii	*Chemical composition (% mass)*		
	Silicon dioxide (SiO_2)	46.6–58.1	—
	Aluminum oxide (Al_2O_3)	35.1–45.3	
	Iron oxide (Fe_2O_3)	0.38–4.64	
	Oxides of silicon, aluminum, iron ($SiO_2 + Al_2O_3 + Fe_2O_3$, min %)	>82.08	70 (85)
	Calcium oxide (CaO)	0.02–2.71	
	Magnesium oxide (MgO)	0.03–1.02	
	Potassium oxide (K_2O)	0.1–3.17	
	Sodium oxide (Na_2O)	<0.01–0.4	
	Available alkalis as Na_2O, (max %)[b]	0.08–2.49	1.5 (1)
	Titanium oxide (TiO_2)	0.01–2.27	
	Sulfur trioxide (SO_3, max %)	0.07–0.99	4.0
	Moisture content (max %)		3.0
	Loss on ignition (max %)	0.51–2.52	10.0 (3)

[a] The brackets in the "Particulars" column refer to the units or the minimum/maximum limits as specified for the ASTM C618 Class N pozzolan.

[b] The equivalent alkali content, expressed as Na_2O, is obtained as $Na_2O + 0.658K_2O$. Applicable only when specifically required by the purchaser for mineral admixture to be used in concrete containing reactive aggregate and cement to meet a limitation of the alkali content.

[c] The brackets in "Characteristic Values" column refer to the typical requirements as specified by the Texas Department of Transportation, United States.

[d] All values have been rounded off at the second decimal point.

N Pozzolan. The range has been obtained from the values as reported in the literature; however, in some cases the characteristic values may be more or less than the upper or lower limits given in Table 5.2, respectively.

In the market, MK is available in two forms:

a. *Clinker*: The MK clinker (size 19–25 mm) has good handling ability and resistance to deterioration. It can be shipped in bulk and is ideal for intergrinding with PC clinker for the production of ASTM C595 Type IP(X) blended cement.

b. *Powder*: Available in packages, big sacks, or in bulk as an ASTM C618 Type N pozzolan.

As shown in Table 5.2, MK is a powder with off-white color, having TAPPI (GE) brightness generally in the range of 79–86 (TAPPI [see abbreviations] brightness is sometimes called GE brightness as the instrument like GE-Photovolt was used first for its measurement). The specific gravity and the bulk density lie in the range of 2.5–2.6 and 0.3–0.4 ton/m^3, respectively. It resembles SF in some respects, particularly in terms of its very large surface area and also that it is a silica based product, which on reaction with CH produces C-S-H gel. The material has a specific surface in the range of 9.5–18 m^2/g and an average particle size of 1–9.5 μm, with 99.8% of the particles passing 45 μm sieve, smaller than ordinary cement (<10 μm) but larger than SF (>0.1 μm).

Whereas MK is off-white in color, SF is typically dark grey or black, making the former particularly attractive in color matching and other architectural applications. The MK powder is quite consistent in appearance and performance, due to the controlled nature of its processing [63].

While kaolin is crystalline, MK has a highly disordered structure. Natural kaolin has different degrees of crystallization according to its origin and classified as ordered (well crystallized) or disordered (poorly crystallized). The kaolin crystal possesses a two-layered structure, where a sheet of octahedrally coordinated aluminum is bonded to a tetrahedrally coordinated silicon sheet. Upon calcination, the bonds break and the kaolin crystal structure collapses. It is observed that the Si–O network remains largely intact and the Al–O network reorganizes itself. MK reacts particularly well with lime and forms, in the presence of water, hydrate compounds of Ca and Al silicates. Therefore, it is considered to be a good synthetic pozzolan [64].

The chemical composition (expressed as oxides) consists mainly of silicon dioxide and aluminum oxide and a small quantity of calcium oxide and alkali. It does not have a capacity to hydrate by itself. Unlike SF, MK also contains alumina, which, on reaction with CH, produces additional alumina containing phases some of which are crystalline. These include tetracalcium aluminate hydrate (C_4AH_{13}), stratlingite (C_2ASH), and hydrogarnet (C_3AH_6) [65–67]. The scanning electron micrograph (Plate 5.3) of MK shows that the particles are angular, like that of cement. The x-ray diffraction analysis

PLATE 5.3
Typical scanning electron micrograph of MK manufactured by thermal activation.

indicates that MK is mainly an amorphous material with only a small quantity of crystallized phases. The main crystallized phases identified include anatase (TiO_2), cristobalite (SiO_2), and quartz (SiO_2) [15].

The pozzolanic activity of MK, determined by the Chapelle test,* is found to be better than FA and SF [68]. The amount of alkali-soluble Al and Si compounds, which reflect the content of active aluminosilicates in a material ready to react with lime, increases after calcination of kaolin [66].

5.4 Quality Control

ASTM C618 [2] and CSA A23.5-03 [69] cover coal FA and natural pozzolans for use as a mineral admixture in concrete. The natural pozzolans in the raw or calcined state are designated as Class N pozzolans and MK falls in that category. Some of the requirements of ASTM Class N pozzolans are given in Table 5.2. The table also mentions, as an example, typical requirements of the physical characteristics and the chemical composition, as specified by the Texas Department of Transportation, United States.

5.5 Addition of Metakaolin to Cement and Concrete

The new ASTM Standard Specification on blended cement, C595-06 [70], was released in August 2006 with some important changes to the nomenclature. The specification was simplified with the intent of making it easier to use.

---

* The Chapelle test is a measure of quantity of lime that reacts with metakaolin. One gram of metakaolin is made up into a suspension in distilled heated water at 100°C in the presence of 2 g of lime and left for 16 h. The remaining quantity of lime in the solution is then determined. The results are expressed in terms of gram of lime reacted per gram of metakaolin.

The new Portland-pozzolan cement, Type IP(X,) can include between 0% and 40% pozzolan, replacing old Types IP and I(PM). The letter "X" stands for the nominal percentage of the supplementary cementitious material (SCM) included in the blended cement. Although blended cement with MK is not commonly in use, the new standard facilitates communication of the total amount of SCM in concrete mixtures, which may also contain other added SCM (like MK) and may be subject to the restrictions on the total amount of SCM in the concrete.

MK enhances the strength and durability of concrete through three primary actions: (a) the filler effect, (b) the acceleration of PC hydration, and (c) the pozzolanic reaction with CH [16]. It is seen that the filler effect is immediate, the acceleration of cement hydration has its major impact within the first 24 h, and the maximum effect of pozzolanic reaction occurs between 7 and 14 days [16]. The filler effect is particularly significant in the interfacial transition zone between the cement paste and aggregate particle, where it produces a denser, more homogeneous, and narrower transition zone due to better packing, leading to reduced bleeding and improved strength of concrete.

The CH gets quickly removed during the hydration of cement with MK. The optimum replacement level of cement depends upon the proportion of the products of the hydration reaction (dependent upon the composition of cement and MK), temperature, and the reaction time in the cement–MK system. The increase in fineness of MK results in the increase in the level of cement replacement [52,71,72].

The addition of MK influences the rheological properties (therefore workability) of the fresh concrete mix. It is due to the high specific surface and chemical activity of MK and also due to its influence on cement–superplasticizer (SP) interaction [73]. The interaction of the cement–SP system in the presence of MK was studied using two point workability test (TPWT) (principle of the method available in Ref. [74]). The results establish that the addition of MK significantly influences the rheological properties of concrete mix. The character and range of this influence depends on the properties of the cement, properties of SP and MK content. The compatibility of cement–SP system must be tested taking into account the presence of mineral additives. When used to replace cement at levels of 5%–10% by weight, the concrete produced is generally more cohesive and less likely to bleed. As a result, pumping and finishing processes require less effort.

The test results on self compacting concrete show that the ternary blends of cement with FA and MK improved the fresh mortar properties and the rheology of the mixtures, when compared to those containing binary blends of FA or MK [75].

The quick removal of CH from the reaction products actually accelerates the hydration reaction, with consequent increase in the temperature [15]. The early strength and the water demand for standard consistency increase and the setting time reduces, with increasing replacement level [15–17,76].

Using Chapelle test, which consists in measuring the amount of CH consumed in reaction with MK, it has been shown that the pozzolanic activity of MK is better than that of the FA, BFS, and SF [77].

Although the cement replacement level up to 40% has been reported [72], 10%–25% replacement is more common. Khatib found that at a low water-to-binder ratio of 0.3, the optimum replacement level to give maximum strength enhancement is 15% MK. The optimum level was lower than that obtained at a higher water-to-binder ratio of 0.45 [58].

It has been demonstrated that the cement-based materials containing MK, within the 10%–25% replacement range, exhibit improved performance in terms of the strength and durability, in comparison with reference materials containing no MK [60,78–82]. Although concrete with MK requires a higher dosage of the superplasticizer and air-entraining admixture, in comparison with that of the control concrete with no MK, satisfactory slump, air content, and setting time can be obtained. The concrete strength can also be improved by partially replacing sand with MK [83,84].

It is observed that the early strength of concrete containing a high proportion of GGBS (>50%) can be improved by adding MK as a cement replacement [85]. However, the durability aspects of such concrete needs to be investigated. Similarly, it has been shown that the loss of workability of concrete, due to the presence of MK, can be compensated for by the incorporation of FA. The degree of restoration of workability, provided by FA, is influenced significantly by the level of cement replacement [86]. Using neural network (NN) technique, models have been developed for workability measured by slump, compacting factor, and Vebe time, for concrete incorporating FA and MK. The NN models provide an efficient, quantitative, and rapid means of obtaining optimal solutions to workability of concrete mixtures using PC–FA–MK blends as binder. Such models would prove extremely useful for deployment in computer based applications. The work indicates that the models are usable in practice to predict the workability of PC–FA–MK blends [87]. Attempts may be made to develop similar, universally applicable models for other tertiary or quaternary blends, incorporating FA, BFS, SF, and RHA with cement. It is reported [88] that replacing cement with MK has an expansive effect, in contrast to the shrinkage effect of FA. In the ternary blends, when both FA and MK replace cement, these influences combine in a complex and approximately compensating manner. Thus, using this technique, it should be possible to significantly reduce cracking, designing concrete with negligible shrinkage or expansion. MK has been found effective in immobilization of toxic materials [89,90].

Samet et al. [76] formulated blended cement using MK and studied its properties. It was observed that the mechanical properties of the blended cement are mainly governed by the level of replacement and the fineness of MK.

Aesthetics is the primary factor driving the popularity of white cement. The material's reflectivity also contributes to the functional advantages such as energy efficiency and safety at night. Using admixtures such as MK, it is possible to achieve bright concrete surface, along with improved workability and durability [91]. With the current shortage of SF and high-quality BFS in some countries, the attitude of concrete producers toward MK may change in the near future.

5.6 Summary

MK, or the so-called HRM, is a mineral admixture, which is relatively new to the cement and concrete industry. It is a natural pozzolan and conforms to ASTM C618, Class N pozzolan specifications. MK is unique in that it is not the byproduct of an industrial process nor is it entirely natural; it is derived from a naturally occurring mineral and is manufactured specifically for cementing applications. It is an engineered, high-strength, pozzolanic material.

MK can be produced by the thermal and mechanical activation of kaolin and the calcination of waste paper sludge in the incinerator. The thermal activation processes, using kaolin clay as raw material, are commonly used on the industrial scale. These processes refer to calcination or dehydroxylation of kaolinite constituent of the kaolin clay, carried out in a processor, under controlled conditions. The key to producing high-quality MK is to achieve a near-complete dehydroxylation, without overheating. The temperature in the processor is typically maintained in the range of 650°C–900°C; the actual temperature depends upon the type of processor, as the process parameters are different. The development of pozzolanic properties in calcined kaolin mainly depends on the nature and abundance of kaolinite in the raw material, on the calcination conditions, and on the fineness of the final product. The calcination of waste paper sludge also yields highly reactive MK. At 600°C–750°C, the calcination produces a reactive pozzolanic material containing MK and calcite. The calcined wastepaper sludge as a source of MK offers a route to utilize the waste material and also an alternative to the production of MK from natural kaolin clay and the associated environmental burden.

In the market, MK is available in two forms suitable for addition to cement, or concrete as cement replacement: (a) clinker, suitable for manufacture of ASTM C595 Type IP(X) blended cement and (b) powder, conforming to ASTM C618 Type N pozzolan. MK powder is off-white in color, having TAPPI (GE) brightness generally in the range of 79–86. The specific gravity and the bulk density lie in the range of 2.5–2.6 and 0.3–0.4 ton/m^3, respectively. The specific surface is in the range of 9.5–18 m^2/g with an average particle size of 1–9.5 μm; 99.8% of the particles pass 45 μm sieve. The particles are smaller than ordinary cement (<10 μm) but larger than SF (>0.1 μm). It is found particularly attractive in color matching and other architectural applications. MK powder is quite consistent in appearance and performance, due to the controlled nature of its processing. The chemical composition (expressed as oxides) consists mainly of silicon dioxide and aluminum oxide and a small quantity of calcium oxide and alkali. The pozzolanic activity of MK, determined by the Chapelle test, is found to be better than FA and SF.

The new Portland-pozzolan cement, Type IP(X), as per ASTM C595-06, can include between 0% and 40% pozzolan. MK enhances the strength and durability of concrete through three primary actions: (a) the filler effect, (b) the acceleration of PC hydration, and (c) the pozzolanic reaction with CH. The increase in fineness of MK results in the increase in the level of cement replacement. The addition of MK significantly influences the rheological properties of concrete mix. The character and range of this influence depends on the properties of the cement and properties of SP and MK content. The quick removal of CH accelerates the hydration reaction, with consequent increase in the temperature. The early strength and the water demand for standard consistency increase and the setting time reduces, with increasing replacement level. It has been demonstrated that the cement-based materials containing MK, within 10%–25% replacement range, exhibit improved performance in terms of strength and durability, in comparison with reference materials containing no MK. It has been shown that the loss of workability of concrete, due to the presence of MK, can be compensated for by the incorporation of FA. It is possible to achieve a bright concrete surface using MK along with improved workability and durability.

6

Hydration

6.1 Introduction

Portland cement (PC) is a product of intergrinding/blending of cement clinker with gypsum and minor additional constituents/additives. The blended cement is manufactured in the cement plant, partially replacing PC with the mineral admixtures. The mineral admixtures are also added during concrete making, as a partial replacement for cement, along with other constituents, like chemical admixtures, aggregate (coarse and fine), and water. The manufacture of PC, varieties of blended cement, as well as concrete are subject to the provisions of respective national standards. The hydration of cement is the combination of all chemical and physical processes taking place after contact of the anhydrous solid with water. The chemical reactions of clinker minerals and mineral admixtures play a major role but other aspects such as agglomeration, adsorption, evaporation, and release of thermal energy also need to be considered. It begins with the addition of water while making concrete, and nearly 60%–65% of the hydration is complete in 28 days. Although the cement minerals like calcium aluminates and silicates and the calcium sulfate (gypsum) take part in the hydration reactions, the process is affected by all other constituents of concrete. There is evidence to show that, under certain conditions, the hydration reactions may continue up to 5 years. The nature and the progress of cement hydration are responsible for the strength and the durability of concrete. The mineral admixtures added to cement take part in the hydration reactions due to their pozzolanic/cementitious properties. They affect the early-age properties, like rheology and therefore workability, setting, shrinkage, etc., besides the strength and the durability of concrete. Therefore, in order to understand the difference, the understanding of hydration process of cement is important, both without and with mineral admixtures.

Although the process of hydration is yet to be fully understood, several modern analytical equipment are being used for the investigation, like transmission electron microscope (TEM), scanning electron microscope (SEM) to observe the particle morphology, quantitative x-ray diffraction (QXRD) for cement phase analysis, microprobe analysis of composite cement phases including backscattered electron image analysis and x-ray mapping, thermogravimetric (TG) techniques, conduction calorimetry, magic angle spinning nuclear magnetic resonance (MAS-NMR) technique, and so on. This chapter reviews the current understanding of the hydration of cement with mineral admixtures and its impact on the early-age properties of concrete, which coincide with the progress of hydration.

Section 6.2 gives a brief depiction of the periods of hydration and the corresponding changes that occur in the concrete during the progress of hydration. These periods are the workability period (Section 6.2.1), setting period (Section 6.2.2), and hardening period (Section 6.2.3). Section 6.3 discusses the nature of the reactants, both from PC and mineral admixtures, that take part in the hydration reaction: the reactive cement compounds (C_3S, C_2S, C_3A, and C_4AF in Section 6.3.1), gypsum (Section 6.3.2), reactive compounds in mineral admixtures (Section 6.3.3), and calcium hydroxide (CH) (Section 6.3.4). Section 6.4 briefs on the voids in hydrated cement paste, which affect its strength, durability, and volume stability. Section 6.5 summarizes how the hydration properties of cement and that of concrete are interrelated. Section 6.6 briefly discusses the properties of concrete during early stages of hydration: workability, yield stress and viscosity, bleeding and laitance, setting time, concrete temperature, and volume changes. Section 6.7 discusses the chemical reactions during hydration and the corresponding changes in early-age concrete properties. It begins with the Section 6.7.1 briefing on the hydration reactions of cementitious materials other than PC. Section 6.7.2 depicts hydration reactions of PC. Section 6.7.3 discusses hydration reactions of cement with mineral admixtures. Section 6.8 summarizes this chapter.

6.2 Progress of Hydration with Time (Hydration Periods)

The use of modern sophisticated analytical equipment, like TEM, QXRD, TG, and conduction calorimetry, has led to considerable progress in understanding of the hydration reaction. The progress of cement hydration can be arbitrarily divided into five stages or time periods, based on the observations, as shown in Figure 6.1 [1–4]. These stages of hydration are observed in all varieties of PC, including those blended with mineral admixtures. The understanding of these stages has significance from the point of view of building good structures, so they have been grouped under three different titles, familiar to the practicing engineer: workability, setting, and hardening periods.

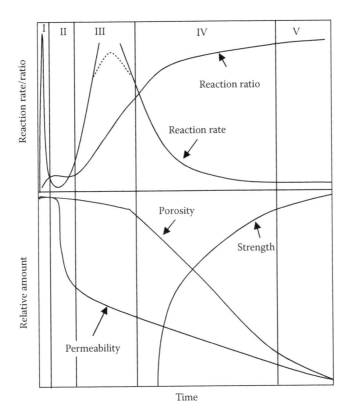

FIGURE 6.1
Arbitrary hydration periods and trends in the change of concrete properties.

6.2.1 Workability Period

The workability period can be divided in two: preinduction (Stage I) and induction (Stage II) periods (Figure 6.1).

Stage I—Pre-induction period: Short but rapid reaction immediately after water addition

Stage II—Induction or dormant period: Signifies very low rate of hydration reaction

The workability period lasts for 1–3 h, depending upon the type of cement and the concrete mix design. The wetting, mixing, transporting, agitating, placing, and finishing of concrete must take place in this period. The period can be extended using mineral admixtures and chemical admixtures. That is important in case of ready mixed (or ready mix) concrete (RMC), particularly for relatively high working temperature and long distance of transportation. Undesirable phenomena, like false set (early loss of consistency, poor heat evolution) and flash set (instantaneous setting, large heat evolution, and poor ultimate strength) are observed in this period.

6.2.2 Setting Period or Stage III: Active Reaction Period

Acceleration of reactions, mostly between the reactive cement compounds and water, takes place during setting. The boundaries of the setting period can be broadly marked by the initial and final setting time, for practical purpose. The cement varieties currently available in market mostly have a minimum initial setting time above 60 min and the maximum final setting time in the range of 3–5 h. The limits of initial (minimum) and final (maximum) setting time are specified separately by the national cement standards. The period is marked by loss of workability, beginning with the initial setting and culminating with the final setting, when the workability is completely lost. The curing of concrete should begin in this period and extend through the hardening period, as required by the standards. The construction engineer wants the initial setting time to be reasonably long so that all operations up to the placing and finishing of concrete are completed before it ends and the final setting time not to be too long so that they can resume the construction activity within a reasonable time after the placement of concrete. The concrete after final set has little or no strength.

6.2.3 Hardening Period

It covers two stages of hydration.

Stage IV—Deceleration period: Rate of hydration reaction substantially decreases with time.

Stage V—Slow reaction period: Rate of hydration reaction is low and does not change significantly with time.

This is a period when strength development takes place in concrete. The porosity and the permeability of the concrete matrix decrease, as shown in the trend in Figure 6.1. All the national standards consider 28 day compressive strength as an indicator of the final (more than 90% of the specified) compressive strength. Whilst the ultimate compressive strength may take some years to develop, at least 60%–65% hydration reaction is complete during this period, as signified by the reaction ratio in this period. It should be noted that almost the entire strength development takes place only after the final set, that is, after the culmination of stage III.

The start, end, and the duration of the workability, setting, and hardening periods are affected by the phase composition and fineness of cement, mineral, and chemical admixtures and the hydration temperature.

6.3 Reactants in Hydration Process

The following four main reactants take part in the hydration process of cement containing mineral admixtures:

a. Reactive compounds in cement clinker: C_3S, C_2S, C_3A, and C_4AF

b. Calcium sulfate (gypsum)

c. Reactive compounds in mineral admixtures (depending upon the type): silicates, aluminosilicates, calciumaluminosilicates, and amorphous silica

d. Calcium hydroxide

There are other reactants also, which take part or assist in the reaction, such as chemical admixtures and minor cement constituents like alkali sulfates; however, their role has not been discussed in this chapter. The readers interested to study details of the chemistry of cement and concrete may refer the book by Lea [5]. The brief summary of the nature and the reactivity of these reactants is given in the following paragraphs. The Section 6.3.1 covers the reactive compounds in cement clinker: tricalciumsilicate (C_3S in 6.3.1.1), dicalciumsilicate (C_2S in 6.3.1.2), tricalciumaluminate (C_3A in 6.3.1.3), and tetracalciumaluminoferrite (C_4AF in 6.3.1.4). Section 6.3.2 characterizes gypsum or hydrated calcium sulfate, which is added to cement as a set regulator. Section 6.3.3 deals with the reactive compounds in mineral admixtures, namely, those obtained in the amorphous phase. Section 6.3.4 discusses the role of CH, a product of C_3S and C_2S hydration, in secondary hydration.

6.3.1 Reactive Compounds in Cement Clinker

PC consists of four major reactive chemical compounds: tricalciumsilicate or alite (C_3S), dicalciumsilicate or belite (C_2S), tricalciumaluminate (C_3A) or aluminate, and tetracalciumaluminoferrite or ferrite (C_4AF)—as shown in Table 6.1, together with a small amount of minor compounds. Plate 6.1 is a reflected light micrograph showing these reactive phases [6]. In the test reports and technical literature, the chemical composition of clinker is expressed in terms of oxides. The range of oxide composition in PC is also given in Table 6.1. The potential compound analysis may be obtained from the oxide analysis, using Bogue formulae [4]. The four phases shown in Table 6.1 are not pure, as the chemical formulae may suggest, but they contain impurities in solid solution, like Na, K, Mg, Fe, Al, Si, Ti, Mn, S, and P. The secondary phases that may occur are (a) periclase (crystalline MgO); the hydration of periclase to magnesium hydroxide is a slow and expansive reaction and may, in certain conditions, cause unsoundness in cement;

TABLE 6.1

Major Clinker Phases and Approximate Range of Oxide
Composition in PC

Major Clinker Phases		Oxide Composition (% Range)	
Notation and Name	Chemical Formula	Oxide	Range
C_3S, tricalciumsilicate (alite)	$3CaO \cdot SiO_2$	CaO	60–67
C_2S, dicalciumsilicate (belite)	$2CaO \cdot SiO_2$	SiO_2	17–25
C_3A, tricalciumaluminate (aluminate)	$3CaO \cdot Al_2O_3$	Al_2O_3	3–8
		Fe_2O_3	0.5–6.0
C_4AF, tetracalciumaluminoferrite (ferrite)	$4CaO \cdot Al_2O_3 \cdot Fe_2O_3$	MgO	0.1–5.5
		$K_2O + Na_2O$	0.5–1.3
		SO_3	1–3

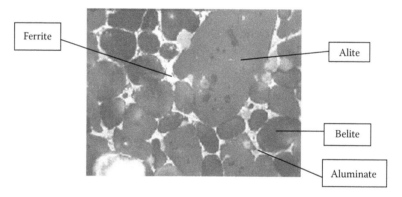

PLATE 6.1

Reflected light micrograph of clinker showing alite, belite, ferrite, and aluminate phases.
Note: Alite occurs as subhedral (having some crystal faces and some grainy surfaces) to anhe-
dral crystals (crystals without a defined external shape) approximately 25 μm in size. Belite
occurs in large clusters with an approximate crystal size of 15 μm. A medium- to fine-grained
lath-like ferrite, with aluminate filling the inter-lath voids, form the interstitial constituents.
Field width: 250 μm. (From Stutzman, P.E. and Leigh, S., Compositional analysis of NIST reference material clinker 8486, in *Accuracy in Powder Diffraction III, Proceedings*, National Institute
of Standards and Technology, Poster #2, Gaithersburg, MD, April 22–25, 2001.)

(b) free lime (CaO) occurs when the raw mix in cement manufacture contains lime (in the form of calcium carbonate) in excess of that which can
be chemically combined or due to improper grinding, homogenization and
burning of raw mix; and (c) alkali sulfates (Na_2SO_4, K_2SO_4). SO_3 as given in
Table 6.1 is mainly in the form of gypsum and alkali sulfates. Most national
cement standards specify upper limits for MgO (magnesia), free lime (CaO),
and SO_3 content. The alkalies (K_2O and Na_2O) are often expressed in terms
of Na_2O equivalent ($Na_2O + 0.658K_2O$). Even a small change in the oxide

composition may actually be a result of relatively big change in the phase composition of clinker. The PC with specific characteristics with regard to hydration, strength, and performance may be manufactured depending upon the distribution of four main phases in the clinker, particle fineness, and size distribution.

6.3.1.1 Tricalciumsilicate

Tricalciumsilicate (C_3S) or alite is the most important component of cement, responsible for the engineering properties. The PC clinker may contain about 40%–60% C_3S. Pure C_3S has a specific gravity of 3.15. The alite crystal occurs in triclinic (having three unequal axes, all intersecting at oblique angles), monoclinic (having three unequal axes, with two perpendicular and one oblique intersection), or rhombohedral (having three equal axes, all intersecting at oblique angles) forms but the one obtained in the PC clinker is mostly monoclinic. These crystal polymorphs are the slight distortions of a simple structure built from SiO_4 tetrahedra (plural of tetrahedron, is a polyhedron composed of four triangular faces, three of which meet at each vertex) and calcium and oxygen ions. The reactivity of polymorphs, at the early stages of hydration, is in the following order: monoclinic < triclinic < rhombohedral. Harada [7] found that the difference in the reactivity reduced with time and at 90 days the degree of hydration of all three polymorphs was about 70%.

The C_3S formed in clinker is not pure and contains various other oxides but mainly MgO, Al_2O_3, and Fe_2O_3, in solid solution. The reactivity of alite is due to the crystal defects created by irregular ionic packing, on account of irregular coordination of Ca and O atoms and the oxide impurities [5]. These oxides have some influence on the strength and the durability properties of cement. The morphology of C-S-H formed after the C_3S hydration depends upon which polymorph is hydrated, as shown in Table 6.2 [7].

The hydration reaction of C_3S is exothermic and can be stoichiometrically* expressed by Equation 6.1 or schematically by Equation 6.3 [8]:

$$3CaO \cdot SiO_2 + 5.3H_2O \rightarrow 1.7CaO \cdot SiO_2 \cdot 4H_2O + 1.3Ca(OH)_2 \qquad (6.1)$$

The composition of C-S-H given in the Equation 6.1 is approximate. The mean C/S ratio in the C-S-H produced may vary between 1.70 and 1.75.

---

* Stoichiometry defines the quantitative relationship between reactants and products in a chemical reaction. In a reaction, the amount of reactants and products is frequently expressed as moles or gram moles. One gram mole is mass in grams numerically equal to the molecular weight of a substance. If the mass of a chemical entity is g grams, the gram moles n are defined as: (mass in grams)/(molecular weight of the entity).

TABLE 6.2

Morphology of C-S-H vis-à-vis C_3S

C_3S	C-S-H
Rhombohedral	Thin sheets
Monoclinic	Fibrous
Triclinic	Amorphous

The exothermic heat evolved is about 122 kcal/kg, but the total heat evolution takes long time period [4]. The rate of heat liberation is important from the point of view of setting, strength and durability properties. It is seen that nearly 50% heat is liberated in the first 3 days. The quantity and the rate of heat liberation becomes a hindrance in mass concrete structures due to the development of unwanted temperature gradients but can also be helpful when concreting is done at temperatures too low to provide activation energy for hydration. The hydration is accelerated in the presence of sulfate ions in the solution. It is known that gypsum (calcium sulfate) is added to cement to retard the hydration of aluminates, but the welcome side effect is the acceleration of alite hydration.

6.3.1.2 Dicalciumsilicate

Dicalciumsilicate (C_2S) or belite is another important component of cement clinker, next to alite. The PC clinker may contain about 12%–30% C_2S. Pure belite is found in four polymorphs: β, γ, α', and α. Out of these, PC contains mainly β and occasionally α' and γ forms. The reactivity of β polymorph is due to its defective crystal structure on account of irregular coordination of Ca ions. The structure contains interstitial spaces but smaller than those obtained in the C_3S structure, which makes it less reactive than the latter. The metastable polymorph* is stabilized by the presence of impurities. On the contrary, the structure of γC_2S is regular, which makes it nonreactive. The specific gravity of βC_2S is 3.28 [5]. The hydration rate of belite is low in comparison to that of alite; hence, it does not take part in the development of concrete properties in the early stages of cement hydration.

Like C_3S, the hydration reaction of C_2S is also exothermic and can be stoichiometrically expressed by Equation 6.2 or schematically by Equation 6.3 [8]. Like in C_3S hydration, the composition of C-S-H given in the Equation 6.2 is approximate. The mean C/S ratio in the C-S-H produced may vary between 1.70 and 1.75.

$$2CaO \cdot SiO_2 + 4.3H_2O \rightarrow 1.7CaO \cdot SiO_2 \cdot 4H_2O + 0.3Ca(OH)_2 \qquad (6.2)$$

---

* Polymorphism is often characterized by the ability of a substance to exist in two or more crystalline phases that have different arrangements and/or conformations of the molecules in the crystal lattice. A metastable solid form can change a crystalline structure in response to the change in environmental conditions, in processing, or over a time period.

or

Alite/belite + water → calcium silicate hydrate (C-S-H) + calcium hydroxide

(6.3)

The exothermic heat evolved is about 59 kcal/kg, but the total heat evolution takes a long time period. Like alite, the hydration of belite is also accelerated in the presence of sulfate ions (gypsum) in the solution.

6.3.1.3 Tricalciumaluminate

Tricalciumaluminate (C_3A) or aluminate constitutes the most important alumina bearing phase in PC. The PC clinker may contain 3%–14% aluminate. According to Lea [5], the aluminate in the PC clinker occurs in equant grains (crystal having all axes of equal length). It belongs to the cubic or isometric system and has a specific gravity of 3.04. C_3A is formed as a result of reaction between calcium oxide and alumina (Al_2O_3), at a high temperature in the rotary kiln, during cement manufacture. Finely ground C_3A rapidly reacts with water, because of its high solubility. There are many kinds of hydrate in the $CaO-Al_2O_3-H_2O$ system but most of them are metastable and the only stable hydrate is C_3AH_6. Large amount of heat is evolved in the hydration, at a very rapid rate. The cement will be useless for any construction purpose unless the rate of heat liberation is slowed down by some means. That is achieved by addition of gypsum to cement.

The hydration of C_3A in the presence of calcium sulfate and CH, as obtained in cement, is quite different. The hydration of C_3A is depressed in the presence of sulfate ions in the solution, due to reduction in its solubility. Gypsum is used as a set-regulating additive in PC, due to this property. In cement hydration, C_3AH_6 is not formed but instead a trisulfate (ettringite, $C_6A\hat{S}_3H_{32}$) and a monosulfate ($C_4A\hat{S}H_{12-18}$) are formed, in the presence of gypsum. Upon depletion of gypsum in the hydration reactions, renewed hydration of C_3A (when C_3A in cement >5%) and C_4AF takes place and the ettringite gets converted to monosulfate. In PC, the heat of hydration of C_3A is reported to be 324 cal/g and nearly 65% of that is evolved in the first 3 days. The ratio $A/\hat{S}$ determines the setting behavior of cement and has practical importance. A detailed discussion on this aspect is given by Mehta and Monteiro [4].

The C_3A in cement can get prehydrated under humid conditions. The prehydration can change the course of cement hydration, disturbing the sulfate (gypsum) balance and also affect the interaction of chemical admixtures with cement. Breval found that if the bound water is less than 3% (m/m) the prehydration products do not affect the subsequent hydration in water. However, if the bound water is more, it has a profound effect on the subsequent hydration, leading primarily to the formation of C_3AH_6 [9].

6.3.1.4 Tetracalciumaluminoferrite

Tetracalciumaluminoferrite (C_4AF) or ferrite is a ternary phase (three component system) occurring in cement clinker, resulting from the solid solution of calcium aluminate and calcium ferrite (C_2A-C_2F). The PC clinker may contain around 7%–16% C_4AF. In its pure form, it has a specific gravity of 3.77. The hydration of C_3A and C_4AF is similar, both in the presence and absence of sulfates, although the retardation effect of sulfate ions on C_4AF is stronger in comparison to that on C_3A. The ferrite reacts rapidly with water, next only to aluminate. The rate of reaction increases when, during the cement manufacturing process, the A/F ratio in the raw mix is increased and the temperature of formation is decreased [4]. At temperatures above 15°C, pure ferrite hydrates to $C_3AH_6-C_3FH_6$, which is a cubic solid solution. However, in the presence of gypsum and lime (i.e., in cement), a solid solution of sulfoaluminate and an analogous sulfoferrite hydrate is formed [5]. The heat of hydration of the ferrite phase in PC is 102 cal/g out of which nearly 68% is liberated in the first 3 days.

6.3.2 Calcium Sulfate (Gypsum)

Calcium sulfate or gypsum is added to cement, primarily to regulate (retard) the hydration (setting) of C_3A. It is obtained in three forms:

Dihydrate gypsum ($CaSO_4 \cdot 2H_2O$): Monoclinic prismatic structure; molecular weight 172 with the specific gravity of 2.3; slightly soluble in water

Hemihydrate gypsum ($CaSO_4 \cdot \frac{1}{2}H_2O$): Basic hexagonal structure; molecular weight 145 with the specific gravity of 2.74, also called plaster of Paris; good water solubility, when obtained after gypsum dehydration in cement grinding mill

Anhydrite gypsum: (a) Soluble anhydrite ($CaSO_4 \cdot 0.001-0.5H_2O$), quasi-zeolitic variant of hemihydrate with hexagonal-trapezohedric structure and a specific gravity of 2.587; good water solubility, when obtained after gypsum dehydration in cement grinding mill; and (b) Insoluble anhydrite ($CaSO_4$) with rhombic, pyramidal structure; molecular weight of 136 and the specific gravity of 2.985; more soluble than dihydrate gypsum but has lower rate of solution in water

The hemihydrate, soluble anhydrite and the so-called insoluble anhydrite is obtained when dihydrate is heated to the temperature of above 97°C, 120°C and 200°C–300°C, respectively. The high-temperature anhydrites are also available and more information on the subject is available at Ref. [10].

In the cement grinding mill, clinker is ground with dihydrate gypsum. When the temperature inside the mill chamber exceeds the dehydration limit (e.g., ball mill), the dihydrate gypsum gets converted to a mixture of dihydrate, hemihydrate, and soluble anhydrite and the proportions may change according to the temperature and the residence time. As the solubility of different dehydrated forms differs substantially, the phenomenon affects the setting characteristics of product cement, namely, the formation of ettringite during hydration [11]. There is an optimum gypsum quantity that renders desired set regulation, compressive strength, and lowest drying shrinkage characteristics. The optimum depends upon C_3A and C_4AF concentration in clinker [12] and their reactivity (function of process conditions, trace impurities), solubility of gypsum, and the temperature inside grinding mill.

A number of gypsum varieties, obtained industrially as a byproduct, are currently used or have potential for use as a set-regulating additive in the production of PC. These are desulfogypsum from flue gas desulfurization (FGD) plants in coal fired power stations, phosphogypsum from phosphoric acid manufacture, fluorogypsum from hydrofluoric acid manufacture, titangypsum from titanium manufacture, borogypsum from boric acid manufacture, citrogypsum and phenologypsum from citric acid and phenol manufacture, respectively, sodagypsum from Solvay process to manufacture soda, and salt-gypsum from salt (NaCl) manufacture from sea water [13]. The compositions of these byproducts differ substantially, with regard to the impurity content. The contaminants include the raw materials used or the products of the chemical process. Even in case of a particular byproduct gypsum, the composition has been found to vary from one consignment to another, depending upon its processing and history before delivery. These contaminants affect the properties of the byproduct gypsum and subsequently formed binders. The proportion of the dihydrate, hemihydrate and anhydrite of calcium sulfate obtained after cement grinding can vary significantly among different gypsum types, obtained from different process routes, plants and operating conditions that influence the dehydration behavior under mill operating conditions. The set-regulating and hydration characteristics of byproduct gypsum are thus affected. In some cases, byproduct gypsum has been reported to enhance the ettringite formation 2–3 times more than the pure natural gypsum. The enhanced ettringite formation is normally at the expense of early C-S-H formation. Therefore the dehydration behavior of the gypsum variety used in cement should be studied separately to obtain an indication of the composition of gypsum in ground cement [14]. The clinker reactivity is likely to have an effect on the rate of gypsum dehydration in the grinding mill [15].

Only a small quantity of gypsum, less than 30%, reacts with C_3A during hydration to form ettringite ($C_6A\hat{S}_3H_{32}$) or monosulfate ($C_4A\hat{S}H_{12-18}$). Majority of the sulfate, 70%–80%, gets incorporated in C-S-H phase helping to raise the compressive strength of cement [10].

The reactivity of gypsum in cement depends upon how quickly it goes into solution and makes available SO_4^{2-} ions for the set-regulating reactions. The three forms of gypsum differ in terms of their solubility in pure water and that containing lime. The presence of lime in the solution is found to assist the reactivity of gypsum. The hemihydrate and the anhydrite (both soluble as well as insoluble) are more soluble than the dihydrate. If the solubility is measured in terms of the $CaSO_4$ going into the solution, at 30°C, the solubility of dihydrate is 0.191 g/100 mL water, that of insoluble anhydrite is

0.209 (however the rate of solution is lesser in comparison to dihydrate) and the hemihydrate/soluble anhydrite is nearly 2–3 times more soluble. Thus in terms of solubility, the four main forms can be arranged in the descending order as hemihydrate/soluble anhydrite > insoluble anhydrite > dihydrate. The insoluble anhydrite, natural or produced by heating gypsum, is relatively less reactive, that is, its rate of solution is lesser in comparison to dihydrate, although the overall solubility lies in the same order as given above.

The hemihydrate in small quantities is useful, as it takes care of easy availability of SO_4^{2-} ions in the initial stages of cement hydration. Hence, hemihydrate is sometimes added to ground cement when the inside temperature of grinding mill does not exceed the dehydration limit. However, larger dehydration to hemihydrate further to soluble anhydrite, may result in false set, which is typically marked by low concentration of aluminate ions but high concentration of calcium and sulfate ions resulting in the formation of large crystals of gypsum and corresponding loss of consistency of the cement paste. The insoluble anhydrite is sometimes added to blast furnace cement, due to its slower rate of solution, as the lower content of C_3A in that cement (resulting from replacement of cement with blast furnace slag [BFS]) does not require much SO_4^{2-} ions in the solution in the initial stages and the greater availability of SO_4^{2-} ions at the later stages helps activation of slag.

The maximum limit of gypsum addition allowed in most national standards vary from 2.5% SO_3 (about 5% dihydrate) for the cements with low C_3A content, to 4.0% (about 8.5% dihydrate) for cements with high C_3A. According to Lea [5] the set-regulating characteristics of gypsum are not proportional to the amount added but show abrupt changes beyond certain proportions.

The set-regulating action of gypsum involves two mechanisms: (i) the precipitation of insoluble sulfoaluminate hydrates, namely, ettringite and monosulfate, and (ii) subsequent retardation of the hydration of C_3A due to the coating of sulfoaluminate film around it. The second action of gypsum is the acceleration of the hydration of alite and belite, due to the increased calcium and consequently reduced hydroxyl ion concentration in the solution.

6.3.3 Reactive Compounds in Mineral Admixtures

The glass forms a major part of mineralogical composition and the reactive phase in PFA and BFS. Under normal operating conditions, modern thermal power plants produce PFA containing more than 70% glass. Mehta [16] classified the mineral admixtures for their pozzolanic/cementitious activity, on the basis of their mineralogical and particle characteristics, for incorporation in the concrete. The PFA obtained from thermal power stations is dry and powdery, with 70%–90% particles <45μ (<45 × 10⁻³ mm size, surface area 2000–4000 cm²/g, Blaine). It may contain up to 4% unburned carbon. The glass particles may also contain cenospheres (hollow spherical particles) and plerospheres (spherical particles filled with smaller particles). The low calcium PFA (ASTM Type F), obtained from the combustion of anthracite and bituminous coal, contains aluminosilicate glass; that seems to be somewhat less reactive than the calcium aluminosilicate glass present in the high calcium PFA (ASTM Type C), obtained from the combustion of lignite and subbituminous coal. The crystalline components, like quartz, mullite, sillimanite, hematite, and manganite, present in the low-calcium PFA do not contribute to its pozzolanic properties. On the other hand, the crystalline phases obtained in high-calcium PFA, like tricalciumaluminate, calciumaluminosulfate, calcium sulfate anhydrite, and free CaO, are reactive except quartz and periclase (free MgO). The high-calcium PFA shows both pozzolanic and cementitious properties [17]. Tenoutasse et al. [18] suggested the following three-step pozzolanic reaction model, based on the microscopic studies.

i. Migration of ions from aqueous phase to reaction sites at the surface of the particles
ii. Precipitation of $Ca(OH)_2$
iii. Reaction between $Ca(OH)_2$ and vitreous phase of PFA leading to formation of C-S-H and stratlingite (C_2ASH_8)

The PFA glass is expected to display an auto-catalytic behavior when it dissolves, which means dissolution of alkaline PFA will cause relatively high pore water pH, further enhancing the PFA glass breakdown. The dissolution of PFA glass in the alkaline medium is expected to follow first order* kinetics [19], represented by the Equation 6.4:

$$\left(\frac{dC_A}{dt}\right) = k(C^* - C_A) \tag{6.4}$$

where
C_A is the measured concentration of glass
C^* is the maximum concentration of glass
k is the rate constant
t is the time

The granulated BFS, obtained from the blast furnace in the integrated steel plants, is dried and ground to the fineness of <45 μ (<45 × 10^{-3} mm size, surface area 4000–5000 cm²/g, Blaine) before use. Like PFA, the hydration and the activity of BFS in cement has also been studied extensively. The factors that affect the reactivity of BFS glass are particle size and size distribution, content and the structure of glass, and the condensation degree of silicate ions (level of combination of silicate with other elements, like Ca, Al) in glass phase [20]. The glass content of slag, used for blending with PC and

---

* In chemical kinetics, the order of reaction with respect to a certain reactant is defined as the power to which its concentration term is raised in the rate equation.

concrete, is generally more than 85%. Besides the chemical composition, the glass content depends upon the processing of BFS from the point it is removed in molten state from the blast furnace up to the granulation stage. The study of phase diagram of SiO_2-Al_2O_3-CaO-MgO system helps in better understanding of BFS glass. The glass phases with melilite composition ($C_2A_xM_{1-x}S_{2-x}$, x varies from 0 to 1) have been found to give the highest hydraulic activity. The glass-modifying elements like Al and Mg have been found to increase the disorder (and entropy) in glass structure and facilitate its dissolution during hydration (to achieve the state of minimum entropy). A more detailed discussion on the subject is given in Ref. [20].

It has been observed that at the time of hydration, a distinct outward migration of Si, Ca and Al ions occurs from the slag grain to the slag-cement paste. Thus the slag grains get enriched in Mg. Pietersen [19] calculated the thermodynamically stable phases using a mathematical model. According to the model, at high percentages of slag (55%–75%), as obtained in most commercial slag cements and also in many concretes, thermodynamically stable phase assemblages consist of about 50% C-S-H with a very low C/S ratio of about 1.0, about 20% siliceous hydrogarnet ($C_3AS_{0-1}H_{6-4}$), about 20% stratlingite (C_2ASH_8), and about 8%–10% hydrotalcite ($M_{4-6}AH_{10-13}$). The limitations of model in predicting the hydrated phases in comparison to those actually observed in practice, have also been given by the author.

The chapters related to the admixtures provide discussion on the chemical characteristics of glass in PFA (Chapter 1) and BFS (Chapter 2), amorphous silica in SF (Chapter 3), and the aluminosilicates in MK (Chapter 5).

6.3.4 Calcium Hydroxide

CH is obtained as a product of hydration of tri- and dicalcium silicates in the cement (Equations 6.1 through 6.3). It plays an active role in the so-called secondary hydration of the reactive compounds present in the mineral admixtures added to the cement and concrete and forms C-S-H with low C/S ratio. The pure CH has a specific gravity of 2.30. It crystallizes in the hexagonal form as plates or short hexagonal prisms. Plate 6.2 shows SEM image of platy CH and the ettringite needles formed during the hydration of cement [21]. As reported by Lea [5], the solubility of CH decreases with temperature. At 25°C, the saturated solution of CH has a pH of 12.45 and contains 1.14 g CaO/L of water. This solubility is much reduced in the presence of alkali hydroxides, like NaOH and KOH. According to Mehta [4], the hydrated paste of ordinary PC contains 20%–25% v/v of CH crystals (also called portlandite). The CH crystals have considerably large size (2–3 times larger than C-S-H particles) and lower surface area, hence the strength contribution, which is mainly due to van der Waals forces, is lower. The presence of considerable amount of CH in the hydrated PC has an adverse effect on the chemical durability of concrete to acidic solutions (like chlorides and sulfates).

PLATE 6.2
SEM image showing hexagonal CH and ettringite needles. (From Petrographic methods of examining hardened concrete: A petrographic manual, FHWA-HRT-04-150, Federal Highway Administration, U.S. Department of Transportation, Washington, DC, 2006.)

The foregoing discussion on the reactants reveals that the four principal reactants (6.3.1 through 6.3.4) participate in the hydration of cement with mineral admixture and mostly contribute toward the formation of the principal strength giving compound, that is, C-S-H. The relative proportion of these reactants and the condition at the time of hydration control the properties of fresh concrete and that of hardened concrete in terms microstructure, strength and the durability. The cement clinker compounds (6.3.1.1 through 6.3.1.4) have different reactivity in terms of the rate of reaction as well as the heat of reaction and they can be arranged in the descending order as: $C_3A > C_4AF > C_3S > C_2S$. They are obtained in different proportions in the clinker, the highest being the C_3S or the alite, followed by C_2S or belite, and the others. The hydration products of these reactants in pure form are different from those in presence of gypsum and lime obtaining in PC. Their reactivity is mainly due to defective crystal structure, impurities and solubility in water. The gypsum (Section 6.3.2) is added to cement, primarily as set-regulating agent. The four main varieties—dihydrate, hemihydrate, soluble anhydrite, and insoluble anhydrite—have different crystal structure, physical, and chemical properties including solubility. The reactivity of gypsum in cement depends upon how quickly it goes into the solution and makes available sulfate (SO_4^{2-}) ions for the set-regulation reaction. The four gypsum varieties have different solubility (and the rate of solubility) in water and water containing CH. The hemihydrate and soluble anhydrite are 2–3 times more soluble, whereas the insoluble anhydrite has more solubility but lower rate of solution, as that of dihydrate. The difference in their solubility characteristics can be used to impart desired setting and hardening properties to cement. The gypsum obtained as a byproduct from different chemical industries, the so-called chemical gypsum, may be used as a set-regulating agent in cement. The impurities in gypsum affect its set-regulating and hydration characteristics. Nearly 70%–80% of it gets incorporated in the C-S-H phase, helping to raise the compressive strength of cement. The reactive compounds such as silicates, aluminosilicates, calciumaluminosilicates, amorphous silica form the major portion of the composition of mineral admixtures. The silicate glass (Section 6.3.3) forms a major part of PFA and BFS composition. The SF and RHA contain 85%–98% silica in the amorphous form and the MK contains more than 80% reactive aluminosilicates. Under normal operating conditions, PFA contains more than 70% and BFS more than 85% glass. Amongst the two, BFS reactivity is more sensitive toward the glass structure and the content. Besides the chemical composition, glass content of BFS depends upon its processing from the point it is removed in molten state from the blast furnace up to the granulation stage. The modifying elements like Al and Mg have been found to increase the disorder in the solid structure and facilitate its hydration. CH (Section 6.3.4), obtained as a product of hydration of tri- and dicalcium silicates, plays a major role in the so-called secondary hydration, along with the reactive compounds present in the mineral admixtures. The presence of excess amount of CH, remaining after the hydration, has an adverse effect on the resistance of concrete to the acidic solutions (like chlorides and sulfates).

6.4 Voids in Hydrated Cement Paste

The concrete strength, durability, and volume stability are greatly influenced by voids in the hydrated cement paste. Besides those due to entrained or entrapped air, two types of voids are formed in hydrated cement paste:

a. *Interlayer hydration space (gel pores)*: It refers to the space between layers of C-S-H with thickness between 0.5 and 2.5 nm (nanometer is 10^{-9} m) and can contribute 28% of paste porosity. It has little impact on strength, permeability, or shrinkage.

b. *Capillary voids*: Depend on initial separation of cement particles, controlled by water-to-cement ratio (w/c). The size is of the order of 10–50 nm, although larger for higher w/c. Larger voids affect strength and durability (permeability), whereas smaller voids impact shrinkage.

6.5 Interrelation of the Hydration Properties of Cement and Concrete

The use of cement in practice, that is, in concrete, requires it to be mixed with aggregate (coarse and fine), water, chemical, and mineral admixtures; thoroughly blended; transported; placed; compacted; and cured properly. The quality of workmanship in any of these areas will affect the strength and durability of concrete. On the other hand, while studying the properties of cement in the laboratory, small specimens of neat cement paste or sand cement mortar are often used. Thus, the method of test eliminates one major component of concrete, namely, the aggregate especially the coarse aggregate. The properties of cement tested in this way cannot be always extrapolated to those of concrete in practice, for the following main reasons:

a. Unlike cement paste or mortar, the concrete comprises of three distinct phases: the binder, the aggregate, and the interfacial transition zone (ITZ) between the binder and the aggregate. The properties of concrete are a cumulative result of these major phases.

b. In cement paste or mortar, the quantity of entrained air is very little. On the other hand, normal concrete may contain 3%–6% (v/v) air entrapped during the process of mixing, placing and compaction or it may be artificially entrained, using a suitable air-entraining agent (AEA). The quantity of entrained air, including the size of air bubbles and their distribution in the concrete matrix, affect the properties of fresh concrete (workability and bleeding), and that of hardened concrete (strength and durability under freezing and thawing conditions). At a given degree of hydration, the porosity and strength of concrete is decided not only by the water/cement ratio but also by the content of air, when air voids are incorporated into the system. As a rule of thumb, in medium- and high-strength concrete, every 1% increase in air content reduces the strength of concrete by about 5% [4].

c. The course aggregates introduce inhomogeneity in the cement paste. The properties of aggregates differ from those of cement paste in respect of the coefficient of thermal expansion (or contraction), modulus of elasticity and Poissons ratio. The inhomogeneity affects the behavior of concrete not only under load but also under drying conditions. The coarse aggregate may propagate or arrest the microcracks, depending on its interaction with the cement paste.

d. The microstructure of ITZ is considerably different from that of the bulk cement matrix. In comparison, ITZ is less compact (more porous) and contains higher proportion of CH. The cement hydration products at the vicinity of aggregate particles have been found loosely bonded to one another and to the aggregate particles. In comparison to the aggregate particles and the bulk cement paste, the ITZ has the lowest hardness value [22]. As a result, the density of fractures and microcracks is higher in ITZ. Thus, in a normal concrete, the ITZ may be considered as a weakest link.

6.6 Properties of Concrete during Early Stages of Hydration

The presence of mineral admixtures affects the properties of concrete during early stages of hydration. These properties are required to be taken into consideration while transporting, placing, consolidating and curing operations. Therefore, it will be worthwhile briefing on the properties of concrete at early age or during the early stages of hydration. The definition of early age as given by Mehta and Monteiro [4] is found to be the most appropriate for the purpose of this chapter. It includes, on the one hand, the freshly mixed concrete of plastic consistency and, on the other, 1–2 day old hardened concrete that is strong enough to be left unattended (except continuation of moist curing). In general, the early-age concrete properties include the following:

6.6.1 Workability

The term "workability" is only broadly defined by different authors and the national standards. The American Concrete Institute [23] describes workability as "that property of freshly mixed concrete or mortar that determines the ease with which it can be mixed, placed, consolidated and finished to a homogenous condition." The requirement of concrete workability depends on the nature of its application; pavement quality concrete, for example, requires low workability, as appropriate for the application. The characteristics of admixtures, both mineral and chemical, those of the aggregate and the content of water and cement affect workability of concrete. The literature mentions number of methods but no single test method measures all aspects of workability [24]. The determination of workability by "compaction factor" and "slump" is common and also recommended by many national standards, the increased knowledge on concrete rheology notwithstanding.

The loss of workability or the slump loss, is defined as the loss of consistency of concrete with time [25]. It is directly related to the depletion of the free water in fresh concrete. The depletion begins as soon as water is added and the mixing operation of the concrete ingredients initiates. The free water in fresh concrete is used up mainly by the processes of cement hydration and the evaporation [26]. The slump loss creates difficulty in handling and finishing of concrete during placement and compaction and may result in reduced ultimate strength and the loss of durability [27]. The mineral as well as chemical admixtures affect slump loss.

6.6.2 Yield Stress and Viscosity

These are the rheological properties of concrete. Concrete is often described as Bingham fluid with a plastic viscosity μ_p (Pa s). That means there exists a minimum value of the stress (τ), known as yield stress (τ_0, Pa), applied to the material for irreversible deformation and flow to occur. The behavior of fresh concrete in steady state is approximated by a yield stress model of the general form, as given in the Equations 6.5 through 6.7 [27]:

$$\gamma = 0 \quad \text{when } \tau < \tau_0 \tag{6.5}$$

$$\gamma \neq 0 \quad \text{when } \tau = \tau_0 + f(\gamma) \tag{6.6}$$

where
  γ is the shear rate (s^{-1})
  $f(\gamma)$ is a positive, continuously increasing, function of the shear rate

$$f(0) = 0 \quad \text{and} \quad f(\gamma) = \mu_p \gamma \tag{6.7}$$

The readers interested in learning more about the rheological properties of concrete may refer to the publication by Tatersall and Banfill [28].

Kovler and Roussel took excellent review of the developments in the understanding of fresh concrete properties. The most common test for the workability of fresh concrete, "slump test," can be correlated, under specific conditions, to the yield stress of a given concrete [29,30]. Whereas the measurement of plastic viscosity has several practical applications, like pumping and casting rate, yield stress is an important parameter for formwork filling in practice [31]. During an industrial casting process, a purely viscous fluid (i.e., no yield stress) would self level under the effect of gravity and the viscosity of the material will dictate the time needed to obtain a horizontal surface. In the case of concrete, if the shear stress generated by gravity during casting, which is a complex function of formwork shape and local density of steel reinforcements, becomes lower than the yield stress of the concrete (τ_0), flow may stop before the concrete self levels or before the formwork is entirely filled.

It is expected that the understanding on the viscosity of concrete shall gain more importance, in future. The trends in mix design show a reduction in the clinker content and the use of mineral admixtures as well as recycled and crushed aggregate in concrete, for environmental reasons. In order to obtain good properties in terms of durability, setting time and mechanical strength, the amount of water in the system is also reduced. The viscosity increases with the relative increase in solid content, as the contact between particles plays an increasing role. The superplasticizer affects yield stress but the knowledge on how to reduce viscosity is far less developed.

6.6.3 Bleeding and Laitance

The bleeding is appearance of water on the surface, after concrete is placed and compacted but before the setting process starts. It is a result of segregation of water from the solids. Such segregation prevents full compaction of concrete, essential to reach its maximum strength potential. Concrete seldom bleeds uniformly. The bleeding is predominantly observed in a highly wet, badly proportioned and insufficiently mixed concrete. It is observed more in thin members like roof slab or road slabs. The bleed water, instead of appearing on the surface, gets often entrapped under the coarse aggregate or the steel reinforcement, especially in the upper portion, making it weaker in comparison with the lower portion of concrete. The laitance is the residue of weak and nondurable material consisting of cement, fines from aggregate or impurities brought to the surface of wet concrete by the bleeding water. Early bleeding, when the concrete mass is fully plastic, may not cause much harm as concrete in that condition gets subsided and compacted. It is the delayed bleeding, when concrete loses its plasticity, that causes harm to concrete. The controlled revibration, while concrete is in plastic state, may be adopted to overcome the bad effects of bleeding. The use of finely divided mineral admixtures reduces bleeding, creating a longer path for the water to traverse. The ASTM C232 [32] gives standard test for the measurement of rate of bleeding and the total bleeding capacity of a concrete mixture.

6.6.4 Setting Time

The setting of concrete is defined as the onset of solidification in fresh concrete mixture. It refers to stiffening of concrete without significant increase in strength. The ASTM C403 [33] arbitrarily defines the initial and final setting time of concrete, using penetration resistance method. The setting time of cement and concrete do not coincide.

The initial setting time defines the limit of handling and the final setting time defines the beginning of the development of mechanical strength for concrete [4]. In respect of pavement construction, to cite an example of the application of setting time concept, Schindler [34] observed that with the knowledge of initial setting time contractors will be able plan measures to finish, texture and saw cut concrete pavements on time. The principal factors controlling the setting time are cement composition, water/cement ratio, temperature, and the admixtures, both chemical and mineral.

6.6.5 Concrete Temperature

The temperature of concrete is affected due to the heat generated from the hydration reactions and the ambient condition. Under adiabatic condition (no heat gain from or loss to the surrounding), the heat of cement hydration is sufficient to maintain satisfactory curing temperature, provided that concrete is delivered at proper temperature (defined by the ambient condition). In hot weather, concrete may be subjected to plastic shrinkage and cracking, while in cold weather the rate of strength development may be lowered. The national standards define the curing practices for concrete placed under different conditions [35,36].

6.6.6 Volume Changes

The volume of concrete changes when it goes from fresh to hardened state. The change occurring in the early age is mainly due to the temperature fluctuations, drying, and chemical reactions.

The temperature rises when the heat is released during hydration reactions and the concrete expands and later shrinks, when it gets cooled due to heat loss. Some of the thermal expansion is elastic, since the concrete returns back to its original dimensions, upon subsequent cooling; however nonelastic behavior in some cases may result in early age shrinkage. The thermal expansion causes problems, when the rate of temperature change is too severe and when large gradients exist over the concrete cross-section.

The volume change or shrinkage due to drying occurs in plastic stage, when concrete loses water due to evaporation. It is also known as the plastic shrinkage. In the initial stage, bleeding water rises to the surface. The evaporation of water also continues. The shrinkage occurs when the rate of evaporation exceeds the rate of bleeding. The magnitude of drying shrinkage is highly dependent on the amount of water lost and the rate of evaporation. When the bleeding rate exceeds the evaporation rate, excess water acts as a curing blanket. In that case, there will be no drying shrinkage, as water is available on the surface to allow for evaporation, without drawing extra water from the internal capillary pores.

The hydration reactions consume water and the volume of hydration products is also lower. When concrete is plastic in the early stages, it results in overall volume reduction, known as autogenous shrinkage. It is defined as the macroscopic volume change occurring with no moisture transferred to the exterior surrounding environment [37]. The chemical shrinkage continues over a longer period, even after the concrete is hardened.

The chemical shrinkage and the heat of hydration are both valid indicators of early-age hydration. The studies conducted on ASTM Type I and Type II cement [38], which differed widely in fineness, showed that the hydration and therefore the early-age properties of coarse and finely ground cement differ. While the coarser cement exhibits compressive strength well below that of the finer cement at all ages, it also releases less heat and results in a substantially lower temperature rise. The coarse cement system exhibits lower risk of cracking at early ages as compared to the fine cement system. The high-early-age-strength cement will generally increase early-age cracking due to both the thermal and autogenous deformation. The partial substitution of cement with mineral admixtures generally reduces heat generation at the early ages. However the addition of silica fume (SF) to cement gives increased heat output during early hydration, due to the rapid and strong lime-pozzolana reaction [39].

6.7 Hydration Reactions and Changes in Early-Age Concrete Properties

In order to get a proper understanding of the reactions that occur during hydration of cement with mineral admixtures, it will be appropriate to begin with understanding the hydration reactions of the other building materials like the mixture of lime, sand, and pozzolan as well as that of the PC, without these admixtures. Section 6.7.1 takes a very brief review of the hydration reactions of building materials other than PC and Section 6.7.2 describes the hydration reactions of PC. Section 6.7.2.1 focuses on the microstructure of calcium silicate hydrate or the C-S-H, which is the principal hydration product as well as the main strength giving phase in the hardened concrete. The major reactions that occur during hydration have been grouped in three time periods, as discussed earlier. Section 6.7.2.2 covers the workability period and explains two important phenomena sometimes observed: the false set and the quick or flash set. Section 6.7.2.3 discusses the main reactions occurring in setting period, that is, the hydration of C_3S and C_2S, signifying the loss of workability. Section 6.7.2.4 deals with the hardening period, signifying the development of strength in concrete. The formation of C-S-H and CH, which begins in the setting period, is completed and the trisulfate hydrate (Af_t) or the ettringite gets converted to the monosulfate hydrate (Af_m) phase in this period. Section 6.7.2.5 compares the hydration of alite (C_3S) and belite (C_2S) in terms of the formation of C-S-H, CH and the heat liberation. Section 6.7.3 discusses hydration reactions of cement with mineral admixtures, and five varieties are covered: pulverized fuel ash (PFA), BFS, SF, rice husk ash (RHA), and metakaolin (MK). Section 6.7.3.1 covers hydration of cement with PFA. It explains how C-S-H and monosulfate is formed through secondary hydration of active silica and alumina (obtaining in PFA) with CH and gypsum. The hydration mechanism in high volume fly ash concrete (HVFAC) is also discussed. It dwells upon the effect of cement-PFA hydration on the early-age properties of concrete. Section 6.7.3.2 incorporates hydration of cement with BFS. The role of Al_2O_3 has been discussed. Sections 6.7.3.3 and 6.7.3.4 cover the hydration of cement with SF and RHA, respectively. Both the mineral admixtures essentially consist of amorphous silica. Section 6.7.3.5 deals with the hydration of cement with MK. The principal reaction is between MK, that is, $Al_2O_3 \cdot 2SiO_2$ or AS_2 and CH.

6.7.1 Hydration Reactions of Cementitious Materials other than Portland Cement

The archeological evidence shows that a cementitious material prepared by intimate mixing of slaked lime, sand, and pozzolan (such as volcanic tuffs, powdered tiles or pottery) was used in the ancient structures. The lime was the binder. The use of pozzolan enhanced the strength and durability properties. The reaction of CH (slaked lime) with atmospheric carbon dioxide produced calcium carbonate (Equation 6.8), which is stable in wide range of environmental conditions. The dissolution of carbon dioxide in water forms carbonic acid, H_2CO_3 (Equation 6.9). It reacts with CH to form more calcium carbonate (Equations 6.10 and 6.12). The calcium carbonate also dissolves in it (Equation 6.11) and recrystalizes after further interaction (Equation 6.12). The process of dissolution and recrystallization produces larger crystals of the preferred orientation.

$$Ca(OH)_2 + CO_2 \rightarrow CaCO_3 \qquad (6.8)$$

$$H_2O + CO_2 \rightarrow H_2CO_3 \qquad (6.9)$$

$$Ca(OH)_2 + 2H_2CO_3 \rightarrow Ca(HCO_3)_2 + 2H_2O \qquad (6.10)$$

$$CaCO_3 + H_2CO_3 \rightarrow Ca(HCO_3) \qquad (6.11)$$

$$Ca(OH)_2 + Ca(HCO_3)_2 \rightarrow 2CaCO_3 + 2H_2O \qquad (6.12)$$

The hydration products observed with pozzolan are calcite, analcimes ($NaAlSi_2O_6 \cdot H_2O$) and zeolites (microporous aluminosilicate material), which are also natural and stable materials [40]. The thoroughness of mixing and long-continued ramming after the placement was ensured. It is now confirmed that the strength and durability of ancient structures was not only due to the composition of the cementitious mixture, formation of stable minerals through carbonization and pozzolanic reactions, but also due to then prevalent construction practice of ensuring thoroughness of mixing and long-continued ramming after placement, which ensured impermeability of those structures [5]. This is in contrast to the modern structures built with PC, where permeability increases and the strength decreases with time. That is because most hydration products of PC (except calcium carbonate) are unstable and over a period of time keep interacting amongst themselves or with the external chemically active agents, causing volumetric changes and ultimately distress in the structure.

6.7.2 Hydration Reactions of Portland Cement

PC is an intimate mixture of reactive phases, mainly C_3A, C_3S, C_2S, and C_4AF (Table 6.1), formed during the high temperature processing in the rotary kiln and subsequent cooling operations, with gypsum (dihydrate, hemihydrate, anhydrite) added during the comminution (size reduction or grinding) process in the cement plant. A small quantity of free lime may also be present. All these components are soluble in water, except the iron compound in C_4AF phase but their solubility and the rate of dissolution are different. When water is added to concrete, the first physical process occurring is the dissolution of cement constituents in water followed by the chemical reactions (hydration), leading to the setting and hardening. The reactions of pure cement components individually with water are different from those that occur when water is added to PC, where all the components are present together. The nature, the rate and the continuity of these reactions are decided by the particle fineness and size distribution, the extent of dispersion and mixing of reactants, temperature, availability of water and gypsum (SO_4^{2-} ions) at the reaction site, pH of the solution, and the mobility of ions taking part in the reaction. In terms of their reactivity, the principal clinker phases may be arranged in the descending order as $C_3A > C_4AF > C_3S > C_2S$, which is almost in line with the rate at which these phases dissolve in water. The properties of fresh concrete change, commonly denoted by the slump, depending upon the nature of products formed, as the hydration progresses. In general, the dissolution of cement components in water followed by the chemical reactions, lead to the precipitation of much less soluble products and as the reactions proceed further, cementitious binding takes place, resulting in the reduction in the porosity and the permeability of the matrix and the concrete hardens (Figure 6.1). The major reactions occurring under different hydration periods are discussed in the following sections. The principal aim is to present these physicochemical changes in such a way that the practicing engineer develops better understanding and appreciation toward their influence on the workability, strength, and durability characteristics of concrete. It is thought that the best way to do so is to correlate these changes with the hydration periods with which the practicing engineer is familiar. While attempting this, some of the reactions may appear rather oversimplified, at places.

The typical hydration products of PC are calcium silicate hydrates (C-S-H), calcium aluminate hydrates (C_xAH_y), and calcium sulfoaluminate hydrates ($C_3A\hat{C}\hat{S}_xH_y$), where the subscripts x, y indicate number of molecules of the particular compound forming the hydrate. The silicate hydrates and the aluminate hydrates are produced in the hydration of silicate and aluminate from cement, respectively, and the sulfoaluminate hydrates in the hydration of aluminate in the presence of gypsum (calcium sulfate). Out of these, C-S-H, is the principal strength-giving phase in the hardened concrete.

The principal chemical reactions that occur during the PC hydration have been broadly grouped according to the hydration periods, in the following sections. The hydration reactions occur simultaneously but at different rates. The physicochemical changes discussed under the respective hydration periods are only the major ones.

6.7.2.1 C-S-H: The Principal Reaction Product and the Strength-Giving Phase

The calcium-silicate-hydrate or the so-called C-S-H is the principal hydration product as well as the principal strength-giving phase of cement, with or without mineral admixture, at the sub-particle level. The structure of C-S-H has been studied widely [41–50]. It has a disordered layer structure. The individual layers are formed from calcium ions to which silicate ions and hydroxyl groups are attached. The structure also has interlayer calcium ions and water molecules. Apart from small amount of monomer, the silicate ions are chains with length of $(3n - 1)$ tetrahedra (where $n = 1, 2, 3, \ldots$). In most cases, it is five tetrahedra. The chemical composition of C-S-H is related (but not exact) to tobermorite ($C_5S_6H_9$ approximately) or C-S-H (I) and jennite ($C_9S_6H_{11}$) or C-S-H (II). The C-S-H (I) refers to C/S < 1.0 and C-S-H (II) refers to C/S > 1.0, as evident from their molecular formulae [51]. With age, a more stable material is formed with an intermediate composition. The variation in C/S ratio during hydration probably occurs due to the variation in concentration of $Ca(OH)_2$ in the pore solution [48]. Although the structure remains same, the C/S ratios differ in C-S-H, with and without mineral admixture. In the PC hydration, mean C/S ratio is 1.7–1.75, whereas with mineral admixtures C/S is in the range of 1–1.6. Some minor components are found as substituents or in phases admixed with C-S-H at nanometer scale. The NMR and other studies indicate that most of the Al and mostly all the Fe and Mg are present in the admixed regions; sulfate is probably also present in such regions or sorbed; and alkali cations probably substitute in interlayer sites. The x-ray microanalysis provides qualitative and semiquantitative spot chemical analysis as well as maps of elemental distribution of hydration products. Figure 6.2 [52] shows a typical spectrum of C-S-H obtained by energy-dispersive x-ray microanalysis.

The ability of C-S-H to incorporate phases like monosulfatehydrate (AFm) and hydrotalcite (M_5AH_{13}) types may be due to the fact that the charges on the layers are of opposite sign [45]. The bonding of C-S-H to other products of hydration is generally good from the strength and durability point of view. The C-S-H has fibrous, foil or rolled sheet-like structure. Plate 6.3 shows the typical microstructure of C-S-H [21]. In slag cement hydration, the fibrillar morphology of C-S-H is gradually replaced by a foil-like morphology, as the slag replacement is increased. This change in morphology is probably responsible for the improved durability performance of slag cement [47].

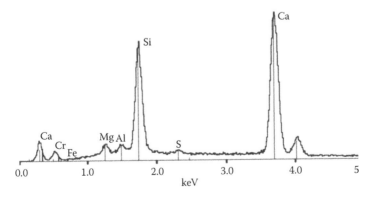

FIGURE 6.2
Energy-dispersive x-ray spectrum of C-S-H indicating presence of calcium, silicon, magnesium, aluminum, and sulfur. (From Ferraris, C.F. et al., Sulfate resistance of concrete: A new approach, PCA R&D Serial No. 2486, Portland Cement Association, Skokie, IL, 2006.)

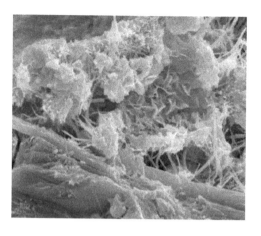

PLATE 6.3
SEM image showing platy or foil-like C-S-H, fine bundles of C-S-H fibers, and platy CH (top). (From Petrographic methods of examining hardened concrete: A petrographic manual, FHWA-HRT-04-150, Federal Highway Administration, U.S. Department of Transportation, Washington, DC, 2006.)

6.7.2.2 Major Reactions Occurring in the Workability Period

As discussed in Section 6.2.1, the workability period lasts up to 1–3 h. The mixing, transporting and placing of concrete must take place in this period. The slump or the workability of concrete does not change significantly in this period. In the first few minutes of wetting and mixing of concrete, rapid dissolution of cement components in water takes place. The principal chemical change taking place during this period is the reaction of aluminate (C_3A) with gypsum and water resulting in the formation of trisulfatehydrate ($C_6A\hat{S}_3H_{32}$) known as ettringite or Af_t phase. The reaction can be schematically and stoichiometrically expressed as in Equations 6.13 and 6.14. The molar mass of the reactants taking part in the reaction is given under the chemical formula

$$\text{Tricalciumaluminate + Gypsum + Water = Ettringite} \qquad (6.13)$$

$$\underset{270.2}{3CaO \cdot Al_2O_3} + \underset{516.5}{3(CaSO_4 \cdot 2H_2O)} + \underset{468.4}{26H_2O} \rightarrow \underset{1255.1}{3CaO \cdot Al_2O_3 \cdot 3CaSO_4 \cdot 32H_2O}$$

$$(6.14)$$

The chemical formula of ettringite is not exact and the number of water molecules incorporated may be between 30 and 32. The molar volume of ettringite is 735 cm³/mol, which is comparable with the total molar volume of the reactants and no expansion is observed. However the formation of ettringite can result in substantial expansion and distress in concrete, when formed in hardened state (low pore volume), due to the reaction of CH and lower sulfates or hydrates of tricalciumaluminate with sulfates within or those diffusing from the surroundings.

Simultaneously with the formation of trisulfate hydrate, (i) nucleation of C-S-H takes place as a result of hydration of C_3S and (ii) the solution gets supersaturated with CH released from the hydration of C_3S and the crystals of CH (portlandite) nucleate.

As can be seen in the reaction (Equation 6.14), formation of trisulfate hydrate requires constant supply of sulfate ions in the solution, produced by the dissolution of gypsum. It is reported that the trisulfate or Af_t phase, formed in the initial stage of C_3A hydration, later converts to the monosulfate or Af_m phase, as the quantity of SO_4^{2-} ions in the solution reduces with time. Thus, for proper hydration, it is essential to control the quantity of gypsum added vis-à-vis the quantity of tricalciumaluminate present in cement. The solubility and the rate of dissolution of gypsum is also important. The imbalance in the supply of SO_4^{2-} ions in the solution can lead to the following undesirable situations:

a. *False set*: The phenomenon occurs when improperly stored cement contains C_3A of low reactivity (due to partial hydration or carbonation) on the one hand and a large proportion of more soluble calcium sulfate hemihydrate (produced due to excessive dehydration of gypsum during grinding operation), on the other [4]. Under this situation, solution will contain low concentration of aluminum hydroxide $[Al(OH)_4]^-$ ions, produced by the dissolution of C_3A, but gets saturated quickly with calcium and sulfate ions. The trisulfate hydrate will not be formed but instead large crystals of dihydrate gypsum will be formed with corresponding loss of consistency (Equation 6.15):

$$CaSO_4 \cdot xH_2O + (2-x)H_2O \rightarrow CaSO_4 \cdot 2H_2O \ (x = 0.001 - 0.67) \quad (6.15)$$

The phenomenon is not associated with large heat evolution. It can be remedied by remixing.

b. *Quick set and flash set*: When cement contains high proportion of reactive C_3A (typically more than 8% with surface area more than $3500 \, cm^2/g$ Blaine) but the soluble sulfate content is less than that required for normal hydration. The reaction with water results in quick formation of the crystals of monosulfatehydrate (also called sulfate-Af_m) (Equation 6.16) and calcium-aluminate-hydrate (Equations 6.17 and 6.18) in large amounts, leading to setting in less than 45 min [4,19]:

$$3CaO \cdot Al_2O_3 + CaSO_4 \cdot 12H_2O + 10H_2O \rightarrow 3CaO \cdot Al_2O_3 \cdot CaSO_4 \cdot 12H_2O$$
$$(6.16)$$

The monosulfate molecule in Equation 6.16 may contain 12–18 water molecules:

$$2(3CaO \cdot Al_2O_3) + 21H_2O \rightarrow 4CaO \cdot Al_2O_3 \cdot 13H_2O + 2CaO \cdot Al_2O_3 \cdot 8H_2O \tag{6.17}$$

The hydrates in Equation 6.17 are not stable and get converted to more stable varieties as given in Equation 6.18:

$$4CaO \cdot Al_2O_3 \cdot 13H_2O + 2CaO \cdot Al_2O_3 \cdot 8H_2O \rightarrow 2(3CaO \cdot Al_2O_3 \cdot 6H_2O) + 9H_2O \tag{6.18}$$

The retarding action of gypsum on C_3A hydration will be effective only when trisulfatehydrate is formed in sufficient quantity in the initial stages and effectively coats the C_3A particles. Conversely, blended cement (containing mineral admixtures) may not require rapid supply of sulfate ions in the initial stages of hydration (due to its low C_3A content) but may require sulfate ions later during the secondary hydration stages, to activate the mineral admixture like BFS. Such cement may require addition of anhydrite gypsum, due to its lower dissolution rate.

6.7.2.3 Major Reactions Occurring in the Setting Period

6.7.2.3.1 Calcium Silicates (beginning)

The calcium silicates (about 30% and mainly C_3S) from cement react with water to form C-S-H and CH:

$$3CaO \cdot SiO_2 + (2.5 + n)H_2O \rightarrow (1.5 + m)CaO \cdot SiO_2(1 + m + n)H_2O$$
$$+ (1.5 - m)Ca(OH)_2 \tag{6.19}$$

$$2CaO \cdot SiO_2 + (1.5 + n)H_2O \rightarrow (1.5 + m)CaO \cdot SiO_2(1 + m + n)H_2O$$
$$+ (0.5 - m)Ca(OH)_2 \tag{6.20}$$

The reactions produce C-S-H with the mean C/S ratio of 1.70–1.75 [5]. The rapid formation of hydrates leads to loss of plasticity and the decrease in porosity. The heat evolution is also at high rate. Reactions 6.19 and 6.20 reduce to Reactions 6.1 and 6.2 when $m = 0.2$ and $n = 2.8$. In terms of the reactivity, C_3S reacts much faster in comparison to C_2S.

6.7.2.3.2 Calcium Aluminoferrite

The reaction of calcium aluminoferrite and gypsum (from cement) with water results in the formation of hydrated calcium-sulfate-aluminoferrite solid solution, with different compositions.

6.7.2.4 Major Reactions Occurring in the Hardening Period

6.7.2.4.1 Calcium Silicates (completion)

The reaction of remaining calcium silicates (mainly C_2S and remaining C_3S) with water results in the formation of C-S-H and CH, as discussed.

6.7.2.4.2 Trisulfatehydrate

The rapid conversion of ettringite or trisulfatehydrate (Af_t) to form monosulfatehydrate (Af_m or sulfate-Af_m), that is, $3CaO \cdot Al_2O_3 \cdot CaSO_4 \cdot 12H_2O$, as C_3A gets further hydrated and the concentration of sulfate ions in the solution gets low (gypsum being exhausted in the reactions):

$$2[3CaO \cdot Al_2O_3 \cdot 6H_2O] + 3CaO \cdot Al_2O_3 \cdot 3CaSO_4 \cdot 32H_2O$$
$$\rightarrow 3[3CaO \cdot Al_2O_3 \cdot CaSO_4 \cdot 12H_2O] + 8H_2O \qquad (6.21)$$

In case, the concentration of sulfate ions is inadequate to form monosulfate, a new hydrate compound $C_3A \cdot Ca(OH)_2 \cdot 12H_2O$ or C_4AH_{13} is formed and remains in solid solution with the monosulfate. The monosulfate and the C_3A hydrate may contain 12–18 water molecules. In these reactions, aluminum can be completely replaced by iron and the hydration of calciumaluminoferrite (C_4AF) also results in the monosulfatehydrate or the solid solution [1].

The sulfate (gypsum) provided in the cement is never sufficient for complete trisulfate (ettringite) formation (19% by mass of gypsum will be required for 10% by mass of C_3A). Therefore, monosulfate and sulfate-free calcium aluminate hydrates are formed depending upon the following internal conditions: dissolved sulfate, pH value, available moisture and external (temperature) conditions. It should be noted that the hydrates of C_3A may eventually get converted to the stable composition C_3AH_6 or form hydrogarnet, combining with silica, if the surrounding temperature and other conditions permit. The ettringite sheaths formed initially are broken up during the secondary reactions so that the C_3A is able to react further, with reduced concentration of sulfate, to form monosulfate [53].

While discussing the hydration reactions taking place in the workability, setting, and hardening periods, all cement phases were assumed pure, although the assumption is seldom correct in practice. In fact, each of the cement phases contains a large number of other compounds in solid solution, which influence the hydration behavior, for example, alkalies and sulfates. In the cement manufacture, due to differences in raw materials processing, kiln conditions, fuels, etc., there is wide scope for the differences in the composition of alite, belite aluminate and ferrite phases, all influencing hydration. The products—C-S-H, CH, ettringite, monosulfate, among others—are also impure. The C-S-H in particular may act like a "sponge," absorbing other ions. The fully hydrated mass of cement normally contains more than 70% C-S-H [3]. Plate 6.4 shows a typical backscattered electron image of hardened cement paste [21].

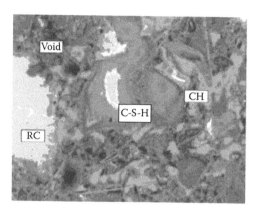

PLATE 6.4

SEM backscattered electron (BE) image of hardened cement paste. *Note*: Constituent phases show residual cement (RC), CH, C-S-H, and other hydration products. (From Petrographic methods of examining hardened concrete: A petrographic manual, FHWA-HRT-04-150, Federal Highway Administration, U.S. Department of Transportation, Washington, DC, 2006.)

6.7.2.5 Hydration of Alite (C₃S) and Belite (C₂S) Compared

Tricalciumsilicate (C_3S) and dicalciumsilicate (C_2S) react with water to form C-S-H and CH ($Ca(OH)_2$). The stoichiometric calculations reveal certain important aspects. Consider a period of first 3 days of hydration. According to Reaction 6.1, the hydration of 100 g alite requires 41.82 g water to produce 141.82 g mixture of products containing 99.63 g (70.25%) C-S-H and 42.19 g (29.75%) CH and the exothermic heat liberated is about 5800 cal. On the other hand, according to Reaction 6.2, the hydration of 100 g belite requires 44.98 g water to produce a mixture of 144.98 g products containing 132.07 g (91.1%) C-S-H and 12.91 g (8.9%) CH and the exothermic heat liberated is about 1200 cal [4]. The stoichiometric mass of reactants and products may vary according to the type of reaction considered. When these hydration processes are compared, for the period of first 3 days, it is seen that C_2S produces 32.6% more C-S-H and 69.4% less CH and also liberates nearly 4.8 times less heat, in comparison with C_3S. Thus, the ultimate strength of high-C_3S cement will be lower than that of high-C_2S cement, considering that the C-S-H is the main strength giving phase in the hydrated cement. The durability of concrete to acidic and sulfate waters is reduced in the presence of excess CH. The cement containing high proportion of C_2S is expected to perform better in acidic and sulfate surroundings than that containing high proportion of C_3S. The higher rate of heat liberation in case of C_3S can become an important consideration, when high-C_3S cement is used for plastering or in mass concreting. The higher heat generation in the initial stages can result in surface cracks (especially under improper curing conditions) and thermal shrinkage. Similarly in mass concreting it may lead to higher temperature gradients. As may be seen from the rate of heat liberation in the initial stages, the C_3S reacts about five times faster in comparison to C_2S. The faster rate of hydration of C_3S is an important factor considered in the design of high-early-strength PC. Some cement standards limit the maximum permissible C_3S and some recommend addition of mineral admixtures to remove the excess CH from the hydrated cement paste.

6.7.3 Hydration Reactions of Cement with Mineral Admixtures

The five principal mineral admixtures, the application of which is widely reported in the literature, are fly ash or PFA, BFS, SF, RHA, and MK. The mineral admixtures are either pozzolanic like PFA (ASTM Type F), SF, RHA, and MK or cementitious (also called latent hydraulic) like BFS. The pozzolans are regarded as substances that do not contain enough calcium in their chemical composition to form cementitious products but combine with CH (lime) at ordinary temperature, in aqueous media, to form cementitious products. The BFS is called cementitious and latent hydraulic because it contains adequate calcium in the chemical composition but cannot form cementitious products at ordinary temperature due to surface film. The CH released during cement hydration provides alkaline medium to break the surface film. It also reacts with both pozzolanic and cementitious admixtures to form cementitious products.

The hydration, representing mass conversion of different phases in the cement paste, develops the microstructure of concrete, which in turn relates to important physical properties such as strength, elastic modulus, toughness, diffusivity, and permeability. Unlike the hydration process in ordinary PC, the hydration process in blended cement is considerably more complex. It involves reactions of mineral additives or pozzolanic/cementitious materials in addition to the hydration of PC. The CH is consumed in the reactions to produce more C-S-H, which leads to higher strength. It is seen that when PC is blended with SF or BFS, the morphology of C-S-H, the principal strength-giving compound in the hardened concrete, changes from fibrillar to foil-like. The change in the C-S-H morphology results in a less well-interconnected capillary pore structure leading to the lower permeability and so the enhanced durability of such systems [54].

6.7.3.1 Hydration of Cement with Pulverized Fuel Ash

The low-calcium PFA, produced by combustion of anthracite or bituminous coal in the thermal power plants, categorized as a normal pozzolan, consists of silicate glass, modified with aluminum and iron [55]. The CaO content is <10%. When used in combination with PC, it forms strength-giving compounds, reacting with CH, due to its pozzolanic activity. The high-calcium PFA, produced by the combustion of lignite or subbituminous coal in the thermal power plants, consists of aluminosilicate glass modified by large amounts of calcium and magnesium. It can be categorized as cementitious material when CaO is greater than 20% or as cementitious and pozzolanic material when CaO varies between 10% and 20% [56].

The low-calcium PFA is more commonly used in cement and concrete. Reactions 6.22 through 6.26 are with regard to the low-calcium PFA; however, they are also common with high-calcium PFA. The readers may read the technical articles by Papadakis at Refs. [57,58], for more details.

The PFA is primarily composed of SiO_2, Al_2O_3, and Fe_2O_3. It also contains a small amount of CaO and very small amounts of other compounds (Chapter 1). Although the glass and crystalline compounds found in PFA as well as the hydration products in the presence of CH are generally non-stoichiometric, simplifying assumptions are made while writing the chemical reactions, for the sake of understanding. The formation of C-S-H in Reaction 6.23 may be seen from that point of view, particularly the C/S ratio, which may actually vary depending upon the microstructure. The alumino-silicate (A-S) glass is the most reactive compound in PFA. The S of A-S reacts with CH [57], without additional water binding, to form C-S-H (Reactions 6.22 and 6.23). The S present as quartz or crystalline A-S is inert and does not take part in the reaction. The ferrous phase is found mostly in crystalline form as hematite (Fe_2O_3) or magnetite (Fe_3O_4) and thus does not participate in the pozzolanic reactions.

$$\text{(Reactive silica in PFA)} + \text{(calcium hydroxide)} \rightarrow \text{C-S-H} \qquad (6.22)$$

$$2SiO_2 + 3Ca(OH)_2 \rightarrow 3CaO \cdot 2SiO_2 \cdot 3H_2O \qquad (6.23)$$

The reactive (noncrystalline) alumina reacts with gypsum and CH to form ettringite, which soon gets converted to the monosulfate, after gypsum is consumed (Reactions 6.24 and 6.25):

$$\text{(Reactive alumina in PFA)} + \text{(gypsum)} + \text{(calcium hydroxide)} + \text{(water)}$$
$$\rightarrow \text{(ettringite)} \rightarrow \text{(monosulfate)} \qquad (6.24)$$

$$Al_2O_3 + CaSO_4 \cdot 2H_2O + 3Ca(OH)_2 + 7H_2O \rightarrow 3CaO \cdot Al_2O_3 \cdot CaSO_4 \cdot 12H_2O \qquad (6.25)$$

The hexagonal tetracalcium aluminate hydrate C_4AH_{13} is formed at low concentrations or the absence of gypsum (Reaction 6.26):

$$Al_2O_3 + 4Ca(OH)_2 + 9H_2O \rightarrow 4CaO \cdot Al_2O_3 \cdot 13H_2O \qquad (6.26)$$

As mentioned earlier, the ferrous phase does not participate in the pozzolanic reaction. In the excess of gypsum, the total pozzolanic reaction of active alumina is described by Reaction 6.25.

The fineness of PFA is generally accepted to be one of the key factors influencing its reactivity [59]. The hydration of PFA generally does not begin until about 7 days, regardless of fineness, glass content and composition. That is because the dissolution of glass phase in PFA is dependent upon the alkalinity of the pore solution, which increases strongly only after some days [60]. It is seen that the mechanism that triggers the beginning of pozzolanic reaction, after the so-called incubation period, is through the availability of OH⁻ ions in sufficient concentration in the pore solution, majority of which come from the alkali (Na and K) hydroxides [61,62].

The hydration of PFA may not be taken into consideration for the heat liberation during initial hydration, which is important for mass concreting. The rate of hydration of cement minerals is affected by PFA. The long-term hydration of alite is accelerated and that of belite and C_4AF is retarded. The addition of PFA results in a low C/S ratio of C-S-H in hardened samples [63]. There is a difference in the precipitation pattern of the hydration products, mainly C-S-H, of PC and PFA. Whereas the hydration products of PC precipitate near the cement particles and do not block the pores, those of PFA precipitate between the small space between the particles, blocking the pores, resulting in the finer pore structure and relatively less number of pores. This process leads to the reduction of permeability of hardened concrete structure and better durability [59].

In high volume fly ash concrete (HVFAC), cement is replaced by PFA in proportions greater than 50% by mass. The mechanisms proposed by Berry et al. for the hydration and strength gain and the early-age properties of HVFAC are given as follows [64–66]:

a. Normal hydration of PC, with slight acceleration of rate at the early age.

b. The spherical shape of PFA particles and superplasticizer added to concrete lead to better particle packing and produce dense concrete matrix.

c. Reactions of PFA particles at later age that produce insoluble silicate and aluminatehydrates at the particle boundary region.

d. Hydration of PFA particles, those remaining physically intact and largely unchanged in morphology, leading to further refinement of pore structure.

e. Very little bleeding due to the low water content requires that curing starts as soon as possible.

f. Longer setting time than for PC concrete. Applications in hot weather may benefit from this characteristic as it allows more time to cast the concrete. Increased setting time and lower bleeding, makes finishing complicated.

g. Less temperature rise due to minimized hydration of the cement helps reduce early-age cracking and enhances durability.

h. Very low slump due to the low water-to-binder ratio and the low water content. Most applications require high-range water-reducing admixture (HRWRA) to increase slump.

The workability is influenced, among other factors, by the volume of paste. The specific gravity of PFA is lower hence it occupies more volume in comparison to PC, for equal mass. When PFA replaces cement on mass-to-mass basis, the paste volume increases and consequently the workability improves and the water requirement decreases, for the given workability. The lubricating effect of spherical PFA particles also plays an important role. The concrete containing PFA (ASTM Class F and Class C) shows similar or reduced shrinkage, in comparison to that with no PFA, under identical curing conditions [67].

The concrete setting refers to the change in consistency of mix from fluid to rigid state, resulting from the hydration of cementitious compounds in the early age. When PFA replaces cement, "initial" and "final" setting time of concrete increase with the level of replacement. However, the rate of strength development remains appropriate for most construction applications, up to a replacement level of 60%. Therefore, time of setting should not be taken as a sole parameter for selecting PFA for a particular application [68–70]. Besides prolonged setting time, addition of PFA may also reduce bleeding and segregation of concrete. It can therefore be difficult for finishing crew to judge the proper time to start the finishing operations. Adequate care may be taken in that regard [65]. It is essential to provide the proper curing temperature and moisture for up to a week to reduce shrinkage cracking and help develop pozzolanic action.

Besides the replacement ratio, the curing temperature also has an effect. At a commonly encountered curing temperature (not higher than 35°C), the presence of PFA at all replacement ratios accelerates hydration of cement due to the cement dilution effect. However, at higher curing temperature (50°C and above) and high PFA replacement ratios, the pozzolanic reaction of PFA becomes important, as it competes with the cement hydration reactions in consuming water and producing large amounts of reaction products, from early ages and that may counteract the acceleration due to the dilution effect. As a result, the hydration is impeded [71].

Although the effect of PFA on hydration has been established experimentally, the quantitative influence of PFA on the kinetics of cement hydration is not well understood. In particular, for modern high-performance concrete with low w/b ratio, the effect of PFA on the cement hydration may be different. In order to predict the performance of PFA concrete accurately throughout its service life, a more quantitative understanding of the effect of PFA on cement hydration in low w/b ratio cementitious mixtures is needed.

6.7.3.2 Hydration of Cement with Blast Furnace Slag

It is the glass component of BFS that reacts with the CH and the gypsum (C$\hat{S}$) in cement, during hydration. The BFS glass mainly consists of minerals composed of lime, silica, alumina, magnesia, and minor quantity of oxide of iron (C-S-A-M-F). It also contains elemental sulfur in the form of calcium sulfide.

The hydration begins when the pH of pore solution exceeds 11.5. Reaction 6.27 showing the reactants and the products (not stoichiometrically balanced) is based on the mechanism as discussed in Refs. [72,73]:

$$C\text{-}S\text{-}A\text{-}M\text{-}F + CH + C\hat{S} + H \rightarrow C\text{-}S\text{-}A\text{-}H + M_5AH_{13} + C_6AFS_2H_8 + C_6A\hat{S}_3H_{32}$$
$$+ C_2ASH_8 + C_4AH_{13}, \ldots \qquad (6.27)$$

The calcium sulfide in slag reacts with water to give $Ca(OH)_2$ and $Ca(SH)_2$ but the quantity is very small [5]. The ferrous phase is found in crystalline form (Fe_2O_3), does not form a part of glass and hence does not take part in the reaction. It gets incorporated in hydrogarnet solid solutions ($C_6AFS_2H_8$). The reactive alumina (A) present in the slag, first combines with M, F, gypsum to form the hydrotalcite (M_5AH_{13}), hydrogarnet ($C_6AFS_2H_8$) and ettringite ($C_6A\hat{S}_3H_{32}$). The remaining A partially substitutes S in C-S-H to form C-S-A-H. The limit of A substitution in C-S-H is related to the C/S ratio of the raw slag glass. When there is sufficient A from the slag to achieve the maximum substitution, the remaining A reacts to form tetracalciumaluminatehydrate (C_4AH_{13}) and/or stratlingite (C_2ASH_8). The C-S-A-H formed during slag hydration has a lower C/S-ratio (approximately 1.1) than that from the PC.

An Al_2O_3 content of up to 15% has a positive effect on the reactivity of BFS. Reactivity increases with rising Al_2O_3 content. Due to the interaction between Al_2O_3 and CaO content, the effectiveness depends on the CaO content and is stronger in lime-rich slags than in lime-poor slags. In order to optimize the chemical composition, Al_2O_3 content should be no lower than 10.5% and the CaO content no lower than 40% by mass, if possible [74,75]. Whereas the hydration products of PC precipitate near the cement particles and cannot block the pores, those of ground BFS precipitate between the small space between the particles (similar to PFA), resulting in the finer pore structure and relatively less number of pores. This process leads to the reduction of permeability of hardened concrete structure and better durability [60].

6.7.3.3 Hydration of Cement with Silica Fume

The SF contains 85%–98% silica in the amorphous form and the crystalline matter may contain quartz, cristobalite, and silicon carbide. It also contains carbon in very small quantities (Chapter 3). It is assumed that the crystalline matter and carbon does not take part in the hydration reaction, which could be written in a generalized form [76] as in Reaction 6.28:

$$xS + yCH + zH \rightarrow C_yS_xH_{y+z} \qquad (6.28)$$

The water content and porosity results show that the reaction takes place without any additional water binding, more than that contained in the CH molecules (i.e., $z = 0$) and the reaction product has the stoichiometry of $C_yS_xH_y$. Also, the values of long-term consumption of CH show that a C/S ratio of 1.5 may be considered. Thus Reaction 6.28 reduces to Reaction 6.29 [56,75]:

$$2SiO_2(amorphous) + 3Ca(OH)_2 \rightarrow 3CaO \cdot 2SiO_2 \cdot 3H_2O \qquad (6.29)$$

The effect of SF on cement system during the early age of hydration can be viewed broadly in two ways: in terms of the rate of cement hydration and the pozzolanic reaction. The addition of SF to cement accelerates the rate of hydration in the first few hours, as a result of enhanced precipitation of hydration products on the fine and initially inert SF particles (nucleation effect). The pozzolanic reaction between cement and SF starts at a relatively early age and proceeds at a faster rate than that with the other mineral admixtures such as PFA and BFS [77].

6.7.3.4 Hydration of Cement with Rice Husk Ash

RHA contains up to 96% silica in amorphous form (Chapter 4). The mechanism of cement hydration with RHA is proposed, based on the two technical articles by James and Subba Rao [78,79] (Reactions 6.30 and 6.31):

$$C_3S + C_2S(Portland\ cement) + H_2O \rightarrow C\text{-}S\text{-}H + Ca(OH)_2 \qquad (6.30)$$

$$SiO_2(RHA) + Ca(OH)_2(Portland\ cement) + nH_2O \rightarrow C\text{-}S\text{-}H + SiO_2(unreacted) \qquad (6.31)$$

The authors, based on the experiments conducted on RHA-lime cement, recommend that no unreacted silica should be left and there should be slight excess of lime, after the hydration is complete. It has been observed that replacing cement with RHA up to 30% (mass basis) leads to increased strength and durability of concrete (Chapter 4).

6.7.3.5 Hydration of Cement with Metakaolin

The principal reaction is between MK ($Al_2O_3 \cdot 2SiO_2$ or AS_2) and CH ($Ca(OH)_2$ or CH) derived from cement hydration, in the presence of water. This reaction forms additional cementitious C-S-H gel, together with crystalline products, which include calcium aluminatehydrates and aluminosilicatehydrates (C_4AH_{13}, C_3AH_6, C_2ASH_8):

$$Al_2O_3 \cdot 2SiO_2 + 6Ca(OH)_2 + 9H_2O \rightarrow 4CaO \cdot Al_2O_3 \cdot 13H_2O + 2CaO \cdot 2SiO_2 \cdot 2H_2O \qquad (6.32)$$

$$Al_2O_3 \cdot 2SiO_2 + 5Ca(OH)_2 + 3H_2O \rightarrow 3CaO \cdot Al_2O_3 \cdot 6H_2O + 2CaO \cdot 2SiO_2 \cdot 2H_2O$$
(6.33)

$$Al_2O_3 \cdot 2SiO_2 + 3Ca(OH)_2 + 6H_2O \rightarrow 2CaO \cdot Al_2O_3 \cdot SiO_2 \cdot 8H_2O + CaO \cdot SiO_2 \cdot H_2O$$
(6.34)

The crystalline products depend principally on the AS_2/CH ratio and reaction temperature. Although Reactions 6.32 through 6.34 show a C/S ratio of 1.0, in actual practice, the ratio varies between 0.8 and 1.5, depending upon the microstructure of the C-S-H produced (amorphous, crystalline or a mixture) [80–83]. The hydration reaction depends upon the level of reactivity of MK, which in turn depends upon the processing conditions and purity of feed clay during manufacturing. The concrete containing MK is likely to exhibit higher rate of heat generation in comparison to the PC, on account of higher pozzolanic activity [84,85]. The high pozzolanic activity also results in reduction in slump (fluidity) and increase in water demand of concrete [86–88]. This effect, to some extent, can be countered making ternary mixtures of PC with MK and PFA.

The studies on the hydration of cement with mineral admixtures continue and a lot is still required to be understood. Some areas for future studies could be (a) understanding of the factors controlling the rate of reaction, using new analytical techniques, and (b) understanding of the changes that occur in C-S-H and the influence of these changes on the properties of concrete [89].

6.8 Summary

The hydration of cement is the combination of all chemical and physical processes taking place after contact of the anhydrous solid with water. It begins with the addition of water, while making concrete and nearly 60%–65% of the hydration is complete in 28 days. Although the cement minerals like calcium aluminates and silicates and the calcium sulfate (gypsum) take part in the hydration reactions, the process is affected by all other constituents of concrete. The nature and the progress of cement hydration are responsible for the strength and the durability of concrete. The mineral admixtures added to cement, take part in the hydration reactions, due to their pozzolanic/ cementitious properties. They affect the early-age properties, like rheology and therefore workability, setting, shrinkage, etc., besides the strength and the durability of concrete.

The progress of hydration with time can be arbitrarily divided in three stages, familiar to the practicing engineer: workability, setting, and hardening. Although hydration reactions continue, concrete remains in plastic condition during workability period, which lasts for 1–3 h, depending upon the type of cement and the concrete mix design. The wetting, mixing, transporting, agitating, placing, and finishing of concrete must be completed in this period. The boundaries of the setting period, which follow the workability period, also called "active reaction period," are marked by the "initial" and "final" setting time, the former signifies the beginning of the loss of workability and the later the total loss of workability, with no significant development of strength. As the hydration is in full form, curing of concrete should begin in this period and extend through the hardening period, as required by the standards. The strength of concrete develops in the hardening period. The hydration may continue for years but the modern world standards consider the 28 day compressive strength as the final indicator, when more than 90% of the specified strength is attained. The start, end, and the duration of the workability, setting, and hardening period are affected by the phase composition and fineness of cement, characteristics of mineral and chemical admixtures, and the surrounding temperature during hydration. It should be noted that almost entire strength development takes place after the final set.

There are four main types of reactants that participate in the hydration process of cement with mineral admixtures. These are (a) reactive compounds in cement clinker, namely, tricalciumsilicate, called alite or C_3S (40%–60%); dicalciumsilicate, called belite or C_2S (12%–30%); tricalciumaluminate, called aluminate or C_3A (3%–14%); and tetracalciumaluminoferrite or C_4AF (7%–16%); (b) calcium sulfate dihydrate, called gypsum or $C\hat{S}H_2$ (5%–8.5% in cement); (c) reactive compounds in mineral admixtures; and (d) calcium hydroxide. The chemical analysis of cement gives the oxide composition; from that the phase composition (reactive compounds) is estimated using Bogue formula. The cement also contains some other compounds, which may be deleterious for the development of strength and durability, when found in excess. Most national standards put upper limit on their content; these are magnesia (M), free lime (C), alkali sulfates, expressed as SO_3 ($\hat{S}$), and alkalies, expressed as Na_2O equivalent.

The alite (C_3S), more reactive in comparison to belite (C_2S), is responsible for the development of early strength of concrete. The aluminate (C_3A), the most reactive of the constituents of cement, reacts first at very high rate with large heat evolution, when water is added. The gypsum is added as set regulator, in order to avoid undesirable side effects on account of high and rapid heat evolution and to impart desired setting characteristics to cement. Whereas it retards the hydration of C_3A, that of alite and belite is accelerated. The partial dehydration of dihydrate gypsum to hemihydrate and anhydrite varieties takes place, when the temperature inside cement grinding mill goes above 100°C. These gypsum varieties, having different solubility in water and water with lime, control the availability of SO_4^{2-} ions and in turn its set-regulating characteristics. The A/$\hat{S}$ ratio determines the setting behavior of cement and has practical importance. It may be noted that less than 30%

gypsum reacts with C_3A, remaining gets absorbed in C-S-H phase, helping to raise the compressive strength of concrete. The byproduct gypsum, obtained from number of industrial processes, is used in cement. The glass, aluminosilicates, or amorphous silica form the major reactive part of the morphological composition of mineral admixtures. Besides the reactive content, the particle fineness and size distribution also plays an important role in the hydration. The unburned carbon should be kept within the standard limits. The CH obtained as a product of hydration of silicates, plays an active role in the so-called secondary hydration of reactive compounds present in the mineral admixtures. The CH makes low contribution to the strength and higher content in the hydrated cement adversely affects the resistance (durability) of concrete toward acidic solutions (chlorides and sulfates).

Unlike cement paste or mortar, the concrete comprises of three distinct phases: the binder, the aggregate, and the ITZ between the binder and the aggregate. The properties of concrete are a cumulative result of these major phases. In comparison to the aggregate particles and the bulk cement paste, the ITZ has the lowest hardness value. Thus, in normal concrete, the ITZ may be considered as a weakest link.

The early-age properties of concrete during hydration, affected by the mineral admixtures, are workability, yield stress and viscosity, bleeding and laitance, setting time, concrete temperature, and volume changes. The requirement of concrete workability depends on the nature of its application. The most common test for the workability of fresh concrete, "slump test," can be correlated, under specific conditions, to the yield stress of a given concrete. Whereas the measurement of plastic viscosity has several practical applications, like pumping and casting rate, yield stress is an important parameter for formwork filling in practice. The bleeding is a result of segregation of water from solids, after concrete is placed. Such segregation prevents full compaction of concrete, essential to reach its maximum strength potential. The delayed bleeding, when concrete loses its plasticity, causes harm to concrete. The use of finely divided mineral admixtures reduces bleeding. The free water in fresh concrete is used up in the hydration reactions and the evaporation. Its rate of depletion defines the workability period and the loss of workability over a period of time defines the boundaries of setting period: the "initial" and the "final" setting. The beginning of initial setting marks the limit of handling and the final setting time marks the beginning of the development of mechanical strength for concrete. The temperature of concrete, while placing, is important from the point of view of later strength development. The volume changes in concrete at the early age are mainly due to the temperature fluctuations, drying and chemical reactions. The drying or plastic shrinkage occurs, when the rate of free water evaporation from fresh concrete exceeds the rate of surface bleeding. When concrete is in plastic condition, autogenous or chemical shrinkage occurs as a result of water loss in chemical reaction as well as due to the net reduction in the volume of hydration products. The chemical shrinkage and the heat of hydration are both valid indicators of the early-age hydration. The high-early-age-strength cement tends to increase thermal and autogenous deformations; partial substitution of cement with mineral admixtures (except SF and MK) can mitigate the situation.

The nature and continuity of hydration reactions depend upon the particle fineness and the size distribution, extent of dispersion and mixing of reactants, temperature, availability of water and SO_4^{2-} ions at the reaction site, pH of the solution, and the mobility of ions taking part in reaction. The C-S-H is the principal hydration product as well as the principal strength-giving phase of cement, with or without mineral admixture, at the sub-particle level. In terms of the terminology, C-S-H (I) refers to C/S < 1.0 and C-S-H(II) refers to C/S > 1.0. In the PC hydration, mean C/S ratio is 1.7–1.75, whereas with mineral admixtures it is 1–1.6.

The main reaction during workability period is between C_3A and gypsum in aqueous medium, leading to the formation of trisulfatehydrate or ettringite. The ettringite formed during this period does not lead to expansion; however, when formed in the hardened state, it leads to substantial expansion and distress in concrete. The imbalance in the supply of SO_4^{2-} ions in the solution can lead to the "false" or "flash" (quick) set. When cement is partially substituted by the mineral admixture, rapid supply of SO_4^{2-} ions may not be required in the initial stage (due to low C_3A content). Such cement may require addition of anhydrite gypsum, due to its lower dissolution rate.

The setting period is marked by the hydration of silicates, mainly C_3S to C-S-H and CH, along with heat evolution at high rate.

The hydration of silicates, started in the setting period, continues mainly over C_2S and gets completed in the hardening period. The trisulfate (ettringite) gets rapidly converted to monosulfate, due to the depletion of SO_4^{2-} ion (gypsum) concentration in the pore solution. The fully hydrated mass of PC may normally contain about 70% C-S-H. When the hydration of C_3S and C_2S is compared, it is found that C_2S produces more C-S-H and lesser CH and heat, in comparison with that for C_3S. Thus the ultimate strength and durability of high-C_3S PC is likely to be lower in comparison.

The CH produced in the hydration of PC and gypsum react with reactive silica and alumina obtaining in the mineral admixtures to form additional C-S-H, (with lower C/S ratio) and the monosulfate hydrate. However it may be noted that the reactivity of mineral admixtures toward CH differs. The hydration of PFA depends upon the alkalinity of the pore solution, hence does not begin until about 7 days. Thus the liberation of heat on account of PFA hydration may not be considered at the early age, especially for mass concreting. The hydration in the HVFAC is characterized by longer setting time and lesser rate of heat liberation (temperature rise). In general, when PFA replaces cement, the "initial" and the "final" setting time may increase but the rate of strength development remains appropriate for most construction applications, up to a replacement level of 60%. The reactivity of BFS increases with rising A (alumina [Al_2O_3]) content, up to 15%. The optimum chemical composition of BFS has A > 10.5% and C > 40%. The addition of SF accelerates the hydration of cement and the pozzolanic reaction also starts relatively early and proceeds at faster rate, in comparison to other mineral admixtures. It may be noted that, in most cases, SF is added in small quantity as an additional constituent of concrete, unlike other mineral admixtures added as the replacement of cement. The replacement of cement with RHA should be decided in such a way that no excess reactive silica is left after hydration; normally replacement up to 30% leads to increased strength and durability. The concrete containing MK is likely to exhibit higher rate of heat generation in comparison to that with PC, on account of high pozzolanic activity; ternary mixtures with other mineral admixtures may counter this effect to some extent. When mineral admixtures are added to concrete, it is necessary to provide proper curing temperature and moisture regime, as appropriate, to reduce shrinkage cracking and assist the pozzolanic or cementitious reactions. In general, the additional C-S-H or C-S-A-H (in case of BFS) produced precipitate in the void space between particles, leading to the refinement of pore structure, reduction in permeability and improvement in the durability of concrete.

7

Strength and Durability

7.1 Introduction

The strength and durability of concrete structure must go hand in hand. Durability is the ability of a structure to resist weathering action, chemical attack, and abrasion, while maintaining minimum strength and other desired engineering properties. In today's context, designing for strength and durability is synonymous to designing for sustainability. This spirit is appropriately described in CEB/FIP Model Code 1990 [1]:

> Structures shall be designed, constructed and operated in such a way that, under the expected environmental influences, they maintain their safety, serviceability and acceptable appearance during an explicit or implicit period of time without requiring unforeseen high costs for maintenance and repair.

The national standard codes of practice mostly follow prescriptive approach, defining the exposure conditions, design procedure, choice of building materials, and their application during the construction, without assuring the expected minimum service life of the structure, when these specifications are followed. However, in 1996, CEB accepted performance-based approach with explicit attention for the durability-based service life, limit states, and reliability in design. A similar change in approach is also reflected in the Indian Standard Code IS 456: 2000 [2]. However, much needs to be done in that area. The objective of the national structural design codes, as in Eurocode 2 [3], should be to enable the design decision based on the life cycle cost. That will be a major step toward building sustainable structures and in order to do that the mechanisms and the mitigation of structural deterioration due to the attack of deleterious agents need to be understood.

Mehta proposed a holistic model of deterioration of concrete [4,5]. According to the model, a well-constituted, properly consolidated, and cured concrete remains watertight as long as the microcracks and pores in the interior do not form an interconnected network of pathways leading to the surface of the concrete. The deterioration takes place in two stages. In stage 1, due to structure–environment interaction, gradual loss of watertightness takes place, as a result of the development of interconnections between microcracks and pores. Consequently, water and along with it ions playing active role in the deterioration get easily transported to the interior of the concrete, which may get saturated. However, little or no damage will be apparent during this stage. In stage 2, the deterioration begins with the successive cycles of expansion, cracking, loss of mass, and increased permeability. The further loss of watertightness and acceleration of damage takes place on account of two processes that follow: (a) increase in the hydraulic pressure of the pore fluid in saturated concrete due to one or more phenomena of volumetric expansion (e.g., freeze and thaw, corrosion of reinforcement, sulfate attack, or alkali–silica reaction [ASR]), and (b) leaching of chloride and sulfate ions and replacement of hydroxyl ions, and, consequently, loss of calcium-silicate-hydrate (C-S-H) adhesion and strength. The holistic model provides an insight into the deterioration process and is helpful in designing cost-effective strategies to prolong service life of concrete exposed to aggressive environment.

The study of the durability of concrete structures needs a multidisciplinary approach. The durability depends both on the ability of concrete to resist the penetration of aggressive substances from the environment and on its ability to protect the embedded steel reinforcement. The transport of aggressive species may follow different mechanisms, depending on the pore structure of the concrete, the exposure conditions, and the characteristics of the diffusing substances. The penetration of carbon dioxide or oxygen, as well as of many other gaseous substances, for example, may occur only if the pores of concrete are almost dry, while the diffusion of chloride or sulfate ions takes place only under moist conditions. Therefore, all these mechanisms should be studied together, with the evolution of moisture content inside the concrete [6]. The durability of concrete is currently one of the most widely studied subjects world over. A better understanding of this phenomenon will contribute toward improvement of the building standards, ensuring durability of the structures, and specifying and directly assessing the required quality of concrete, namely, its resistance to chemical and physical attacks, which is the origin of damage.

Section 7.2 summarizes the two approaches to concrete structure design: the prescriptive (Section 7.2.1) and the performance-based (Section 7.2.2) approaches. Section 7.3 on concrete strength discusses various related aspects, when mineral admixtures replace cement. Section 7.4 attempts to present the mechanisms, mathematical models, national standards, guidelines, and practices to mitigate concrete deterioration, as related to (a) carbonation (Section 7.4.1), (b) alkali–aggregate reaction (AAR) (Section 7.4.2), (c) chloride attack and corrosion of reinforcement (Section 7.4.3), (d) external sulfate attack (Section 7.4.4), (e) internal sulfate attack (ISA) or delayed ettringite formation (DEF) (Section 7.4.5), (f) decalcification or leaching (Section 7.4.6), and (g) frost or freeze–thaw action (Section 7.4.7). Section 7.5 discusses performance-based design of structures using durability indices. Section 7.6 refocuses on the sustainability of cement and concrete as building materials. The important issues discussed in this chapter are summarized in Section 7.7.

7.2 Designing Structures for Strength and Durability

Presuming that the structural design is carried out as per the norms and standard construction practices are followed, the durability of a reinforced concrete structure depends upon both the environment to which the structure is exposed as well as the quality of concrete. Therefore, identification of the exposure condition for the structure is the first step toward the design for durability. The current trend in the national standards is to provide an extensive classification of the exposure conditions to cover a wide range of situations encountered in practice. Table 7.1 lists some publications and standards on durable concrete.

The next step is to choose a proper design procedure and make a choice of materials, which includes concrete, suitable for the exposure condition. In that regard, two approaches can be followed: prescriptive and performance based. The possible design steps required in both the approaches are summarized in Table 7.2 [7, 8].

One important factor common to both the approaches is the general acceptance of "permeability" as a criterion for durability. Although the term

TABLE 7.1

Publications and Standards on Durable Concrete

Sl No	Country	Number	Title
i	United States	ACI 201.2R	Guide to durable concrete
		EB 221	Specifier's guide to durable concrete (published by PCA)
ii	United Kingdom	BS EN 206-1	Concrete—Part 1: Specification, performance, production and conformity
		BS 8500	Concrete—Complementary British standard to BS EN 206-1
			Part 1: 2002 Method of specifying and guidance for the specifier
			Part 2: 2002 Specification for constituent materials and concrete
iii	Germany	DIN 1045-2	Concrete, reinforced and prestressed concrete structures—Part 2: Concrete—Specification, properties, production and conformity—Application rules for DIN EN 206-1
iv	India	IS 456	Plain and reinforced concrete-code of practice

ACI, American Concrete Institute; PCA, Portland Cement Association; BS EN, British Standard European Norm; DIN, Deutsches Institut für Normung (DIN: in English, the German Institute for Standardization); IS, Indian Standard.

TABLE 7.2

Possible Steps to Designing Structures for Strength and Durability

Prescriptive Approach	Performance-Based Approach
Identification of the exposure conditions for the structure	Specification of the target service life and the design service life
Formulation of the functional requirements to be fulfilled	Identification of the exposure conditions for the structure
Selection of design solution and the choice of materials	Selection of design solution and the choice of materials
Execution in compliance with Standards	Identification of the durability models for degradation mechanisms
Inspection and maintenance of structure or structural elements	Selection of a durability factors and degradation mechanisms (depth of deterioration of concrete and corrosion of reinforcement, concrete cover, and diameter of bars)
	Calculation of durability parameters using available calculation models
	Transfer of the durability parameters into the final design
	Efficient execution in compliance with standards; adequate inspection and maintenance and introduction of measures to prevent potential causes of failure and/or to reduce their consequences, that is, provide the required reliability

"permeability" is strictly related to the flow that occurs under an applied pressure differential, it is also frequently used in a general sense to cover other transport mechanisms including absorption and diffusion [9]. The RILEM technical committee, TC 116-PCD, was formed to evaluate the use of permeability as a criterion of concrete durability. The available test methods were evaluated and reported with regard to their suitability for routine testing of concrete transport parameters [10,11]. The Roads and Transport Authority (RTA) of New South Wales, Australia, adopted the water sorptivity concept as an additional durability requirement for concrete to be used in bridge construction. The RTA simplified the test, for practicality, determining only the depth of water penetration after 24 h of wetting and referred to it as the sorptivity depth [12]. The other two important factors, common to both the approaches, are the quality of construction work and the continuous maintenance of structures during use.

The standard construction practices, when not followed properly, can easily negate the best design provisions given to produce a durable concrete structure. The construction variables that influence durability include concrete placing, compacting, curing, reinforcing bar placement, ducts and tendon placement, grouting procedures, and materials. The two major faults in concrete construction that can lead to the loss of durability are inadequate cover to reinforcement and inadequate compaction of cover concrete. A durability-based design procedure should cover quality control aspects related to the materials of construction and the construction process. While the quality control norms related to the former are well established, those for the later are practically nonexistent. Thus there is a need to pay attention to the quality and reliability of the construction execution process. The provisions for the maintenance of structure or structural elements should be incorporated at the design stage. The two approaches to the design, prescriptive and performance based, are discussed briefly in the following sections.

7.2.1 Prescriptive Approach

The national codes of practice, which mostly follow the prescriptive approach, do not explicitly specify the minimum service life that can be achieved by following the durability provisions specified. The codes offer specifications and guidelines, in a prescriptive manner, to build durable structures. The requirements for the strength and the durability change according to the exposure conditions and the properties desired. The standards define the service environment or exposure conditions for the structure; recommend structural design criteria; specify for concrete ingredients and their application with respect to the service environment, to achieve the desired durability properties and the life of concrete. The structural Euro Codes incorporate adequate maintenance strategy also as a part of design concept [13]. The review of the provisions of different national standards on the subject has been reported by Anoop et al. [14]. The requirements for durability are tied up with the exposure conditions in the standards, in terms of the minimum strength or the grade of concrete, the maximum water-cement ratio and the minimum cement content, the minimum cover thickness, the minimum (percent) air content with regard to the maximum aggregate size along with the special requirements for aggregate, and the permeability expressed as maximum (mm) water penetration. Special mention may be made of Eurocode 2 [13] and CEB Design Guide [15] that provide a detailed list of the nine exposure classes and EN 206 [16], which defines exposure conditions based on different degradation mechanisms. The CEB Design Guide underlines the importance of producing crack-free concrete, specifying the maximum limit of water penetration. The two important aspects, related to the exposure conditions, mostly ignored by the standards are (i) the condition of microenvironment surrounding the site of deterioration and (ii) the influence of temperature [14].

7.2.2 Performance-Based Approach

The structural design based on durability-based service life is the hallmark of performance-based approach. The aggressiveness of the exposure conditions needs to be quantified. Masters and Brandt [17] tried to quantify the aggressiveness of the environment, introducing a concept of corrosion aggressivity index, for example. In order to estimate the durability-based service life, we should have the knowledge of the models of the degradation processes and the characteristics of materials possessing adequate resistance toward the attack of deteriorating agents. The models should be time-tested, based on the reliable data on service life of materials, generated from the field performance tests carried out in a well-planned manner. It is not practicable for one laboratory or even one country to pursue, on its own, the long-term, complex, and costly research needed in the area. It offers the opportunity for continued and increased international interactions and to perform challenging research on building materials and their degradation processes [17].

The limit state design requires the structure to satisfy two principal criteria: the ultimate limit state (ULS) and the serviceability limit state (SLS). A structure is deemed to satisfy the ULS criteria, when all the factored stresses—bending, shear, and tensile or compressive—are below the factored resistance calculated for the section under consideration. A structure is deemed to satisfy the SLS, when the constituent elements do not deflect by more than certain limits laid down in the building codes, the floors fall within predetermined vibration criteria and the crack width in concrete is kept below specified dimensions.

The structure's reliability is the probability of a structure to fulfill the given functions during its service life, that is, to keep the performance characteristics—safety, durability, and serviceability—within the given limits [18]. The readers may also like to go through Section 7.5 on the performance-based design of structures using durability indices.

7.3 Concrete Strength

The series of experiments conducted by Cyr et al. and Lawrence et al. [19–22] on mortars containing PFA revealed that the compressive strength of such mortars ($f_{p(admixture)}$) can be represented by the combination of three distinct, additive, factors as follows:

$$f_{p(admixture)} = f_{(dilution)} + f_{\varphi(physical)} + f_{pz(chemical)} \qquad (7.1)$$

where
$f_{(dilution)}$ represents the strength reduction proportional to the amount of cement in the mixture without considering any physical or chemical effect of mineral admixture

$f_{\varphi(physical)}$ represents the increase in strength due to the physical effect or heterogeneous nucleation due to mineral admixture

$f_{pz(chemical)}$ represents the increase in strength related to the pozzolanic reaction

The authors developed an empirical model to quantify these factors.

7.3.1 Interfacial Transition Zone

The interfacial transition zone (ITZ) is considered as the weakest link in concrete. The reduction in width, porosity, and calcium hydroxide (CH) content of the ITZ between the paste and the aggregate in concrete with mineral admixtures, attributed to the factors mentioned in Equation 7.1, is reported to be responsible for the strength enhancement and the durability. Hence efforts to improve the strength and durability of concrete must also pay attention to improving the ITZ. Plate 7.1 shows the thresholded back-scattered electron (BE) image and the gray-level histogram of ITZ. Stutzman found, when microstructure of ITZ was compared with the bulk of paste, the increase in porosity was more than five times and the decrease in C-S-H gel was nearly twice in the ITZ [23]. Both the aspects have a bearing on the strength and durability of concrete. Plate 7.2 shows, as an example, the SEM BE images of ITZ microstructure of concrete at 28 days, containing (a) 0%, (b) 10%, and (c) 20% silica fume (SF) (mass basis) [24]. It can be seen, as the

0.45 M/G 26 d, 1000x

PLATE 7.1
Thresholded BE image and gray-level histogram of ITZ. *Note*: The four constituents are identified using the gray levels in backscattered electron imaging: residual cement being the brightest followed by calcium hydroxide, C-S-H gel, and pores showing as black. The gray-level histogram to the right of the image plots the number of image pixels across the gray scale. The image is segmented into the four phases on the basis of gray level, and the area fractions of constituents are estimated as a function of distance from the interface. It provides a graphical display of the changes in paste microstructure with distance. The paste/aggregate interfacial transition zone (ITZ) often shows increased porosity (dark) and calcium hydroxide, and less residual cement in comparison to the bulk cement paste (field width 150 μm). (From Stutzman, P.E., Scanning electron microscopy in concrete petrography, in *Materials Science of Concrete Special Volume: Calcium Hydroxide in Concrete (Workshop on the Role of Calcium Hydroxide in Concrete). Proceedings*, Skalny, J., Gebauer, J., and Odler, I., eds., The American Ceramic Society, Anna Maria Island, FL, pp. 59–72, November 1–3, 2000.)

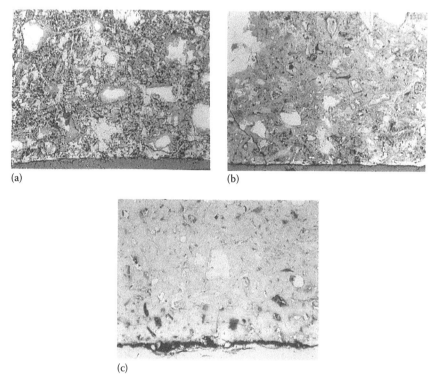

(a)   (b)

(c)

PLATE 7.2

SEM BE images of interfacial zone microstructure of concrete at 28 days containing (a) 0%, (b) 10%, and (c) 20% by weight SF. *Note*: In the micrographs, cement particles are bright white; CH, silica fume, and C-S-H are light gray; primary C-S-H is dark gray; and pores are black. The scale is such that the full image height corresponds to about 160 μm. The flat dark surface at the bottom of each image is aggregate. (From Bentz, D.P. et al., *Cem. Concr. Res.*, 22, 891, 1992.)

cement replacement with SF increases, the microstructure of ITZ gets denser and homogenized. It has been observed by other researchers also that the replacement of cement by mineral admixtures improves ITZ microstructure, leading to the improvement in strength and durability of concrete [25–28].

7.3.2 High-Performance Concrete

The structures satisfying the requirement of cost, service life, strength, and durability require the use of high-performance concrete (HPC). Meeks and Carino [29] gave a definition of HPC, discussed in NIST/ACI Workshop. As per the definition, HPC meets special performance and uniformity requirements that cannot always be achieved routinely by using only conventional materials and normal mixing, placing, and curing practices. The requirements may involve enhancements of characteristics such as placement and compaction without segregation, long-term mechanical properties, early-age strength, volume stability, or service life in severe environments.

The concrete possessing many of these characteristics often achieves higher strength. Therefore, HPC is often of high strength, but high-strength concrete may not necessarily be of high performance. The special performance requirements using conventional materials can be achieved only by adopting low water-to-binder (w/b), which necessitates the use of high cement content. However, judicious choice of chemical and mineral admixtures reduces the cement content and that results in economical HPC.

HPC is used today to build structures in aggressive environments, highway bridges and pavements, tunnels, nuclear structures, and precast units. The published literature on subject [30–38] shows that it is possible to make HPC using mineral admixtures. However, many researchers report that the conventional mix design methods, primarily developed keeping PC in mind, require modification. In order to take into account the contribution of mineral admixtures to the mechanical properties of concrete, many researchers developed the concept of "efficiency factor," which is empirically determined for the given set of materials and the exposure conditions [39–45]. The cementing efficiency factor of a mineral admixture may be defined as the number of parts of cement in a concrete mixture that could be replaced by one part of mineral admixture, without changing the property being investigated, which is usually the compressive strength. The work published by Bharatkumar et al. can be taken as an example [35]. The authors took a review of the general philosophy of different concrete mix proportioning methods obtained in the literature and proposed a new mix proportioning method, based on the ACI method of normal concrete mix design [46], which utilizes optimum water content and the efficiency factor for mineral admixture.

High-strength concrete is a design requirement in many infrastructure constructions, such as bridges, dams, and high-rise buildings. The designer has to take into account the aspects related to high heat generation as well as sharp temperature gradients in concrete (especially in mass concrete), as a result of high cement content and increased cement fineness (consequence of modern cement manufacturing process). The situation can lead to thermal cracking. The experimental results reported in the literature reveal that, incorporating mineral admixtures, it is possible to make concrete with high early-age strength as well as low heat liberation [47–50].

Yazici et al. [51] reported that the production of reactive powder concrete (RPC) with high compressive strength is possible using high volume (up to 60% by mass) of GGBS. The concrete contained relatively coarse (1–3 mm) granite or bauxite as aggregate and a certain quantity of SF. The cement content of these mixtures was considerably lower than the cement content of conventional RPC (800–1000 kg/m³). Besides the reduced heat of hydration and shrinkage, these mixtures also yield important environmental and economical benefits. The SF content can be decreased with increasing GGBS replacement. The modification also reduced the superplasticizer demand considerably.

The study carried out by Hill and Sharp [52] demonstrated that cementitious systems with 75%–90% replacement of OPC by BFS or PFA are sound, with potential application as grout. Although it may require longer to set with much reduced compressive strength, in applications where that is unimportant, other properties may be advantageous, such as low heat evolution and low level of CH.

7.3.3 Importance of Concrete Curing

The importance of concrete curing for the development of strength and durability properties, especially when mineral admixtures are used, needs special mention. Curing is a process of preventing loss of moisture from concrete, while maintaining a satisfactory temperature regime. The early and long term curing is beneficial for the development of concrete properties [53,54]. The prevention of moisture loss from concrete is particularly important, when the water-to-cement/binder ratio is low, when the cement has a high rate of strength development and when the concrete contains mineral admixtures.

The hydration reaction in concrete containing mineral admixtures continues longer, through the reaction of CH with the reactive silica and alumina present in the admixtures, in presence of water. This reaction, known as secondary hydration, produces additional C-S-H, which is the principal strength giving compound in concrete. The secondary C-S-H is also responsible for the refinement of pore structure and thereby reducing the permeability of concrete. Thus adequate and continuous supply of water is necessary for the long-term strength development and improved durability performance of concrete with mineral admixtures, in as much as the inadequacy of curing, in some cases, may result in a situation worse than that without the addition of mineral admixtures. The rate of moisture loss from the concrete surface, more so for the thin concrete elements, is relatively high in hot, dry climate and adequate measures should be undertaken for its prevention.

Meeks and Carino took an excellent review of the state-of-the-art on the curing of HPC and listed parameters given by Hilsdorf that must be considered while deciding the minimum curing duration [29]. These parameters are as follows:

a. *Concrete composition and water-to-cement ratio*: Strength development properties are affected by this parameter.

b. *Concrete temperature*: The rate of hydration and in turn, strength development, and reduction in porosity are affected greatly by this parameter.

c. *Ambient conditions during and after curing*: This parameter affects the severity of drying of the surface layer.

d. *Exposure condition of the structure in service*: This parameter affects the required near-surface or skin properties for adequate service life.

It is further mentioned that the minimum duration of curing may be estimated based on the requirements of (a) depth of carbonation (C-Concept),

(b) permeability (P-Concept), (c) maturity or degree of hydration (M-Concept), and (d) compressive strength (R-Concept). The assessment of field situation determines which of these requirements, as enunciated by Hilsdorf, needs to be considered.

It should be noted that concrete compressive strength, measured under controlled conditions of laboratory, is not an adequate indicator of concrete durability because it does not account for the construction process variables such as placing, compaction, and curing. These variables significantly affect the quality of the surface zone of the concrete and in that way, have a direct influence on the durability, controlling the movement of deleterious substances from the environment into concrete. The important factors that control concrete deterioration, besides strength, are therefore the near-surface quality of the finished concrete and the aggressiveness of the environment. There is a little that could be done to control the environment and, hence, the strategies for improving the service life of structures have to focus on the quality of construction including that of the materials. The durability specifications, therefore, increasingly rely on the measurement of the fluid flow properties of the surface or cover zone of concrete, usually 28 days after casting.

7.4 Mechanisms, Models, Standards, and Mitigation of Concrete Deterioration

When concrete is exposed to the environment, the commonly observed processes causing deterioration are (a) carbonation, (b) alkali-aggregate reactions, (c) corrosion of reinforcement, (d) sulfate action, (e) decalcification or leaching, and (f) frost action. Several experimental studies carried out as well as the theoretical models developed aim at analyzing different mechanisms of material deterioration. The mathematical models are useful in determining parameters required for design of structures, for durability against aggressive environment. A brief review of the mechanisms and the models of these processes is presented in this section. The formulation and understanding of the mathematical models require understanding of the mechanisms of micro- and macroscopic transport of species occurring in concrete. An excellent account, including the basic mathematical formulations for such transport processes, has been presented by Samson et al. [55] as well as Glasser et al. [56]. As far as possible, simple but widely applied mathematical models have been presented in this section, in order to make the practicing engineer appreciate their importance in analyzing the practical phenomena. However, the verification of mathematical models should be carried out on fairly large scale specimens, using reliable test methods, over a long period of time. The tests must be carried out under conditions simulated to represent those obtained in the actual practice.

This section also reviews the provisions of different national standards, guidelines, as well as means to prevent or mitigate the deterioration of concrete due to the attack of deleterious agents.

It is generally accepted that under optimum conditions of effective blending of components, transportation, placing and curing, the addition of mineral admixtures to cement and concrete improves its performance, including high ultimate strength, low heat of hydration, low permeability, and the resistance to deteriorating agents. As such, cement replacement up to about 35% PFA and up to about 65% BFS are now well established in the construction industry and fully characterized in the scientific literature [57,58]. Higher levels of replacement are less fully documented, but may have advantages in specific applications, for example, when very low heat evolution is required. One or more of the following factors are responsible for the improvement in performance [53]:

a. Low C/S ratio in C-S-H fixes alkalis through adsorption or solid solution, thereby decreasing the alkali ion concentration in pore solution. About 95% of the total alkali content can be retained by blended cement, whereas only 15% is blocked in PC.

b. The C-S-H formed by the secondary hydration of mineral admixtures fills up the pores in hardened cement paste, thus making the structure finer and denser. The resultant reduction in permeability suppresses the movement of pore solution within and the diffusion of deteriorating agents from outside.

c. The surplus C_3A, which remains after reaction with gypsum during hydration, promotes the attack of sulfates (Na_2SO_4 and $MgSO_4$) on the hardened cement paste, at later stage. The replacement of cement with mineral admixtures reduces the C_3A content.

d. The pozzolanic/cementitious reactions occurring during the hydration of mineral admixtures consume CH produced during Portland cement hydration. In the sulfate attack, alkali sulfates react with CH to produce additional $C\hat{S}$, which further leads to expansive ettringite formation. The reduction in CH during hydration of mineral admixtures leads to reduction in $C\hat{S}$ generation and helps mitigate sulfate attack.

e. The packing effect created by the unreacted particles of mineral admixture contributes toward densification and consequent reduction in the permeability of concrete structure.

f. It is found that some mineral admixtures have greater chloride-binding capacity in comparison to PC. This property limits the ingress of chloride ions, responsible for corrosion, into the hardened concrete.

It should be underlined that water-to-binder ratio plays a major role in deciding the ability of concrete to resist the attack of deteriorating agents, as it affects the permeability of concrete toward the ionic transport [59]. At the same time, low water-to-binder ratio should always be associated with good workmanship—workable mixes, good consolidation, hard finish, and good curing—along with the use of a relatively rich mix.

The author of this book subscribes to the view on "attack" expressed by Adam Neville and also quotes what he has to say on developing concrete resistance toward external deteriorating agents. Although the opinion is with regard to external sulfate attack, it is equally applicable to other deteriorating agents [60]:

> … From an engineer's point of view, what matters is what has happened to the concrete: an action that does not result in deterioration or in a loss of durability is not an attack. … There is, however, an overriding requirement, necessary but not sufficient, for good sulfate resistance of concrete, and that is its density in the sense of a very low permeability. This is achieved by good proportioning of the mix ingredients, such that full compaction can be obtained, by actual achievement of full compaction, and by effective curing to maximize the degree of hydration of the cement paste. These requirements are obvious to an engineer, but often not fully considered by laboratory experimenters. Hence, my repeated plea for consideration of field concrete.

7.4.1 Carbonation

One of the processes that takes place in the pores of concrete and which may limit the service life of reinforced concrete structures is the carbonation of materials. Carbonation refers to the precipitation of calcite ($CaCO_3$) as well as other CO_2-based solid phases, through the reaction of penetrating atmospheric CO_2 with the calcium ions in the pore solution. The carbonation process by itself does not have a negative effect on concrete. In some cases, it can even result in the reduction of the porosity and favor formation of a protective layer on the surface of concrete. However, the consumption of calcium in the pore solution leads to increased dissolution of portlandite (CH). The main consequence of carbonation is the drop of the pH of the pore solution of concrete from the standard values of 12.5–13.5, to a value of about 8.3 in the fully carbonated zones, so that the passive layer that usually covers and protects the reinforcing steel against corrosion becomes unstable. The next sections summarize different aspects of the carbonation process.

7.4.1.1 Mechanism of Carbonation

The carbonation process begins when gaseous carbon dioxide (CO_2) penetrates into the cement matrix, dissolving in the pore solution, it dissociates into bicarbonate ions (HCO_3^-) and carbonate ions (CO_3^{2-}), which react with calcium ions (Ca^{2+}) from CH, C-S-H, and the hydrated calcium aluminates and ferro-aluminates, to precipitate as various forms of calcium carbonate ($CaCO_3$), silica gel, and hydrated aluminum, and iron oxides.

$$CO_2(g) \rightarrow CO_2(aq) \tag{7.2}$$

$$CO_2(aq) + H_2O \rightarrow H^+ + HCO_3^- \tag{7.3}$$

$$HCO_3^- \rightarrow H^+ + CO_3^{2-} \tag{7.4}$$

$$CO_2(aq) + 2OH^- \rightarrow CO_3^{2-} + H_2O \tag{7.5}$$

$$Ca^{2+}(aq) + CO_3^{2-} \rightarrow CaCO_3 \tag{7.6}$$

$$C\text{-}S\text{-}H + CO_2(aq) \rightarrow CaCO_3 + SiO_2 + H_2O \tag{7.7}$$

The experimental evidence of CH reduction upon calcite formation has been reported [61,62]. The pH drop associated with these reactions leads to further dissolution of CH. While aragonite and valerite polymorphs of $CaCO_3$ have also been found, calcite is generally identified as the main reaction product of carbonation [6,63]. The volume of the carbonate (calcite) formed is 11%–12% greater than the volume of CH. Therefore, it is frequent to find carbonated samples with increased weight, lower porosity, and higher compressive strength, at early ages of carbonation [64]. However, when the porosity is sufficiently high to permit constant CO_2 diffusion inside, the CH gets further depleted and the calcium from C-S-H also reacts with carbon dioxide, leading first to the decalcification (removal of calcium) and later to the formation of silica gel (amorphous and highly porous form of silica) [65–67]. The formation of silica gel in this manner may cause loss of strength, volumetric decrease (shrinkage) and cracking, and increase in the porosity of concrete. In concrete with mineral admixture, where the amount of CH is reduced due to pozzolanic or cementitious reaction, the carbonation of C-S-H is dependent on the permeability and lower permeability hinders the ingress of CO_2.

7.4.1.2 Mathematical Models for Carbonation

The mechanism of carbonation is complex and involves different steps that are mutually interdependent. The numerical models presented in literature [6,68–71] have proven to be powerful tools, allowing us to treat these complex interactive processes in a quantitative way. The parameters used in the simulation must be linked with concrete characteristics and with exposure conditions, by means of suitable experimental studies.

PLATE 7.3
Damage caused to structure due to carbonation and subsequent corrosion of reinforcement.

The carbonation rate is essentially controlled physically by the carbon dioxide diffusion process and chemically by the reserve of CH of the concrete. Therefore, an effective numerical model to simulate the whole phenomenon should consider the interaction between many processes: the CO_2 diffusion, the moisture and heat transfer, the mechanism of $CaCO_3$ formation, and the availability of CH in the pore solution.

The carbonation should be studied in context with other processes, such as chloride penetration, sulfate attack, crack formation, freezing of pore solution, oxygen diffusion, and water sorptivity, which may affect the durability of the structure. Plate 7.3 shows a typical structure, where carbonation of the concrete cover is the principal cause of the reinforcement corrosion. In particular, carbonation is usually associated with volume change, which may lead to cracking and in turn can change the effective permeability. Therefore, carbonation probably causes changes in permeability and diffusivity directly. All these phenomena would affect the diffusion equations and should be taken into account in the development of the model.

The ingress of $CO_2(g)$ in concrete is expressed using a diffusion-based equation (Equation 7.8) in most models [56]:

$$\frac{\partial(\phi - w)[CO_2(g)]}{\partial t} - \text{div}((\phi - w)D \, \text{grad}[CO_2(g)]) - f_c = 0 \qquad (7.8)$$

where
  ϕ is the porosity of the material
  w is the volumetric water content
  $[CO_2(g)]$ is the gaseous carbon dioxide concentration
  D is the gas diffusion coefficient

The sink term f_c accounts for the transfer of carbon dioxide from the gaseous phase to the pore solution. The concentration of CO_2 in the pore solution can be estimated using Henry's law (Equation 7.9), which is applicable to sufficiently dilute solutions and when the solvent does not chemically react with the solute [72]:

$$[CO_2(aq)] = K_h P_{CO_2} \ldots \tag{7.9}$$

where
  $[CO_2$ (aq)] is the concentration of carbon dioxide in aqueous solution
  K_h is Henry's law constant
  P_{CO_2} is the partial pressure of carbon dioxide in gas phase, in relevant units

Reactions 7.2 through 7.6 are summarized in Reaction 7.10:

$$Ca(OH)_2 + CO_2(aq) \rightarrow CaCO_3 + H_2O \ldots \tag{7.10}$$

Simple rate equations are then used to calculate the formation of calcite or the loss of portlandite (Equation 7.11):

$$\frac{\partial [CaCO_3]}{\partial t} = f(w, T, [Ca(OH)_2(s)], [CO_2(aq)]) \tag{7.11}$$

The penetration of the carbonation front is a result of a diffusion process with chemical reaction. The carbonation depth is an important index to estimate both the damage and durability of reinforced concrete structures. It is traditionally estimated using a phenolphthalein indicator. It is an indirect measure, since the pink indicator actually shows where the pH drops below 9, by de-colorizing (Plate 7.4).

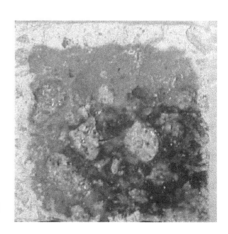

PLATE 7.4
Carbonation depth measurement using de-colorizing phenolphthalein indicator.

Liang et al. obtained an expression for the maximum depth of carbonation (x), analytically solving the one-dimensional diffusion equation, with a chemical reaction (Equation 7.12) [73]:

$$\frac{\partial C}{\partial t} = D\frac{\partial^2 C}{\partial x^2} - kC \tag{7.12}$$

where C is the concentration of $CO_2(g)$. It may also be written as $C'e^{-kt}$ [74], where C' is the new variable term for the CO_2 concentration. The first term on the right-hand side of Equation 7.12 represents the diffusion and the second term represents consumption of CO_2 in the carbonation reaction, respectively. Solving Equation 7.12 using Laplace Transform and substituting the boundary conditions, we get Equation 7.13:

$$C(x,t) = \left[C_i + \left(C_s e^{kt} - C_i \right) \text{erfc}\left(\frac{x}{\sqrt{4tD}} \right) \right] e^{-kt} \tag{7.13}$$

where $C(x, t) = C$ is concentration at depth x and time t. At $t = 0$, $C(x, t) = C_i$ and at the exposed surface or at $x = 0$, $C(x, t) = C_s$. The erfc represents the complementary error function. At time t, the depth of carbonation, x, may be written, using Equation 7.14:

$$x = 2\sqrt{D} \ \text{erfc}^{-1}\left[\frac{C e^{kt} - C_i}{C_s e^{kt} - C_i} \right] \sqrt{t} \tag{7.14}$$

or

$$x = A\sqrt{t} \tag{7.15}$$

where, carbonation coefficient A is

$$A = 2\sqrt{D} \ \text{erfc}^{-1}\left[\frac{C e^{kt} - C_i}{C_s e^{kt} - C_i} \right] \tag{7.16}$$

The authors further predicted that the carbonation depth at corners is $\sqrt{2}$ times larger than the general concrete surface. Papadakis et al. [63] developed a model for the prediction of carbonation depth (Equation 7.17):

$$x = \sqrt{\frac{2D[CO_2]_s t}{[CH]_i + 3[C\text{-}S\text{-}H]_i + 3[C_3S]_i + 2[C_2S]_i}} \tag{7.17}$$

where
   D denotes the effective diffusivity of carbon dioxide
   $[CO_2]_s$ is the concentration of carbon dioxide at the surface, the brackets in the denominator indicate the initial concentration of CH, C-S-H, C_3S, and C_2S

Equation 7.17 may also be written as the empirical expression of the form $x = A\sqrt{t}$, identical to Equation 7.15.

The exposure conditions that help carbonation are the concentration of CO_2 and relative humidity of the surroundings. The carbonation rate increases with the increase in carbon dioxide concentration in the surroundings. The carbonation-induced corrosion in concrete may often occur in a high carbon dioxide environment [75].

The relative humidity alters the internal moisture content or the saturation degree of pores in concrete. Concrete under low humidity does not react with carbon dioxide, as the quantity of water is insufficient to form carbonic acid. Similarly penetration of carbon dioxide into saturated concrete is difficult, under the condition of high relative humidity. Under normal atmospheric conditions, the relative humidity in the range of 50%–80% is optimum for carbonation to progress [76]. The main factors affecting concrete carbonation are the type and the content of binder, the water-to-binder ratio, the degree of hydration, the concentration of CO_2 and the relative humidity of the surroundings [77].

7.4.1.3 National Standards and Guidelines on Carbonation

Table 7.3 lists some standards and guidelines on carbonation. The European Standard EN 13295 [78] specifies an accelerated laboratory method to measure concrete resistance against CO_2 penetration. The test is recommended at low carbon dioxide concentrations to avoid unreal coefficients of carbonation

TABLE 7.3

Standards and Guidelines on Carbonation

Sl No	Standard	Organization	Title
i	EN 13295:2004	European Committee for Standardization (CEN), Brussels, Belgium	Products and systems for the protection and repair of concrete structures. Test methods. Determination of resistance to carbonation
ii	EN 14630:2006		Products and systems for the protection and repair of concrete structures—Test methods—Determination of carbonation depth in hardened concrete by the phenolphthalein method
iii	EN 206-1		Concrete: Specification, performance, production and conformity
iv	CPC 18:1988	International Union of Laboratories and Experts in Construction Materials, Systems, and Structures (RILEM), Bagneux, France	Measurement of hardened concrete carbonation depth

found in higher concentrations. The EN 13295 recommends that the accelerated carbonation test should be conducted at 1% CO_2 environment and at a temperature of 21°C ± 2°C. The concentration of 1% CO_2 in air develops the same reaction products with hydrated cement as in the case of normal atmosphere at 0.03% CO_2. The EN 14630 [79] specifies a test method to determine the carbonation depth in hardened concrete by phenolphthalein method. In the European Standard EN 206-1 [16], exposure classes are defined. The XC exposure classes deal with carbonation. The minimum binder content and maximum water-to-binder ratio are specified for each exposure class. The minimum strength is also mentioned, and in the French complementary national standard, maximum mineral admixture contents are also defined [80]. The new European standard allows performance based specifications through the equivalent performance concept. The equivalent performance is evaluated in comparison with a reference concrete mixture, complying with the prescriptive requirements for a given exposure. The comparison may be done through durability tests, provided that they give a reliable ranking of performances of concrete exposed to a given degradation, such as carbonation. The procedure published by RILEM in 1988 [81] has been widely used for the measurement of carbonation depth by phenolphthalein method. The ASTM and ISO have no test methods to evaluate the carbonation penetration in concrete. Although the accelerated carbonation and the phenolphthalein colorimetric methods to measure carbonation depth are widely used, they are still very diversified in terms of time and type of curing, preconditioning of specimen, surrounding temperature, and specimen dimensions.

7.4.1.4 *Mitigation of Carbonation*

The disadvantage sometimes mentioned in the literature is the increased carbonation in concrete containing mineral admixtures [75,82,83]. It is partly related to the hydration or the strength development and the CH buffer available. In dense and compact concrete, the carbonation initially produces calcite having 11%–12% higher volume than CH, resulting in reduced total porosity and specific surface of cement pastes as well as the permeability [84]. In general, concrete with a low hydration rate will carbonate faster than that with high hydration rate, when exposed to the environment at an early stage. The hydration of PFA or BFS in concrete, responsible for the refinement of pore structure, begins late in comparison with the PC. The concrete also contains less CH. Thus, in the initial stages, CO_2 or carbonation penetrates deeper. This phenomenon occurs in the laboratory, when concrete samples are stored at constant temperature and humidity condition. In practice, the difference is found much less or absent. Jan Bijen et al. reported the results of a large scale study on the durability of concrete structures built along the Dutch coast. The structures included sluices, piers, and quays, ranging in age from 3 to 63 years. In 48 out of 51 structures, where BFS was used in concrete, the carbonation depth was found less than 5 mm [85,86].

The factors controlling carbonation are the diffusivity of CO_2 and the reactivity of CO_2 with the concrete. The diffusivity of CO_2 depends upon the pore structure of hardened concrete and the exposure condition. The pore structure of concrete depends upon the type and the content of binder, water-to-binder ratio, and the degree of hydration. The time duration and the type of curing control the degree of cement hydration and the development of concrete microstructure. The hydration proceeds faster when concrete is cured in water in comparison to air and has been found to reduce the carbonation depth. Besides, concrete with a low water-to-binder ratio needs a longer and more efficient curing process. Early and longer curing produces better carbonation resistance, especially in concrete with mineral admixtures [87–90].

Blended cements containing large replacements (up to 90%) of ordinary Portland cement (OPC) with blast furnace slag (BFS) are currently used for encapsulation of nuclear waste in the United Kingdom. The replacement levels, much higher than those used in the construction industry, are used to help reduce the heat evolution of the mixes and avoid problems associated with thermal cracking and loss of durability and also to ensure that the rheology of the pastes is appropriate for mixing and pouring into the encapsulation drums [91]. Borges et al. [64], based on the accelerated carbonation tests, found that BFS:OPC pastes, with 9:1 blending ratio, suffered carbonation shrinkage and cracking during testing, the increase in the overall density and reduction in overall porosity notwithstanding. The authors recommended that systems with very high OPC replacement levels may not be suitable for encapsulation of nuclear wastes, when carbonation is a concern.

7.4.2 Alkali–Aggregate Reactions

Most aggregates are chemically stable in hydraulic cement concrete and do not have deleterious interactions with other concrete ingredients. The micropores in hardened concrete contain highly alkaline fluid (pH $\geq$ 12.5) with dissolved alkali (K, Na) hydroxides [92]. The aggregates containing certain dolomitic or siliceous minerals react with soluble alkalies in concrete and sometimes result in detrimental expansion, cracking, and premature loss of serviceability of concrete structures affected. These chemical reactions are termed as AAR. AAR related problems were first identified in early 1940s in California, United States [93]. All kinds of concrete structures may be affected, although structures in direct contact with water, such as dams and bridges, are particularly susceptible to AAR, given that the moisture conditions play an important role in the chemical process. The research on the subject has been carried out extensively and the principal areas are (a) mechanisms, (b) alkali reactivity of concrete aggregates and test methods, (c) specification for prevention, and (d) management of affected structures.

7.4.2.1 Mechanisms of AAR

The two types of AAR are (a) alkali–carbonate reaction (ACR) and (b) alkali–silica reaction (ASR). They differ in the type of aggregate mineral phases and reaction mechanisms. Figure 7.1 shows an idealized sketch of cracking pattern in concrete mass caused by internal expansion as a result of AAR [94].

7.4.2.1.1 Alkali–Carbonate Reaction

It is a process of degradation of concrete containing dolomitic (dolomite: $CaMg(CO_3)_2$) aggregate. The dolomite crystals undergo de-dolomitization (separation of calcium and magnesium compounds) process under the attack of alkali hydroxides of the concrete pore fluid, which opens channels through which ions from the pore fluid penetrate deeper into the reacting particle [95]. The reaction was first found at Kingston, Ontario, Canada, and identified by Swenson [96].

$$CaMg(CO_3)_2 \; + \; 2(Na, K)OH \; \rightarrow \; Mg(OH)_2 + CaCO_3 \; + \; (Na, K)_2CO_3$$

Dolomite   Alkali hydroxides   Brucite   Calcite   Alkali carbonates

(7.18)

It is found that the most reactive carbonate aggregates are partially dolomitized. However, all dolomitic limestone aggregate are not prone to ACR. Only the argillaceous (with substantial amount of clay-like components) limestone, with requisite texture and composition, produces expansion and cracking in concrete due to ACR [97,98]. The texture of reactive carbonate aggregates is characterized by rhombic crystals of dolomite, $50\,\mu m$ or smaller in size, sparsely distributed in a matrix of micritic calcite

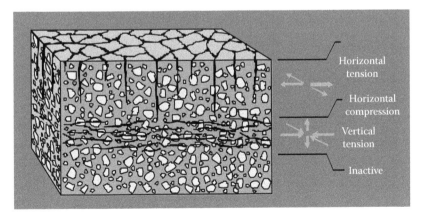

FIGURE 7.1
Idealized sketch of cracking pattern in concrete due to internal expansion caused by AAR. (From Walker, H.N. et al., Petrographic methods of examining hardened concrete: A petrographic manual, FHWA-HRT-04-150, Virginia Department of Transportation, Richmond, VA, 2006.)

(microcrystalline calcite, made of mud or clay-sized particles) and in which there is 5%–25% acid-insoluble residue composed of illite (layered aluminosilicate) and chlorite (micaceous clay mineral with chemical formula: iron-aluminum-magnesium-silicate-hydroxide), with some silt quartz. The ACR expansion is found to be its highest when the calcite/dolomite ratio in the carbonate portion of the rock approaches 1:1. This unique mineral composition and texture has been used to identify reactive carbonate aggregates. However, it should be noted that high ACR expansion is found in other varieties of dolomitic limestone also, besides the dolomitic micritic variety [99].

The expansion and cracking due to ACR occurs due to one or a combination of the following processes [93,100]: (a) hydraulic pressure caused by migration of water and alkali ions into restricted space around dolomite crystals, (b) adsorption of alkali ions and water molecules on the surface of active clay minerals scattered around the dolomite grains, and (c) growth and rearrangement of the products of de-dolomitization, namely, brucite (magnesium hydroxide) and calcite.

The alkali carbonates produced in the de-dolomitization reaction react with CH in concrete to regenerate the alkali hydroxides in the pore solution (Reaction 7.19):

$$(Na, K)_2 CO_3 + Ca(OH)_2 \rightarrow CaCO_3 + 2(Na, K)OH \qquad (7.19)$$

The rate of expansion affects the extent of distress in concrete. If expansion of the aggregate is sufficiently slow, the concrete may not deteriorate presumably because of the long time factors of creep and autogenous healing [101]. Plate 7.5 shows a thin section, under the microscope, of rock fracture

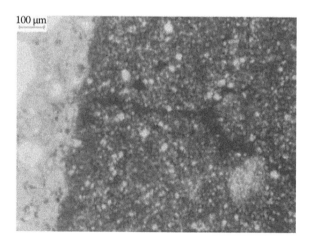

PLATE 7.5
Thin section under microscope of rock fracture caused by ACR. (From Walker, H.N. et al., Petrographic methods of examining hardened concrete: A petrographic manual, FHWA-HRT-04-150, Virginia Department of Transportation, Richmond, VA, 2006.)

caused by ACR [94]. The localized alteration due to carbonation of the paste adjacent to the aggregate particles can be observed in polished slabs and thin sections. This alteration proceeds from reactions associated with de-dolomitization (see Reaction 7.18) rather than the more common carbonation by atmospheric CO_2.

7.4.2.1.2 Alkali–Silica Reaction

ASR refers to the increased reactivity or instability of amorphous, disordered, micro- or crypto-crystalline form of silica found in the aggregate rocks, in the alkaline pore solution in concrete. The following rock types are susceptible to the ASR.

a. *Rocks incorporating poorly crystalline or metastable silica minerals*: These are found in opal, tridymite, cristobalite, and volcanic glasses. The concrete with aggregate incorporating such silica minerals, even in amounts as low as 1%–2% (opal and cristobalite), is susceptible within a few years of construction, when other conditions essential for the initiation and propagation of ASR exist.

b. *Rocks incorporating very fine grained quartz*: ASR in aggregates containing micro- or crypto-crystalline quartz is characterized by delay (5–25 years) in the onset of expansion and cracking in concrete, when other conditions essential for initiation and propagation of ASR exist.

When the aggregate prone to ASR is attacked by the alkali hydroxides present in the concrete pore solution, silica dissolves and alkali–silica gel is formed. The gel expands as water and various ionic species in the surrounding pore fluid flow into it, due to the difference in the free energy.* As the expansion is restrained in the set concrete matrix, tensile stresses build up at the localized sites. When these stresses exceed the tensile strength of aggregate particles and cement paste, it leads to the extensive formation of microcracks, under ASR-conducive conditions. The viscous substance called ASR gel spreads freely though these microcracks. It has been observed to appear as pop-outs or to exude from cracks in concrete structures. Over a period of time, it loses its expansive properties, incorporating calcium through an ion-exchange process with the cement hydrates, CH and CSH. ASR may be observed after 5–25 years of construction and may affect any concrete structure [102,103]. Plate 7.6 shows a typical portion of cracked concrete pavement [104], and Plate 7.7 shows a crack induced in a prestressed concrete beam by ASR [105].

The conditions that initiate and propagate AAR in concrete are as follows: (a) critical quantity of reactive aggregate, (b) sufficient alkali in concrete, and

---

* Helmholtz free energy is a thermodynamic quantity. It is minimized for a system in equilibrium. Reader may refer textbook on *Thermodynamics* to know more on the subject.

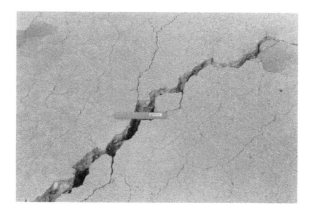

PLATE 7.6
Concrete pavement cracked due to ASR. (From Malvar, L.J. et al., Alkali-silica reaction miti-gation: State-of-the-art, TR-2195-SHR, U.S. Naval Facilities Engineering Service Center, Port Hueneme, CA, 2001.)

PLATE 7.7
Crack induced in prestressed concrete beam due to ASR. (From Ahlstrom, G., FHWA alkali-silica reactivity development and deployment program, *HPC Bridge Views*, 51, September–October 2008.)

(c) sufficient moisture. It should be noted that the attack and the damage due to AAR can occur only when all the three conditions prevail in concrete [106]. AAR is time-dependent and its occurrence is unpredictable, as there is wide variability in the conditions causing it. The variations occur in the degree of aggregate reactivity, type and composition of cement, concrete alkali content, water-to-cement ratio, type of mineral admixture, temperature, and moisture conditions.

Usually, while determining the safe alkali content in concrete made with an alkali-reactive aggregate, only the alkalies derived from the cement are considered. However, alkalis get added to concrete from different sources like aggregates and deicing salts. It is standard practice to express the alkali content in cement and concrete in terms of "sodium oxide equivalent," as given in Equation 7.20:

$$Na_2O_{eq} = Na_2O + 0.658 \, K_2O \text{ (mass percent)} \qquad (7.20)$$

Powers and Steinour [107] suggested that an alkali content of less than 0.6% Na_2O_{eq} (mass basis) in cement is normally harmless with respect to the AAR. Although such a limit will greatly minimize the risk, it should be noted that the alkali content of concrete is determined both by the alkali content of cement as well as the total cement content. Oberholster [108] gave a relationship between the active alkali content of cement, the cement content of concrete and the alkali content of concrete, and the resultant potential of the concrete for AAR. The investigation showed that an alkali content less than $1.8 \, kg/m^3$ of concrete is innocuous.

Water has a threefold function in AAR: first, it dissolves and ionizes alkali; second, it carries alkali to the other parts of concrete; and third, it is the cause of expansion of the gel produced by AAR. As per one PCA report on ASR [109], an internal relative humidity of less than 80% at 23°C (73°F) indicated insufficient moisture in concrete for the expansion. When other conditions conducive to AAR exist, mass concrete of hydraulic structures, such as dams, weirs, locks, and canals is more vulnerable, for the following reasons: (a) sufficient availability of moisture over the service period, (b) large aggregate size (76–150 mm) leading to stress differential and stress concentration [110], and (c) low resistance to AAR-induced cracking due to low compressive strength (10–20 MPa at 90 days) and corresponding low tensile strength. These aspects should be kept in mind, while designing these hydraulic structures.

The ions present in the concrete pore solution—hydroxyl, sodium, potassium, and calcium—penetrate aggregate having poorly crystalline structure. Glasser and Kataoka studied the AAR extensively [111–113] and distinguished four reactions for the silica gel/sodium hydroxide/CH system. Bazant and Steffens [114] gave an excellent review on the chemistry of ASR. The mechanism is summarized as follows:

a. When aggregate with poorly crystallized silica is placed in an alkaline medium of concrete pore solution, the attack of hydroxyl ions breaks the Si-O-Si or siloxane groups to form Si-OH or silanol groups (Reaction 7.21), resulting in loosening of the framework and dissolution of silica:

$$Si\text{-}O\text{-}Si + OH^- \rightarrow SiOH + SiO^- \qquad (7.21)$$

$$\rightarrow Si\text{-}OH \ldots OH\text{-}Si \qquad (7.22)$$

Normally, the surface oxygen of siliceous aggregate is hydroxylated up to a depth of several atoms or even tens of atoms (Reaction 7.22). When the aggregate comes in contact with highly alkaline pore solution, its potential to undergo further hydroxylation is enhanced. It holds good for quartz also but due to the strong bonds of its crystalline framework, the rate of hydroxylation is too slow for being considered on the normal engineering time scale. The poorly crystallized silica reacts much more rapidly.

b. The silanol groups on the surface of silica react with OH^- of the pore solution:

$$SiOH + OH^- \rightarrow SiO^- + H_2O \qquad (7.23)$$

As more siloxane bonds are attacked, a gel-like layer forms on the surface of the aggregate. Some silica may even pass into the solution, where the principal solute species is the monomer $H_2SiO_4^{2-}$, for highly alkaline solution [115].

c. The negatively charged species of this gel attract positive charges present in the pore solution in the form of mobile species such as sodium, potassium, and calcium. They diffuse into the gel in sufficient number to balance the negatively charged groups.

d. The presence of these positively charged ions determines important properties of the gel.

- When the surrounding environment is rich in Ca^{2+}, these ions will be taken up by the gel to form C-S-H. Thus, the gel is transformed into a rigid and unreactive structure. This process may be considered similar to the pozzolanic reaction in concrete.

- When pore solution is low in Ca^{2+}, the gel takes up mainly Na^+ and K^+ ions. That results in more viscous gel. The loosening of the crystal structure and the binding of sodium and potassium ions enhance the tendency to imbibe water. The silica is thus converted into a gelatinous polyelectrolyte. More hydroxyl ions then reach the interior and the process continues inexorably. With the increasing swelling pressure, concrete eventually ruptures [111]. It is difficult to predict the constitution of gel species that may form in a certain concrete. Plate 7.8 shows a thin-section cut from concrete affected by ASR, viewed under transmitted-light microscopy. The ASR gel is shown in the crack forming through the aggregate and extending into the surrounding cement paste. The damage is typical of ASR-induced deterioration at the microstructural level of concrete [116].

e. When the gel gets in contact with atmospheric CO_2, it carbonates. The carbonation is observed on the surface of cut concrete specimens that have undergone ASR.

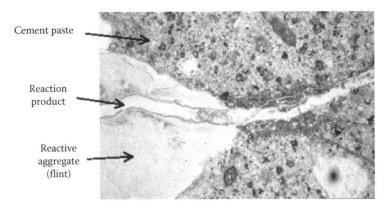

PLATE 7.8
Thin-section of ASR-damaged concrete, viewed under transmitted-light microscopy, showing ASR gel and crack pattern through aggregate and into surrounding matrix. (From Folliard, K.J. et al., FHWAHRT-06-073: Interim recommendations for the use of lithium to mitigate or prevent alkali-silica reaction (ASR), Office of Infrastructure Research and Development, Federal Highway Administration, Lakewood, CO, 2006.)

7.4.2.2 Mathematical Models for AAR

The structural behavior of concrete affected by the AAR is difficult to model due to the number of variable parameters that govern the chemical reactions. As discussed earlier, the occurrence of AAR depends on the simultaneous availability of three factors: critical amount of reactive aggregate, sufficient alkali in concrete, and sufficient moisture. Several micro-structural factors related to concrete are also involved in AAR expansion, such as porosity, amount, and location of reactive regions and permeability. These parameters, added to the heterogeneous nature of concrete make the modeling and simulation of AAR expansion difficult. According to the studies carried out by Glasser and Kataoka [111–113] the AAR pass through two distinct phases, first the formation of gel and next water absorption by gel, causing expansion. A comprehensive mathematical model on AAR must include (a) a model for the kinetics of the chemical and diffusional processes involved and (b) a model for the mechanical damage to concrete. The models on the specific aspects as referred in (a) and (b) as well as the attempts to develop a comprehensive model, to incorporate both, have been reported; still much remains to be done at this stage [117–121].

Bazant and Steffens [114] modeled the kinetics of the chemical and diffusional processes involved in ASR, considering incorporation of ground waste glass (mainly, bottle glass) into concrete. A characteristic unit cubic cell of concrete containing one spherical glass particle is analyzed. The spherical layer of basic ASR gel grows radially inward into the particle, controlled by diffusion of water toward the reaction front. The modification of the mathematical solution, for the case of mineral aggregates with veins of silica, is also indicated. The authors obtained an expression for the mass of gel formed, solving the steady-state diffusion equation. The imbibition of additional water from the adjacent capillary pores, which causes swelling of the gel, is described as a second diffusion process, limited by the development of pressure due to resistance of concrete to expansion.

Ichikawa and Miura [120] studied the mechanism of the development of pressure that leads to cracking in aggregates affected by ASR. The authors found that a hard reaction rim is slowly generated from the alkali silicate that covers the ASR-affected aggregate. The expansive pressure is generated due to the accumulation of the alkali–silica gel that continuously imbibes water from the surroundings inside that rim. Plate 7.9 shows the rim [122]. The reaction rim is a semi-permeable membrane that allows the penetration of alkaline pore solution inside but does not allow the hydrated gel to leak out, thus leading to the generation of expansive pressure inside the aggregate. The authors derived an expression for the maximum pressure that could be generated and illustrated, taking a typical example, that such pressure could be substantially more than that required for cracking. The mechanism of ASR-induced cracking of concrete has been schematically presented. The model claims that the ASR, when completed before the formation of the reaction rim, does not lead to the deterioration of concrete. The reactive but tiny silica-rich particles of PFA or the municipal waste incinerator bottom ash do not induce the deterioration of concrete, since they get completely converted to alkali silicate before the formation of reaction rim.

Grimal et al. [123] presented a model developed on the basis of finite element method, to calculate the mechanical behavior of AAR-damaged structures. The model takes into account all major phenomena occurring in AAR: the concrete creep, the stress induced due to the formation of gel, and the mechanical damage. The AAR gel pressure is calculated, based on the assumption that there is no coupling between the mechanical stress and the chemical

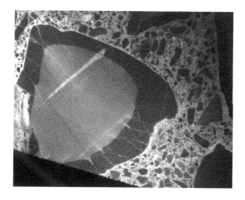

PLATE 7.9
Alkali–silica gel seen as dark rim around aggregate particle. (From Martin, L.C., Engineering Technical Letter (ETL) 06-2: Alkali-aggregate reaction in Portland cement concrete (PCC) airfield pavements, U.S. Department of Air Force, Washington, DC, February 9, 2006.)

phenomena responsible for the gel formation. The long-term behavior of concrete, that is, creep and shrinkage, is verified by a rheological model. The swelling transfer between the restrained and free swelling directions is taken into consideration by the anisotropic (non-isotropic or having properties that differ according to the direction of measurement) damage model, based on the effective stress concept. The model has been tested with the experimental results obtained on various reinforced concrete beams damaged by AAR. The comparison shows the capability of the model to reproduce the mid-span deflection and the crack pattern of the beams with acceptable accuracy. The method is a part of the global methodology already used to predict the past and the future behavior of a French dam affected by AAR.

7.4.2.3 National Standards and Guidelines on AAR

There is no universally accepted standard testing method for all cases of AAR.

The national standards, world over, have several test methods to identify potential reactivity of aggregate and to test the concrete durability against AAR. These aggregate testing methods may be classified into three types: petrographic examinations, expansion tests, and chemical analyses. Table 7.4 lists the ASTM standards to test aggregate reactivity and concrete durability against AAR. ASTM C227 is not suitable for determining the threshold level of alkali required to initiate expansion with a particular

TABLE 7.4

ASTM Standards to Test Aggregate Reactivity and Concrete Durability against AAR

Number	Title
ASTM C227	Standard test method for potential alkali reactivity of cement-aggregate combinations (mortar-bar method)
ASTM C1567	Standard test method for determining the potential alkali-silica reactivity of combinations of cementitious materials and aggregate (accelerated mortar-bar method)
ASTM C1260	Standard test method for potential alkali reactivity of aggregates (mortar-bar method)
ASTM C289	Standard test method for potential alkali-silica reactivity of aggregates (chemical method)
ASTM C586	Standard test method for potential alkali reactivity of carbonate rocks as concrete aggregates (rock-cylinder method)
ASTM C856	Standard practice for petrographic examination of hardened concrete
ASTM C441	Standard test method for effectiveness of pozzolans or ground blast-furnace slag in preventing excessive expansion of concrete due to the alkali-silica reaction
ASTM C1105	Standard test method for length change of concrete due to alkali-carbonate rock reaction
ASTM C1293 REV B	Standard test method for determination of length change of concrete due to alkali-silica reaction

aggregate. It is seen that the alkali content required to produce expansion in this test is much greater than that required in the concrete prism test (CPT) or field-exposed concrete blocks. ASTM C441 has been specifically developed to evaluate the mineral admixtures in construction. The test is essentially the same as ASTM C227 except that Pyrex (borosilicate) glass is used as the reactive aggregate. ASTM C1260, "Standard Test Method for Potential Alkali Reactivity of Aggregate," is a short-term accelerated test to measure expansion of mortar bars and probably the most widely used test method. It is also commonly known as Accelerated Mortar Bar Test (AMBT). The test involves the immersion of mortar bars in 1 M NaOH solution at 80°C (176°F) for 14 days and produces results within 16 days. The accepted maximum expansion for nonreactive aggregates is 0.1% at 14 days after the zero reading or at 16 days after casting; between 0.1%–0.2% it is inconclusive and higher than 0.2% it is deleterious. The experience shows that this test method is best suited to normally reactive aggregate, while some slowly reactive aggregate (example: gneiss, quartzite, metabasalt, and granite) have been found to be deleteriously expansive in field performance, even though their expansion in this test is less than 0.1% at 16 days [124]. ASTM C1293, "Standard Test Method for Determination of Length Change of Concrete Due to Alkali-Silica Reaction," is a long-term testing method using concrete prisms. It is also known as CPT. The aggregate is classified as deleteriously reactive, when the average expansion of the samples tested is ≥0.04%, at 1 year. The 1 year testing duration is considered too long for the structures under construction; moreover the quality of aggregate itself may not remain consistent over the year. The method may be useful for slowly reactive aggregates. The main shortcomings of the CPT as a means to evaluating the efficiency of mineral admixtures in controlling damage due to ASR are (a) the duration of the test and (b) its inability to determine how the minimum level of cement replacement changes with the cement alkali content, as additional alkali is required to compensate for leaching effect, while conducting the test. The extracted samples from the existing structure may be examined according to ASTM C856, "Standard practice for petrographic examination of hardened concrete."

7.4.2.4 Mitigation and Management of AAR

The earlier paragraphs discuss about the three essential prerequisites, which must be obtained simultaneously, for the initiation and sustenance of AAR. The measures to prevent or mitigate AAR involve eliminating one of the three prerequisites and/or changing the nature of the reaction by introducing mineral admixtures. It is difficult to fully avoid water ingress into concrete, especially in hydraulic structures. Limiting the alkali content of cement and concrete is the second alternative and for potentially reactive aggregates, a maximum alkali content of 0.4% in cement is recommended. However, experience shows that it is not an absolutely reliable method to control AAR. The third alternative, the use of nonreactive aggregates, may not be feasible for the reasons of availability and the economy.

As an electrochemical method, lithium compounds, especially lithium nitrate ($LiNO_3$), can be added to the concrete mix to counter and mitigate AAR [95,125,126]. However, special precaution is required with regard to other lithium compounds, as lithium hydroxide (LiOH) and lithium carbonate (Li_2CO_3) have been found to promote ACR. Some lithium compounds, in insufficient quantities, can actually increase the expansion. This is known as the pessimum effect, that is, at a certain lithium level the concrete will expand significantly, whereas at other levels expansion may be negligible. The lithium nitrate does not exhibit a pessimum effect. Although lithium compounds have been used to mitigate AAR expansion in some structural concrete, research is needed to evaluate the effect on mass concrete [127].

The mineral admixtures replacing cement, such as BFS, PFA and natural pozzolans, mitigate or eliminate AAR in concrete. There is no general agreement on the predominant mechanism by which the mineral admixtures reduce the AAR expansion. Glasser reviewed different theories [128]. The total alkali content of the concrete and the level of cement replacement by mineral admixtures are the key factors governing the expansion behavior of concrete. Significant reduction in AAR expansion is achieved for higher replacement level of cement by PFA (>25%) or BFS (>40%) or the combination of both. The ASTM Class F or N PFA should also have a maximum of 1.5% available alkali, a maximum 6% loss on ignition and a maximum of 8% CaO [104,129,130]. However, low volume replacement of even Class F PFA may actually worsen ASR problems, as reported by Malvar et al. [131]. The pozzolanic activity of PFA depends on its fineness and glass content. Whereas the alkali content (expressed as Na_2O equivalent = $Na_2O + 0.658K_2O$) of PFA tends to accelerate the ASR, quartz and mullite in it decrease the expansion and the quantity of alkali in cement determines the minimum amount of PFA needed in the mix to control the expansion within 0.04% in 1 year (ASTM C1293). Kobayashi et al. suggested an empirical formula to assess the effectiveness of PFA in alkali-attack-prone concrete [132], as given in Equation 7.24:

$$\Sigma a_c + 0.83\Sigma a_p - 0.046\Sigma \text{pfa} \leq 4.2 \ \text{kg/m}^3 \tag{7.24}$$

where
Σa_c and Σa_p are the total alkali content from cement and PFA, respectively
Σpfa is the PFA content kg/m^3 concrete

In practice, the quantity of PFA obtained from Equation 7.24 should be used with safety factor, based on the experience. The impact on ASR of using PFA, with various calcium content, as cement replacement is shown in Figure 7.2 [133] and Plate 7.10 illustrates the effect of using low calcium

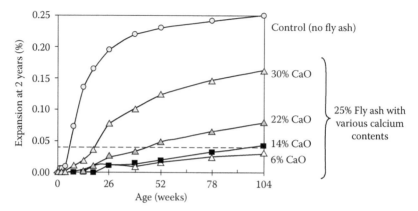

FIGURE 7.2
Expansion of concrete due to ASR using PFA with various calcium content (ASTM C1293). (Courtesy of T.W. Bremner, http://www.unb.ca/civil/bremner/CIRCA/Images/Circa_8_Use_in_b_v.jpg)

(a)   (b)

PLATE 7.10
Effect of ASR on 900mm concrete cubes at 11 years with 100% ordinary PC (a) and with 25% low calcium PFA replacement (b). (From Thomas, M. et al., ICAR 302-1: Test methods for evaluating preventive measures for controlling expansion due to alkali-silica reaction in concrete, International Center for Aggregates Research, The University of Texas at Austin, Austin, TX, 2006.)

PFA [134]. It is observed that the level of cement replacement by mineral admixtures depends upon (i) the composition of the mineral admixture (particularly CaO, SiO_2, and Na_2O equivalent), (ii) the nature and the level of reactivity of the aggregate, and (iii) the alkali content provided by the Portland cement (and other sources such as aggregate and the chemical admixtures) [134]. The high-volume fly ash concrete (HVFAC) is found highly effective in inhibiting expansion due to AAR [135].

The BFS added to cement and concrete has been found effective in controlling ASR, like that in case of PFA. The contribution of alkalis in slag to the ASR determines the total expansion [136,137]. The replacement of cement in large quantity (>60%) is effective in reducing the ASR and in some instances ACR [138].

SF is considered effective in suppressing concrete expansion due to AAR, provided it is used in sufficient quantities. However, the long-term effectiveness of SF against AAR is questioned by some workers [139,140]. The phenomenon is attributed to probable recycling of alkali entrapped in the C-S-H with low C/S, formed during hydration. There is a need for long term field data to determine all the parameters that influence the effectiveness of SF.

The addition of mineral admixtures to concrete, replacing cement, is found as an effective and economically feasible means to control the AAR, in most cases. The effectiveness of the mineral admixtures in reducing the AAR expansion potential is verified by testing, before use. The Texas Department of Transportation (TxDOT), United States, uses concrete in large volume (46×10^6 m^3 in 2006). The department developed the following eight mix design options with industry input, considering all aggregates as potentially reactive [105].

TxDOT Option 1: Replace 20%–35% of the cement with ASTM Class F PFA.

TxDOT Option 2: Replace 35%–50% of the cement with GGBS or ASTM Class F PFA modified to improve early strength gain and setting properties.

TxDOT Option 3: Replace 35%–50% of the cement with a combination of ASTM Class F PFA (35% max), GGBS, ASTM Class F PFA modified to improve early strength gain and setting properties, ultrafine PFA, MK, or SF (10% max).

TxDOT Option 4: Use ASTM C595 Type IP or Type IS cement. Up to 10% of a Type IP or IS cement may be replaced with ASTM Class F PFA, GGBS, or SF.

TxDOT Option 5: Replace 35%–50% of the cement with a combination of ASTM Class C PFA (35% max) and at least 6% of SF (10% max), ultrafine PFA, or MK.

TxDOT Option 6: Use a 30% solution of lithium nitrate admixture at a minimum dosage of 0.55 U.S. gal/lb (or 4.58 L/kg) of alkalies.

TxDOT Option 7: Use straight cement, if the total alkali contribution from the cement in the concrete does not exceed 4.00 lb/yd^3 (or 2.372 kg/m^3) of concrete.

TxDOT Option 8: Performance testing. Test both coarse and fine aggregates separately in accordance with ASTM C1567 and certify that expansion for each aggregate does not exceed 0.10%.

The test results revealed that the permeability of concrete is reduced, when mineral admixtures are used, which means less moisture penetrates concrete to form ASR gel. The mineral admixtures react with CH and lower the alkalinity of concrete and tie up free calcium ions needed to form ASR gel. It should be noted that the TxDOT generally uses prescriptive specification, when specifying HPC, to keep the construction cost down, as the performance specification tends to increase the bid price, in order to cover the contractor's risk, on account of the high volume of concrete consumption.

Fournier et al. [141] have published an excellent review on the management of concrete structures affected by the ASR. It involves overall interpretation of the results of both the field and the laboratory investigations. This is essential to develop long-term monitoring programs and to determine the nature and the extent of the repairs required. According to the authors, the management of ASR consists of one or more of the following:

- No further action necessary.
- Continue regular monitoring and inspections; in situ and periodical laboratory investigations.
- Perform minor or periodic (small scale or maintenance) repairs to minimize moisture ingress and deterioration.
- Perform significant restoration including chemical (e.g., injection of CO_2 and treatment with lithium) and physical (stress relief, strengthening, and post-tensioning) interventions to reduce or restrain the expansion process due to ASR.
- Replace the damaged structure.

It should be noted that it is impossible to interrupt AAR in most cases. The only way to lessen its harmful effects is by taking remedial measures, whose effectiveness depends strongly on an adequate prediction of the stress and strain field development in the built structure. The dams, for instance, face operating problems due to closing of expansion joints and changing of the original geometry of structural components. A frequently used remedial technique is opening or reopening of expansion joints, to release the generated stresses and allow the original geometry to be restored [119].

Reviewing 60 years of research on ASR, Idorn [142] made this observation: "It can be concluded that ASR occurs in field concrete all over the world. The majority of available aggregate materials are alkali-reactive, nevertheless, the majority of ASR occurring with field concrete is harmless."

7.4.3 Chloride Attack and Corrosion of Reinforcement

The reinforced concrete is durable and cost-effective; hence, its use is common in construction. The attack by aggressive agents such as chloride ions, leading to the corrosion of reinforcing steel, may cause a structure to deteriorate. As the cost of structural repairs is high, chloride transport in concrete structures continues to receive great attention. The corrosion resistance of concrete structures becomes important in salt-bearing environment such as in marine or coastal areas, with deicing salts and also when concrete is prepared with contaminated aggregates. Besides general constructions for domestic and industrial use, large number of other reinforced concrete constructions built in marine and coastal areas, such as docks, jetties, bridges, breakwaters, and varieties of sea-defense and off-shore oil exploration work require protection from corrosion.

7.4.3.1 Mechanism of Chloride Corrosion

The reinforcing steel, surrounded by the alkaline environment in concrete structure, is protected from the attack of aggressive agents by the passive iron oxide film formed on its surface. However, when the chloride ions penetrate through porous concrete and build up around the reinforcement and the alkalinity (pH) of the surrounding pore solution falls substantially, the protective iron oxide film depassivates and cracks, exposing the steel. The exposed steel gets corroded in the presence of water and oxygen, resulting in the formation of expansive corrosion products (rust) that occupy several times the volume of the original steel consumed. The expansive corrosion products create tensile stress in the concrete surrounding the corroding steel reinforcing bar. That leads to cracking and spalling of concrete cover, a usual consequence of corrosion of steel in concrete.

The corrosion of reinforcing steel in concrete is an electrochemical process. Table 7.5 [53] mentions such deleterious agents that cause or assist corrosion and summarizes the attendant electrochemical reactions and their effects on the concrete structures.

The corrosion process of steel can be described in a simplified way by an equivalent electrical circuit and the corrosion current (I_{corr}) is written as in Equation 7.25:

$$I_{corr} = \frac{(U_c - U_a)}{(R_a + R_c + R_e)} \tag{7.25}$$

where
U_c is the open circuit potential at cathode
U_a is the open circuit potential at anode
R_a is the anodic polarization resistance
R_c is the cathodic polarization resistance
R_e is the electrolytic resistance of concrete

Equation 7.25 shows that corrosion current (I_{corr}) and consequently the rate of corrosion (generally expressed in micron or 10^{-6} m/year) will increase, if either or both the anodic polarization resistance (R_a) and the electrolytic resistance (R_e) of concrete decrease. The polarization resistance of anode comes due to the formation of passive film of iron hydroxides/oxides during the process of corrosion. This resistance will decrease, if this protective film breaks under certain conditions. As shown in Table 7.5, under marine

TABLE 7.5

Corrosion-Causing or Assisting Chemical Reactions in Concrete

Sl No	Reaction Causing Agent	Reaction	Effects
i	Oxygen[a]	Anode: $2Fe \text{ (metal)} \rightarrow 2Fe^{++} + 4e^-$ Cathode: $2H_2O + O_2 + 4e^- \rightarrow 4OH^-$ $2Fe^{++} + 4OH^- \rightarrow 2Fe(OH)_2$	Corrosion of steel. Formation of protective passive film of nanometer thickness of iron hydroxides/oxides
ii	Chloride[a]	Without oxygen at anode $Fe + 2Cl^- \rightarrow (Fe^{++} + 2Cl^-) + 2e^-$ $(Fe^{++} + 2Cl^-) + 2H_2O \rightarrow Fe(OH)_2 + 2H^+ + 2Cl^-$ In presence of oxygen at anode $6(Fe^{++} + 2Cl^-) + O_2 + 6H_2O \rightarrow 2Fe_3O_4 + 12H^+ + 12Cl^-$ Chloride acts as catalyst in corrosion of steel and becomes free to take part in corrosion reaction again Attack on hydrated paste: $Ca(OH)_2 + MgCl_2 \rightarrow CaCl_2 + Mg(OH)_2$ $3CaO \cdot Al_2O_3 \cdot 6H_2O + CaCl_2 + 4H_2O \rightarrow 3CaO \cdot Al_2O_3 \cdot CaCl_2 \cdot 10H_2O$ (Friedel salt) Friedel salt $\rightarrow$ ettringite (in presence of $CaSO_4$) C-S-H + $MgCl_2 \rightarrow$ C-M-S-H or M-S-H + $CaCl_2$	Chlorine ions break the passive oxide film formed on steel. External penetration causes differential concentration and sets up microcells. Presence of salt increases its electrical conductivity Fe-hydroxide tends to react further with oxygen to form higher oxides
iii	CO_2[a]	$Ca(OH)_2 + H_2O + CO_2 \rightarrow CaCO_3 + 2H_2O$	Reduces alkalinity of microenvironment, increasing the risk of corrosion. Releases more water
iv	Sulfate[a]	$SO_4^{--} + Ca^{++} + 2H_2O \rightarrow CaSO_4 \cdot 2H_2O$ Ettringite formation: $4CaO \cdot Al_2O_3 \cdot 13H_2O + 3CaSO_4 \cdot 2H_2O + 14H_2O \rightarrow 3CaO \cdot Al_2O_3 \cdot 3CaSO_4 \cdot 32H_2O + Ca(OH)_2$ $xCaO \cdot ySiO_2 \cdot zH_2O + xMgSO_4 + (3x + 0.5y - z)H_2O \rightarrow xCaSO_4 \cdot 2H_2O + xMg(OH)_2 + 0.5y(2SiO_2 \cdot H_2O)$ $4Mg(OH)_2 + SiO_2 \cdot nH_2O \rightarrow 4MgO \cdot SiO_2 \cdot 8.5H_2O + (n - 4.5)H_2O$	Ettringite causes expansion in concrete, leading to crack formation. Concrete becomes more susceptible to penetration of external corrosive agents. Reduction in the strength of concrete

[a] All reaction causing agents assist corrosion in presence of water. Hence water should also be considered as one of the corrosion-causing agent.

conditions, chloride ions (from $MgCl_2$) first react with the $CaOH_2$, reducing the alkalinity of the micro-environment surrounding the reinforcement. The $CaCl_2$ thus formed may either get leached out or react with C_3A present in the structure to form expansive product (ettringite) in the presence of gypsum. The $MgCl_2$ may also react with C-S-H and form less cohesive and weak C-M-S-H or M-S-H. The free Mg^{++} ions may react with C-S-H and form less cohesive and weak C-M-S-H. Once the crystalline $Ca(OH)_2$ is removed from the vicinity of reinforcing steel, the pH of pore solution drops below 11 and the iron oxide film loses its protection. In order to explain the deterioration of reinforced concrete structures in the marine environment, Mehta and Gerwick proposed a cracking–corrosion–cracking model [143]. The model postulates that substantially increased permeability of concrete through enlargement and interconnection of microcracks (due to carbonation, chloride and sulfate attack) assists corrosion.

The Pourbaix diagram (named after the scientist Pourbaix who studied and developed the diagram) for iron–water system, gives a "qualitative" understanding of "where" and "when" the corrosion will take place [144]. During the process of corrosion, metallic ions get converted to hydroxides and oxides depending on the pH and the availability of oxygen and water. The volume of these corrosion products is larger than the metallic ions. Their excessive formation results in expansive pressure. The measure of this expansive pressure can be obtained by Pilling and Bedworth Ratio (PBR), which may be written as in Equation 7.26 [145]. On the basis of the PBR, it can be predicted if the metal is likely to passivate in dry air by the creation of a protective oxide layer. It indicates whether the volume of corrosion product is greater or lesser than the volume of metal from which it is formed. When PBR < 1, the metal oxide tends to be porous and nonprotective because it cannot cover the whole metal surface. When PBR $\gg$ 1, the compressive stresses are likely to develop in metal oxide, leading to buckling and spalling. The protective oxide layers generally have a PBR of 1–2. The PBR for iron oxides, FeO and Fe_3O_4, are 1.7 and 2.1, respectively:

$$\text{PBR} = \frac{V_{\text{oxide}}}{V_{\text{metal}}} = \frac{\left[M_{\text{oxide}}/\rho_{\text{oxide}}\right]}{\left[n\left(M_{\text{metal}}/\rho_{\text{metal}}\right)\right]} \tag{7.26}$$

where
  V is the molar volume
  M is the atomic or molecular mass
  ρ is the density
  n is the number of atoms of metal per one molecule of the oxide

The role of chloride ions in initiating and propagating the corrosion of steel in concrete led to the concept of a chloride threshold level (CTL). The CTL can be defined as the content of chloride that is necessary to sustain breakdown

TABLE 7.6

Maximum Chloride Content Specified in ACI and BS Documents

| Sl No | Concrete Particulars | Maximum Chloride Content (% Mass of Cement) | | | |
		BS 8110[a]	ACI 201[b]	ACI 357[c]	ACI 222[d]
i	Prestressed concrete	0.10	—	0.06	0.08
ii	Reinforced concrete: exposed to chloride in service	0.20	0.10	0.10	0.20
iii	Reinforced concrete: dry or protected from moisture in service	0.40	—	—	—
iv	Other reinforced concrete	—	0.15	—	—

[a] BS 8110: Structural use of concrete. Code of practice for design and construction.
[b] ACI 201: Committee on Durability of Concrete.
[c] ACI 357: Committee on Offshore and Marine Concrete structures.
[d] ACI 222: Committee on Corrosion of Metals in Concrete.

of passive oxide film and initiate the corrosion of exposed steel [146]. It is usually presented as the ratio of chloride to hydroxyl ions, the free chloride content, or the percentage of the total chloride content relative to the mass of cement. The representation most widely used for the CTL is total chloride content relative to the cement mass, as it is convenient and takes into account the inhibiting effect of cement as well as the aggressive nature of chloride [147]. The values of maximum chloride content specified by different ACI and BS documents are given in Table 7.6. The time needed to reach the CTL corresponds to the corrosion initiation period [148]. It is determined by series of parameters such as the properties of the concrete cover, its thickness and the exposure conditions. The prediction on corrosion initiation period can be reasonably improved with advanced modeling approach, when reliable CTL is available [56].

The reinforced concrete bridge decks and parking structures deteriorate due to chloride ion induced corrosion. They are exposed to chloride ions and high moisture content that decrease the resistivity of the concrete. The type of corrosion observed is typically macro-cell corrosion, where the top reinforcement is the anode and the bottom reinforcement acts as the cathode. The potential difference between the anode and cathode causes current flow and corrosion.

The ongoing discussion makes it clear that the onset and propagation of the corrosion in the steel reinforcement in concrete structures take place under the following conditions [53]:

a. For the anodic reaction, iron must be available in a metallic (Fe) state on the surface of reinforcing steel. Under marine conditions, chloride facilitates anodic reaction.

b. For the cathodic reaction, oxygen and moisture must be continuously available.

c. Low electrical resistivity of concrete surrounding the reinforcement facilitates the flow of corrosion current.

d. Substantially increased permeability of concrete through enlargement and interconnection of microcracks (due to CO_2, chloride, and sulfate attack) assist corrosion.

7.4.3.2 Mathematical Models for Chloride Attack

The results of chloride analyses on cement, mortar or concrete, subjected to chloride attack, are generally presented in the form of chloride profiles, for the purpose of comparison. The comparison of the chloride resistance of cementitious materials in terms of diffusion coefficients is also useful, particularly in the design of concrete structures for durability against aggressive environment. The diffusion of chloride ions through concrete is a complex and time-dependent phenomenon, controlled by numerous interdependent parameters. Some important influencing factors are the w/b ratio, temperature of surrounding environment, effect of cations coexisting with chloride ions, type of exposure conditions, curing temperature, cement type, and the presence of mineral admixtures in the concrete mixture [149].

The Fick's second law of diffusion has been used to predict the chloride profiles and the time required for the initiation of chloride-induced corrosion in concrete [150–153]. The diffusion is the process by which ions or molecules move from an area of higher concentration to that of lower concentration. It is the primary mechanism of chloride transport in concrete, where there is no applied electric field and the moisture content of the concrete pores is stable. The Fick's second law predicts how diffusion causes the concentration to change with time (Equation 7.27):

$$\frac{\partial C}{\partial t} = D\frac{\partial^2 C}{\partial x^2} \qquad (7.27)$$

where
C is the concentration, mol/m^3 or percent by mass of cement
t is the time, s
D is the diffusion coefficient, m^2/s
x is the position, m

The solution as given in Equation 7.28, in a simple case of diffusion in one dimension (taken as the x-axis), gives chloride concentration $C(x, t)$ at time t and depth x from the boundary located at position $x = 0$, where the concentration is maintained at a value C_s. The erf is error function.*

$$C(x,t) = C_s\left[1 - \text{erf}\,\frac{x}{2\sqrt{Dt}}\right] \tag{7.28}$$

When the amount of chloride ions accumulated at the steel-concrete interface exceeds the hydroxyl ion concentration (responsible for the alkalinity) by a certain proportion, the attack of corrosion on the reinforcement is imminent, unless the interface is devoid of oxygen. The model at Equation 7.28 is commonly used in determining chloride transport in concrete [154].

It should be noted that the validity of diffusion Equation 7.28 rests on a series of simplifying assumptions—semi-infinite domain, constant diffusion coefficient (D), and constant exposure conditions—which are often not met in reality. The diffusion coefficient calculated using chloride profiles also includes the effects of chloride binding and the ingress from other transport modes, namely, sorption and convection. In certain situations (like marine splash and tidal zones) the combined effect of diffusion, sorption and convection may significantly increase the chloride content of concrete and may yield an incorrect value of diffusion coefficient, several times larger than the actual [155]. The standard tests also suffer from these drawbacks.

The appearance of the first corrosion crack is usually used to define the end of functional service life, when rehabilitation of a corroding structural element is required [156]. Maaddawy and Soudki [157] presented a mathematical model that can predict the time from corrosion initiation to corrosion cracking, based on the relationship between the steel mass loss and the internal radial pressure caused by corrosion. The concrete ring is assumed to crack when the tensile stresses in the circumferential direction reach the tensile strength of the concrete. The comparison of model's predictions with experimental results, published in the literature, showed that the model gives reasonable prediction. Lin et al. [158] proposed a model to predict the service life of RC structures, which takes into account the environmental humidity and temperature fluctuations, chloride binding, diffusion and convection, as well as the decay of structural performance. The application of the numerical model was demonstrated by predicting service life of a RC slab exposed to chloride environment.

---

* *Error function*: In mathematics, the error function (also called the Gauss error function) is a special function (non-elementary) of sigmoid shape which occurs in probability, statistics, materials science, and partial differential equations. It is defined as:

$$\text{erf}(x) = \frac{2}{\sqrt{\pi}}\int_0^x e^{-t^2}dt$$

The complementary error function, denoted by erfc, is defined as:

$$\text{erfc}(x) = 1 - \text{erf}(x)$$

7.4.3.3 National Standards and Guidelines on Chloride Corrosion

AASHTO T259 is a 90-day salt-ponding test [159] to evaluate the chloride diffusion coefficient. It suffers from the drawbacks such as the effect of other transport modes and the time dependent changes occurring in the pore structure. The steady-state diffusion tests eliminate the influence of other chloride transport modes by maintaining saturation but still require long exposure periods (long-term ponding in years) to obtain reliable estimates of chloride diffusion coefficient [160]. The accelerated chloride migration or the so-called rapid chloride permeability test (RCPT) is developed to nondestructively obtain the indication on chloride permeability of concrete, in a short time period.

The rapid test method was originally developed by the Portland Cement Association (PCA), under a research program sponsored by the Federal Highway Administration (FHWA), United States [161]. It was modified and adapted by various agencies and standards' organizations. The construction industry accepts such test procedures as means to obtain indication on chloride permeability of concrete. The common procedures include the following.

- AASHTO T277: "Standard method of test for rapid determination of the chloride permeability of concrete"
- ASTM C1202: "Standard test method for electrical indication of concrete's ability to resist chloride ion penetration"

Many concrete structures are built today with specifications that require low-permeability concrete. Table 7.7 lists ASTM/AASHTO standards to test concrete durability against chloride attack.

The accelerated migration of chloride ions in the rapid tests is achieved applying the electrical voltage across the specimen and the passage of charge, that is, coulombs (integral of current-versus-time plot) is measured. The test results are compared to the values in the standard chart, to obtain the indication of chloride permeability. Figure 7.3 shows the typical results of RCPT test conducted on concrete samples, as per ASTM C1202 [162]. The test methods measure concrete resistivity not the permeability. The resistance is calculated as volts divided by current. It has been shown that there is a fair correlation between concrete resistivity and concrete permeability. The standards allow large variability in test results (which indicates the relative inaccuracy of the test methods), while maintaining that concrete samples which lie within the large acceptable range are essentially equal in quality. Notwithstanding the attempts made to correlate chloride diffusion coefficient from ponding test and the transient (unsteady) state diffusion coefficient from the rapid tests [160,163,164], there is still a need to develop a better understanding of the relationship among short-, medium-, and long-term tests to assess the resistance of concrete to chloride penetration.

TABLE 7.7

ASTM/AASHTO Standards to Test Concrete Durability against Sulfate
and Chloride Attack

Number	Title
ASTM C1012/C1012M	Standard test method for length change of hydraulic-cement mortars exposed to a sulfate solution
ASTM C452/C452M	Standard test method for potential expansion of Portland-cement mortars exposed to sulfate
ASTM D516	Standard test method for sulfate ion in water
ASTM C1582/C1582M	Standard specification for admixtures to inhibit chloride-induced corrosion of reinforcing steel in concrete
ASTM C1202	Standard test method for electrical indication of concrete's ability to resist chloride ion penetration
ASTM C1556	Standard test method for determining the apparent chloride diffusion coefficient of cementitious mixtures by bulk diffusion
ASTM C1152/C1152M	Standard test method for acid-soluble chloride in mortar and concrete
ASTM C1218/C1218M	Standard test method for water-soluble chloride in mortar and concrete
ASTM C1524	Standard test method for water-extractable chloride in aggregate (Soxhlet method)
AASHTO T259	Standard method of test for resistance of concrete to chloride ion penetration
AASHTO T260	Standard method of test for sampling and testing for chloride ion in concrete and concrete raw materials

ASTM, American Society for Testing and Materials; AASHTO, American Association of State Highway and Transportation Officials.

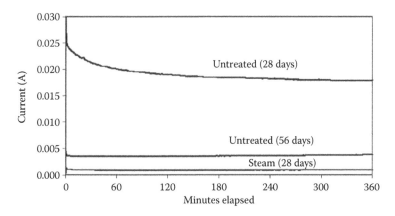

FIGURE 7.3

RCPT results on concrete as per ASTM C1202. Example: If 0.003 A current passed through the concrete for the duration of the test (360 min), then the total charge passed would be 65 coulombs. (From Graybeal, B.A., FHWA-HRT-06-103: Material property characterization of ultra-high performance concrete, Office of Infrastructure Research and Development, Federal Highway Administration (FHWA), U.S. Department of Transportation, Washington, DC, August 2006.)

The Florida method of test for concrete resistivity, FM 5-578 [165], is a nondestructive laboratory test method, which measures the electrical resistivity of the surface of water-saturated concrete and provides an indication of its permeability. The test result is a function of the electrical resistance of the specimen. According to the method, the surface resistivity of <12 kΩ cm is indicative of high chloride permeability. The test document mentions that the method is applicable to concrete with mineral admixtures, such as FA, BFS, SF, or MK, but can produce misleading results when calcium nitrite, reinforcing steel, conductive fibers, or other embedded electrically conductive materials are present in the concrete. The test method is not applicable to cores, as these can be contaminated with conductive chloride ions.

The qualitative assessment on the state of corrosion can be obtained from the criteria given in the American Standard ASTM C876 [166], which gives the half-cell potential of steel reinforcement in RCC structures for saturated colomel electrode (SCE, which is normally saturated copper/copper sulfate ($Cu/CuSO_4$) reference electrode) under standard conditions (Table 7.8). As stated in the standard document, the test method is suitable for in-service evaluation of the concrete members regardless of their size or the depth of concrete cover over the reinforcing steel. The standard test estimates the electrical corrosion potential of uncoated reinforcing steel in field and laboratory concrete, for the purpose of determining the corrosion activity of the reinforcing steel. It provides general principles for evaluating the probability of corrosion of reinforcing steel in concrete. It should be noted that the potential can change in a wide range, depending upon the moisture content of concrete and may sometimes lead to erroneous conclusions [167]. Under such situations the state of corrosion of the reinforcement can be predicted reliably by analyzing the data from both half-cell potential and corrosion rate ($\mu A/cm^2$) measurements and by considering the effects of environmental conditions [168]. The estimate of possible reinforcement corrosion can be made measuring the corrosion current; 9.12 kg iron will corrode (oxidize) each year for each ampere of positive current flowing from the exposed iron surface to the surrounding electrolyte [169].

In a study sponsored by TxDOT, United States, the newly developed rapid methods were evaluated to measure the corrosion performance of

TABLE 7.8

Qualitative Assessment on the State of Reinforcement Corrosion as per ASTM C876

Potential (mV SCE)	State of Corrosion
More negative than −270	Active
More positive than −220	Passive
−220 to −270	Active or passive

reinforcing steel in concrete [170]. The research evaluated four accelerated test procedures: (a) accelerated chloride test (ACT), (b) rapid macrocell test, (c) chloride ion threshold test, and (d) modified ASTM G109 test [171]. The evaluation was done on the basis of reasonableness of the test results, simplicity, cost, and duration. Not considering the one-time equipment cost, tests (a), (b), and (c) decrease the cost by approximately 58%, 75%, and 67%, respectively, in comparison to the standard ASTM G109 test. The rapid macrocell test was determined to be relatively simple, while the other two tests were considered to be more complex to perform.

7.4.3.4 Mitigation of Chloride Corrosion

The mineral admixtures impart enhanced resistance to the diffusion of chloride ions as well as the other deteriorating agents, on account of the refinement of pore structure due to the generation of additional C-S-H, through secondary hydration. The addition of mineral admixtures to concrete inhibits corrosion of reinforcement, improving the resistance toward chloride penetration and reducing the quantity of free (soluble) chloride in concrete. Besides delaying the initiation, the corrosion propagation period is also extended. The wide-scale, long-term studies reported in the literature conclude that the values of diffusion coefficient of chloride ions in concrete or concrete permeability increase with the w/b ratio and shows reduction for blended cement concrete. Figure 7.4 shows a typical trend [172]. The permeability of concrete is affected by the admixture type, content, moist-curing period, and age [149,155,160,173–177]. The additional C-S-H produced by the PFA pozzolanic reaction with the available CH allows concrete to continuously gain strength with time. The concrete mixture designed to produce equivalent strength at an early age (less than 90 days) will ultimately exceed the strength of plain cement concrete mix (Figure 7.5) [172]. The higher long-term strength and improved

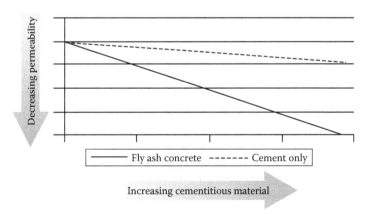

FIGURE 7.4
Permeability of concrete with PFA. (From American Coal Ash Association, FHWA-IF-03-019: Fly ash facts for highway engineers, Federal Highway Administration, Washington, DC, 2003.)

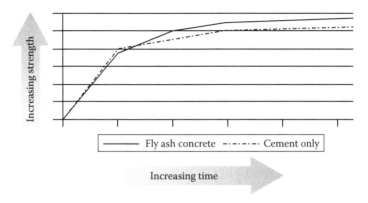

FIGURE 7.5
Typical strength gain of PFA concrete. (From American Coal Ash Association, FHWA-IF-03-019: Fly ash facts for highway engineers, Federal Highway Administration, Washington, DC, 2003.)

PLATE 7.11
Fly ash concrete is used under severe exposure condition such as in the decks and piers of Sunshine Skyway Bridge, Tampa Bay, Florida. (From American Coal Ash Association, FHWA-IF-03-019: Fly ash facts for highway engineers, Federal Highway Administration, Washington, DC, 2003.)

resistance toward corrosion makes PFA concrete suitable for construction of heavy-duty structures under severe marine conditions (Plate 7.11) [172].

The analysis of cement systems exposed to chloride shows that they chemically react with the aluminate phases in the paste to form $3CaO \cdot Al_2O_3 \cdot CaCl_2 \cdot 10H_2O$, that is, Friedel's salt (Equations 7.29 and 7.30). The physical binding or adsorption has also been reported as one of the mechanisms of chloride binding [178]. The mechanisms for Friedel's salt formation have been reported in the literature.

The NMR results [179] suggest two different mechanisms for Friedel's salt formation: dissolution/precipitation and the ionic exchange. Many studies suggest that the main mechanism is ionic exchange [180–181].

The possible mechanisms of chloride binding through ion exchange are given in Equations 7.29 and 7.30.

1. The hydroxyl-Af$_m$, C_4AH_{13} and its derivatives, release OH$^-$ ions from the interlayers and replaces it with Cl$^-$ ion:

$$[\text{X-OH}] + Cl^- \leftrightarrow [\text{X-Cl}] + OH^- \tag{7.29}$$

2. The monosulfate or sulfate-Af$_m$, $C_4A\hat{S}H_{12-18}$, releases SO_4^{2-} ion and replaces it with Cl$^-$ ions:

$$[\text{X-SO}_4] + 2Cl^- \leftrightarrow [\text{X-Cl}_2] + SO_4^{2-} \tag{7.30}$$

where X denotes the ion exchange sites.

The Friedel's salt is stable over a wide range of chloride concentrations. It is found that PFA and BFS added to cement have greater chloride binding capacity in comparison with PC. The superior chloride binding capacity of mineral admixtures enhances the resistance of concrete toward chloride ion transport [37,182,183]. However, some studies suggest that chloride binding and carbonation are intimately related [184]. In presence of $CO_2(g)$, the Friedel's salt tends to react. The reaction liberates Cl$^-$ ions in the pore solution. The chloride ions thus liberated may reach the reinforcement and initiate corrosion, depending upon the concentration gradients in the material. In the presence of excess SO^{4-} ions in the pore solution, Friedel salt may get converted to ettringite.

Samson and Marchand [185] developed a model for chloride transport based on the mass and energy conservation. The chemical interaction of chlorides with the hydrated cement paste was modeled based on an ionic exchange mechanism between monosulfates and Friedel's salt as in Equation 7.30. The predicted total chloride content accounts for chloride ions present in the pore solution and also those found in Friedel's salt.

7.4.4 External Sulfate Attack

The deterioration of concrete due to external sulfate attack is a commonly observed phenomenon when structures are exposed to sulfate solutions or built in sulfate bearing soil and/or ground water. All commonly obtained water soluble sulfates are deleterious (Mg > Na > Ca) to concrete, but the effect is severe when it is associated with Mg cations. Bapat reviewed the mechanisms of sulfate attack [53]. The cement-based materials exposed to sulfate-bearing solutions such as some natural or polluted ground waters (external sulfate attack or ESA) or by the action of sulfates present in the original mix (internal sulfate attack or ISA) can show signs of deterioration. The sulfate ions react with ionic species of the pore solution to precipitate gypsum, ettringite or thaumasite or a mixture of these phases, depending upon the temperature and other favorable conditions. The precipitation of these solid phases can lead to stress within the material, inducing expansion, strength loss, spalling, and severe degradation.

The following sections describe various aspects of the ESA. The delayed ettringite formation (DEF) is considered as the ISA and discussed in the subsequent section.

7.4.4.1 Mechanisms of External Sulfate Attack

The cement-based materials subjected to sulfate attack suffer from two types of damages: loss of strength due to decalcification of cement hydrates, mainly C-S-H, and cracking due to the formation of expansive compounds, mainly gypsum and ettringite. The volume change taking place is of the order of 48%–126%, depending upon the conditions in which sulfate attack takes place [186].

The solubility of calcium sulfate in water is limited at normal temperature (approximately 2400 mg/L at 20°C, dihydrate). Hence higher concentration of sulfate ions is generally due to the presence of magnesium sulfate (water solubility approximately 255 g/L at 20°C, anhydrous) and sodium sulfate (water solubility approximately 161 g/L at 20°C, anhydrous). These salts are abundant, in the arid, saline Sabkha soils along the Arabian Gulf coast, for example [187]. There is some difference in the mechanisms of deterioration caused by the two sulfates.

In the set cement, Na_2SO_4 and $MgSO_4$ react with free CH to form calcium sulfate ($CaSO_4 \cdot 2H_2O$), in the presence of moisture:

$$Ca(OH)_2 + Na_2SO_4 + 2H_2O \rightarrow CaSO_4 \cdot 2H_2O + 2NaOH \qquad (7.31)$$

$$Ca(OH)_2 + MgSO_4 + 2H_2O \rightarrow CaSO_4 \cdot 2H_2O + Mg(OH)_2 \qquad (7.32)$$

Reaction 7.32 goes to completion because the Mg $(OH)_2$ formed has low solubility (0.01 g/L) compared to Ca $(OH)_2$ (1.37 g/L) and low pH, that is, low alkalinity. The calcium sulfate formed in Reactions 7.31 and 7.32 reacts with the tricalcium aluminate (C_3A) as well as the hydrates of calcium aluminate (typically $4CaO \cdot Al_2O_3 \cdot 13H_2O$) and the monosulfate hydrate ($3CaO \cdot Al_2O_3 \cdot CaSO_4 \cdot 12H_2O$) in set cement (see Chapter 6), to form secondary ettringite ($3CaO \cdot Al_2O_3 \cdot 3CaSO_4 \cdot 32H_2O$), which is highly expansive (see Reactions 7.33 through 7.35) [187]:

$$3CaO \cdot Al_2O_3 + 3CaSO_4 \cdot 2H_2O + 26H_2O \rightarrow 3CaO \cdot Al_2O_3 \cdot 3CaSO_4 \cdot 32H_2O$$
$$(7.33)$$

$$4CaO \cdot Al_2O_3 \cdot 13H_2O + 3CaSO_4 \cdot 2H_2O + 14H_2O \rightarrow 3CaO \cdot Al_2O_3 \cdot 3CaSO_4 \cdot 32H_2O$$
$$+ Ca(OH)_2 \qquad (7.34)$$

$$3CaO \cdot Al_2O_3 \cdot CaSO_4 \cdot 12H_2O + 2CaSO_4 \cdot 2H_2O + 16H_2O$$
$$\rightarrow 3CaO \cdot Al_2O_3 \cdot 3CaSO_4 \cdot 32H_2O \qquad (7.35)$$

Magnesium sulfate attacks C-S-H in both PC and blended cements. Massive quantities of calcium sulfate are liberated (Reaction 7.36):

$$x\text{CaO} \cdot y\text{SiO}_2 \cdot z\text{H}_2\text{O} + x\text{MgSO}_4 + (3x + 0.5y - z)\text{H}_2\text{O} \rightarrow x\text{CaSO}_4 \cdot 2\text{H}_2\text{O}$$
$$+ x\text{Mg(OH)}_2 + 0.5y(2\text{SiO}_2 \cdot \text{H}_2\text{O}) \tag{7.36}$$

The low level of alkalinity produced by MH precipitation in Reaction 7.32 destabilizes C-S-H, which tends to liberate lime to establish the equilibrium. However, the liberated lime further reacts with $MgSO_4$ to form $CaSO_4$. The formation of MH in Reactions 7.32 and 7.36 causes the reactions to proceed in a repetitive manner with more gypsum ($CaSO_4 \cdot 2H_2O$) accumulating in pores and the crystals of gypsum separate out [49]. The massive gypsum precipitation facilitates induction of sulfate ions into the C-S-H matrix, leading to the reduction in the strength of concrete.

The silicate hydrate ($2SiO_2 \cdot H_2O$) or silica gel formed in Reaction 7.36 can be considered having a general composition $SiO_2 \cdot nH_2O$. The MH and silica gel can react slowly to form hydrated magnesium silicate (M-S-H) with an approximate composition: $4MgO \cdot SiO_2 \cdot 8.5H_2O$ [188]:

$$4\text{Mg(OH)}_2 + \text{SiO}_2 \cdot n\text{H}_2\text{O} \rightarrow 4\text{MgO} \cdot \text{SiO}_2 \cdot 8.5\text{H}_2\text{O} + (n - 4.5)\text{H}_2\text{O} \tag{7.37}$$

The M-S-H formed in Reaction 7.37 is non-cementitious and fibrous. The formation of M-S-H thus represents the final stage of deterioration of the concrete attacked by magnesium sulfate, although in practice, it is reached after a long time period.

Thus it is seen that $MgSO_4$ is more aggressive in comparison with Na_2SO_4, when the pH of pore solution in the hydrated cement is low (below 11.5); a condition frequently encountered in the laboratory investigations. However, the following findings reported in the literature [60,189,190,191] may be noted:

a. The laboratory test using magnesium sulfate solution is not valid for field conditions, where the pH is not lowered during the sulfate attack [189].

b. The magnesium hydroxide (brucite) forms a protective layer on concrete, unless this is damaged mechanically [190]. That explains the good record of concrete in seawater, which has a high magnesium sulfate content [60].

c. The carbonation of a concrete surface prior to exposure to sulfates reduces the buildup of sulfate within concrete [191].

The volume increase resulting out of the reactions generates expansive pressure and subsequent cracking in concrete. Initially, in the beginning of hydration, C_3A reacts with the gypsum to produce primary ettringite. This ettringite does not generate expansive pressure, despite large volume, as the concrete is not fully set and volume adjustment takes place at that stage [192]. The surplus C_3A, left after the reaction with gypsum, reacts with water to form $4CaO \cdot Al_2O_3 \cdot 13H_2O$, which reacts with $CaSO_4$ at a later stage, as shown earlier.

Unlike $MgSO_4$, Na_2SO_4 does not react with C-S-H to any appreciable extent. The extent of Na_2SO_4 attack on concrete is primarily dependent upon the quantity of free CH available and the C_3A content of cement. Plate 7.12 illustrates that aspect [193]. The changes in microstructure (relative to the

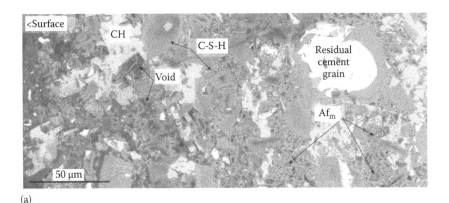

(a)

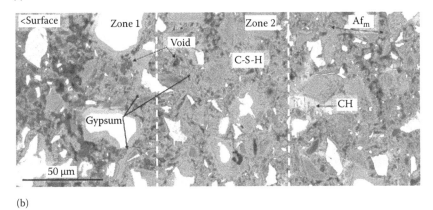

(b)

PLATE 7.12
SEM BE image of (a) hardened cement paste and (b) hardened cement paste exposed to sodium sulfate solution. *Note:* (a) Hardened cement paste in lime-water cured, control mortar specimen (outer surface is oriented left). The residual cement grains appear brightest followed by calcium hydroxide (CH), calcium-silicate-hydrate (C-S-H), and dark voids that are filled with epoxy. The ettringite (Af_t) and monosulfate (Af_m) occur in the bulk paste and are identified based upon their texture and chemical signature. The specimen was exposed to sodium sulfate solution for 105 days. After exposure (b), it shows increased porosity near the surface and loss of CH within 150 μm distance from the surface. The second zone is characterized by replacement of monosulfate with ettringite, densification of inner-product C-S-H (seen as a loss of coarse capillary porosity), and deposition of gypsum in place of CH. The field width in both images is 250 μm. (From Ferraris, C.F. et al., Sulfate resistance of concrete: A new approach, PCA R&D Serial No. 2486, Portland Cement Association, Skokie, IL, 2006.)

control), revealed in the SEM BE image, include increased porosity near the surface, replacement of monosulfate with ettringite, loss of CH and deposition of gypsum in its place and increased sulfate content in the C-S-H, at various depths.

The limestone powder is widely used as a filler in concrete for many years. The calcareous aggregates are also extensively used in many countries. At low temperatures (generally below 15°C), when such concrete is subjected to sulfate attack, especially $MgSO_4$, the formation of thaumasite $(CaSiO_3 \cdot CaCO_3 \cdot CaSO_4 \cdot 15H_2O)$ takes place, within a few months of exposure. It is a nonbinding calcium-carbonate-silicate-sulfate-hydrate, formed at low temperature from C-S-H, under conditions of destructive sulfate attack. The deterioration of concrete as a result of the thaumasite form of sulfate attack (TSA) has been recognized as a separate form of sulfate attack, which has the potential to affect a wide variety of structures. These include concrete foundations and floor slabs, roads and subbases, and tunnel linings and sewer pipes exposed to sulfate bearing surroundings, that is, water or soil with mobile groundwater. The thaumasite has also been found at the interface of lime-gypsum plasters with cement-based renders; in lime gypsum plasters with cement clinker or clay impurities and the mortar and renders of exposed sulfate-bearing brickwork. It is now established that the carbonate required for TSA may come from a source other than aggregates in the concrete. In particular, it can come from bicarbonate dissolved in groundwater. The full extent of this problem is still unknown [194–198].

The degradation can also take place when concrete is exposed to wetting and drying cycles with sulfate containing water (marine conditions). In addition, cyclic migration of water into concrete can be induced by capillary flow and the variations in the atmospheric relative humidity. The result is the concentration of sulfates (and other salts) at the subsurface. That can lead to disintegration of surface layer, when the concentration reaches sufficient level. There is no general agreement on the mechanism of such disintegration [199]. There is a difference of opinion as to whether the "physical attack" or the deposition of sulfates and other marine salts at the subsurface should be called as sulfate attack [60]. However, the physical attack is often followed by the chemical attack and to that extent it requires an important consideration. Santhanam et al. [200] proposed mechanism of sulfate attack according to which the attack due to sodium sulfate solution progresses in stages. The expansion of an outer skin of the specimen leads to the formation of cracks in the interior region, which is chemically unaltered. The surface skin disintegrates with continued immersion and the sulfate solution is able to react with the hydration products in the cracked interior zone leading to the deposition of reaction products in this zone. The interior zone then becomes the expanding zone, leading to further cracking of the mortar. In the case of magnesium sulfate solution, a layer of MH (brucite or magnesium hydroxide) forms on the surface of the mortar specimen. The penetration of the sulfate solution then occurs by diffusion across this surface layer.

As the attack progresses, the formation of attack products such as gypsum and ettringite in the paste under the surface leads to expansion and strength loss. The expansion also causes cracking in the surface brucite layer and that leaves mortar susceptible to direct attack by the magnesium sulfate solution. The conditions favorable for the decalcification of C-S-H are thus created and the ultimate destruction of the mortar occurs as a result of the conversion of C-S-H to the non-cementitious magnesium silicate hydrate (M-S-H).

In spite of considerable research in the past few years, the action of sulfates on concrete is still not properly understood. The understanding on sulfate attack is mostly developed on the basis of the experiments on concrete conducted in laboratory, the situation is more complicated in actual practice in the field [60,201]. The reactions between cement hydration products and the sulfate bearing solutions manifest in various ways and the mechanisms are complicated. Additional research is required to provide adequate means for the selection of materials for concrete subjected to sulfate attack. In particular, the role of cation in the sulfate solution and the effect of reaction products like gypsum, ettringite, and thaumasite on the extent of damage, requires further investigation.

7.4.4.2 Mathematical Models for External Sulfate Attack

Glasser et al. [56] reviewed the various aspects of sulfate attack on concrete and mentioned three types: empirical, mechanistic, and ionic transport (numerical) models. The empirical models estimate the sulfate resistance factor, the expansion under sulfate attack, or the location of the visible degradation zone. The mechanistic models typically attempt to take into account the mechanisms leading to the deterioration of concrete. These models usually predict the rate of sulfate attack and the fractional or volumetric expansion. The ionic transport (numerical) models simulate the chemical reactions occurring during sulfate attack and, in some cases, also estimate the damage caused by expansion. According to the authors, the ability of empirical and mechanistic models to predict the behavior of concrete structures under sulfate attack remains somewhat limited, whereas the ionic transport models offer a more detailed description of the process through dissolution–precipitation reactions, coupled with transport of ions, in concrete matrix. The ionic transport models for sulfate attack become more complex [55,185] for two reasons: (a) sulfate ions react strongly with hydrated cement phases; hence, models are required to include resultant mineralogical transformations; and (b) concrete structures under sulfate attack are usually affected by decalcification also. In order to predict the mechanical consequences of the sulfate ingress, such transport models can be coupled to mechanical models where the effects of macroscopic expansions due to ettringite formation and crystallization pressures can be considered [186,202].

The mechanistic model proposed by Atkinson and Hearne [203] predicts, based on a mathematical expression, increase in degradation rate with the ion diffusivity and concentration of sulfate in the bulk solution as well as with the concentration of ettringite in reacted cement. However, the Atkinson and Hearne model, as well as the other proposed by Clifton and Pommersheim [204], predict that the rate of degradation is not linearly related to the concentration of ettringite. Sufficient void space is generally available to accommodate the maximum volume of ettringite formed. Their findings were found applicable for the normal range of w/c ratio and for the commonly used varieties of cement, i.e. ASTM Type I and Type V. The Atkinson and Hearne model has been applied to reasonably predict the service life of underground concrete vaults [205]. The problem in using the model is the nonavailability of validated data on sulfate ion diffusivity.

Basista and Weglewski developed two micromechanical models to simulate the expansion of cementitious composites exposed to external sulfate attack, through ettringite formation by two mechanisms: through solution and through topochemical reaction [206]. Both the models combine transport of sulfate ions, the stress and strain fields due to expanding ettringite crystals, the microcracking of elastic matrix and the percolation. The predicted axial expansions of the mortar specimens immersed in sodium sulfate solution were compared with the experimental data reported in the literature. The results indicated that the topochemical mechanism explains the experimentally observed amount of expansion. The pressure of expanding ettringite formed in the topochemical reaction may cause formation and propagation of microcracks in the hardened cement paste. The findings support the mechanism proposed earlier by Santhanam et al. [200].

Tixier and Mobasher [186,202] presented a chemo-mechanical mathematical model to simulate the response of concrete exposed to the external sulfate attack. It is based on the interaction of the chemistry of the mechanisms involved in terms of the rate of reaction and the volume change during reaction, the physics in terms of the diffusion with the moving boundary and variable diffusivity, and the mechanics in terms of the stress–strain response of concrete. The final model is derived successively from the solution of the diffusion-reaction equation, then the addition of the moving boundary effect, the change in diffusivity due to cracking and finally the approximate adaptation of the 1D solution to the 2D case of a prismatic specimen. Using the volumetric information, the model predicts the generation of internal stresses, evolution of damage, reduction in stiffness, and expansion of the matrix phase. The theoretical expansion-time responses were obtained and compared with a variety of available data in the literature. The most important parameters are identified as the w/c ratio, internal porosity, diffusivity of the cracked and uncracked material, and available calcium aluminates. The importance of controlling the pH of the test solution was also observed. The model simulations indicate a reasonable agreement with experimental expansion-time data available in the literature. The parametric study revealed that the effect of increase in diffusion coefficient is much more important than the same operation on the rate constant of chemical reaction, indicating that sulfate diffusivity plays a more dominant role than the rate of reaction of sulfates and aluminates. The finding agrees with those reported by researchers earlier [189,203] and important for predicting the possible response of concrete with mineral admixtures.

7.4.4.3 National Standards and Guidelines on External Sulfate Attack

Table 7.7 lists the ASTM/AASHTO standards to test concrete durability against sulfate attack. Neville [60] gives selected tables from American, Canadian, British, and European sources giving a classification of exposure to sulfates. The limitations of the two ASTM Test methods, ASTM C1012 and ASTM C452, mentioned in the literature [189,199,207–211] are as follows:

a. *Applicability and utility*: The accelerated test method should give reliable results within a relatively short time, 4 weeks or less, if it is to be of practical benefit to the cement manufacturers and the users and if it is to be applicable to both Portland and blended cement. ASTM C452 is reasonably rapid, completed within 14 days. However it is not considered to be applicable to blended cement. ASTM C1012 is applicable to both Portland and blended cement but cement meeting the performance criteria will require 6 months of testing [190,199,212]. The reduction in the size of specimens could reduce the test period, without anyway disturbing the mechanism of degradation, as shown by experiments conducted at the National Institute of Standards and Technology (NIST) and others [213,214]. That can make the test more acceptable to the cement manufacturers and users.

b. *Simulation of field conditions*: The continuous immersion of test specimens in the sulfate solution, as stipulated by the ASTM Standards, does not represent the field situation as the pH and the concentration of sulfate in solution continuously change during the testing period. The high sulfate concentration, generally used in such tests, leads to the precipitation of ettringite and gypsum and to significant accumulation of SO_3 near the surface of the sample, while lower sulfate concentrations, as present under field conditions, lead mostly to ettringite but little or no gypsum precipitation and to a less distinct accumulation of SO_3 near the surface. Even the test solutions containing lower sodium sulfate concentrations are significantly different from the solutions predominant under the field conditions, where a number of other cations such as potassium, calcium, and magnesium and anions, mainly bicarbonate, are also present. The occurrence of bicarbonate in sulfate-containing solutions lowers the expansion very strongly as shown experimentally [215–217]. In the field, the concentration of sulfate solution remains almost constant; besides, the specimens are also subjected to atmospheric effects, such as wetting and drying. The studies carried out by the U.S. Bureau of Reclamation, on a large scale, lasting for 20 years [218], indicate that the degradation occurring due to wet and dry cycling is more rapid than that associated with continuous immersion. In view of some observations, Clifton et al. [199] expressed a need for standard test for concrete durability under partial immersion in solution or soil with sulfates, combined with drying. Mehta [210] proposed a new method to evaluate the sulfate resistance of cement. The translation of laboratory results to the field conditions is difficult due to the complex interplay of physical and chemical factors [207,210,211].

  c. *Concentration and temperature effects*: The mechanism of sulfate attack depends upon the concentration of sulfate solution and the temperature of the surroundings. The concentration defines which cation (Na^+ or Mg^{2+}) will play the prominent role and whether gypsum, ettringite, or both will be formed after the sulfate attack. Whereas, at low temperatures ($0°C–5°C$) thaumasite is formed [207], when limestone powder or calcareous aggregate is used in concrete.

The American Concrete Institute Guide to Durable Concrete [219] gives recommendations on water-to-cement ratio and the cement material for durable concrete, exposed to sulfates in soil and solution, in relation with the severity of exposure.

The European Norm EN 206 for concrete durability [16] recommends maximum w/c ratio, minimum strength class, and minimum cement content (kg/m^3 concrete), in relation to the severity of exposure in terms of sulfate content of the soil or water, expressed as Exposure Class. The standard, for a typical case of underground concrete structure (such as foundation or pipe), requires the determination of (a) sulfate content of the soil to select the corresponding exposure subclass (XA1, XA2, XA3), (b) water permeability to consider whether or not the environment should be moved into a lower class depending on the specific soil permeability, and (c) acidity of the soil based on the German Standard test DIN 4030-2 [220] in order to move the exposure subclass from XA1 to XA2, when the acidity is higher than 20° Baumann Gully in soil with a sulfate content in the range of 2000–3000 mg/kg. Troli and Collepardi [221] expressed that the stipulated durability requirement is too complicated for an underground concrete structure.

In all the national standards on concrete durability, the quantity of sulfate is expressed in parts of SO_4 in milligrams per kilogram of water, that is, in ppm (parts per million), generally without consideration of the type of cation in the sulfate. The role of cation, Na^+ and Mg^{2+}, in the severity of sulfate attack is known.

The Building Research Establishment (BRE), United Kingdom, published BRE Special Digest1 (SD1) [222], dealing with concrete in aggressive sulfate environment. The new edition published in 2005, updates and consolidates the first edition published in 2003. It is a long and complex six-part (A–F) document, which provides practical guidance on the specification of concrete for installation in natural ground and in brownfield* locations. According to the guidelines, while building a structure in aggressive sulfate environment, the first step is to determine the "Design Sulfate Class (DS Class)." The DS Class is a five-level classification for sites based principally on the sulfate content, including total potential sulfate of the ground, groundwater, or both. It is dependent on the presence or absence of substances including magnesium ions and pyrite and, for pH less than 5.5, chloride and nitrate ions. The next step is to find the "Aggressive Chemical Environment for Concrete Class (ACEC Class)" for the site location (Part C). Further, determine the "Intended Working Life" of proposed building or structure and the form and use of specific concrete elements. This performance factor brings the Digest in line with BS EN 206-1 and provides for the general building structures to have working life of at least 50 years and that for civil engineering structures of at least 100 years. Based on the ACEC class and some other factors, define the "Design Chemical Class (DC Class)," which decides the qualities of concrete that are required to resist chemical attack. After obtaining the concrete specification determine the options for "Additional Protection Measures (APM)" (Part D). The Digest gives the step-by-step procedure for the general use of (a) cast in situ concrete (Part D), (b) surface-carbonated precast concrete (Part E), and (c) specific precast concrete products (Part F).

The study reported by Ferraris et al. [213] claims to have developed a specimen test to determine the cement resistance to sulfate attack, that is three to five times shorter than current tests and a test to determine the resistance to sulfate attack, when concrete is not totally immersed in the solution.

7.4.4.4 Mitigation of External Sulfate Attack

The pozzolanic/cementitious reactions occurring in blended cement consume CH produced during the hydration of cement. Hence, the quantity of gypsum formed in Reactions 7.31 and 7.32 will be smaller in blended cement in comparison to PC. It is seen that, in contrast to PC hydration where CH or the C-S-H gel precipitates on the cement grain, those produced during the hydration of blended cement, with PC and BFS or PFA, precipitate in the void space between or on the grains of mineral admixture. After the initiation of the pozzolanic or cementitious reaction, the cement paste becomes increasingly denser. Thus, the secondary C-S-H produced in blended cement hydration helps refining the pore structure reducing the permeability of set cement. The reduction in the permeability reduces diffusion of SO_4^{2-} ions into the interior of the structure. Also, the likely precipitation of secondary C-S-H on the aluminate phase would mitigate its role in the formation of the

---

* *Brownfield locations*: A brownfield location is defined as a site or part of a site that has been subject to industrial development, storage of chemicals (including for agricultural use), or deposition of waste and which may contain aggressive chemicals in residual surface materials or in ground penetrated by leachates [222].

secondary ettringite [223–225]. The low water-to-binder ratio, maintains low ionic permeability of concrete and helps hinder destructive sulfate attack such as the formation of thaumasite at low temperatures [196].

The reduction in the alkalinity of pore solution through pozzolanic/ cementitious reaction increases the solubility of hydrated calcium alumi- nate in the pore solution. That promotes the ettringite formation in the solu- tion rather than in the solid state (topochemical), preventing the expansion. The reduced alkalinity would further cause the ettringite to be produced in the form of large lath like crystals, which are not expansive [226].

The Na and Mg sulfates have deleterious action on the concrete, but they produce different effects. The action of Na_2SO_4 is manifested in the expan- sion, whereas that of $MgSO_4$ is manifested in the reduction of the strength of the concrete. Therefore, the resistance of blended cement concrete to a Na_2SO_4 environment can be evaluated on the basis of expansion measure- ments and that toward $MgSO_4$ environment can be evaluated on the basis of compressive strength measurements.

The concrete with blended cement exposed to Na_2SO_4 environment, in gen- eral, shows lower expansion. It is attributed to the lower content of CH and the formation of secondary C-S-H gel due to pozzolanic/cementitious reactions tak- ing place in the blended cements [135]. Figure 7.6 illustrates this aspect. However when concrete is exposed to $MgSO_4$ environment, it shows higher strength reduction, in comparison to that with PC. The phenomenon may be attributed to the lower availability of CH in the hardened matrix of blended cement concrete. The CH forms the first line of reaction, acts as a buffer and retarder (defense) for $MgSO_4$ attack. Its reduction deflects the $MgSO_4$ attack toward the C-S-H binder, thereby enhancing its deterioration. However, the negative effect of mineral admixtures, during magnesium sulfate attack, is often offset by the reduced

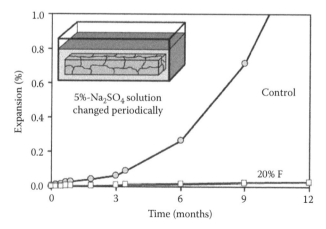

FIGURE 7.6

Expansion in mortar with blended cement exposed to Na_2SO_4 environment. *Note*: F is the pulverized fuel ash (PFA) or fly ash. (Courtesy of Dr. Theodore W. Bremner, PhD, PEng., http://www.unb.ca/civil/bremner/CIRCA/Images/Circa_8_Use_in_b_iv_3.jpg)

permeability and densification caused by their use. This necessitates the determination of the optimum dosage of the mineral admixtures. The application of a protective coating, like epoxy-based coating, is also recommended on the concrete surface exposed to $MgSO_4$ environment [52,200,227–230].

Moderate sulfate resistance can be provided by the appropriate Portland or blended cement or by any hydraulic cement meeting the requirements of ASTM C1157: cement Type MS [231]. High sulfate resistance can be achieved with ASTM C150: Portland cement Type V [232] or any hydraulic cement meeting the requirements of ASTM C1157: cement Type HS [231]. The sulfate-resistant concrete may also be made using other cements such as ASTM C150: Portland cement Type I [232], in combination with sufficient amount of appropriate mineral admixture.

It is seen that cements with high tricalcium silicate content generally have poor durability in sulfate environment. High C_3S content combined with high C_3A content was particularly detrimental to the concrete resistance to sulfate attack [233–236]. It should be noted that, partial replacement of Portland cement with mineral admixtures reduces the content C_3S as well as C_3A in the blend.

The Al_2O_3 (alumina) content of BFS (chemical analysis of BFS gives Al_2O_3 content) is said to be affecting its sulfate resistance. Increasing Al_2O_3 content in the slag may have unfavorable influence on its sulfate resistance [237]. However, there is no direct relationship between the Al_2O_3 content of BFS and its sulfate resistance, as a part of the Al_2O_3 is taken up in C-S-H structure, where it is less prone to sulfate attack. In general, blended cement containing 60% or more of BFS, shows moderate to good sulfate resistance, irrespective of the Al_2O_3 content of slag. With regard to the PFA, a replacement level of 25%–30% of cement has been found to increase the sulfate resistance of concrete [238–241].

However the control of permeability of concrete is more important than the control of chemistry of cement [189,212]. This can be deduced from the fact that the diffusivity of sulfate ions, which controls the rate of degradation by sulfate attack [199], varies by several orders of magnitude in comparison to C_3A, which varies only between 1% and 12%.

The thaumasite formation is delayed in concrete with mineral admixtures. The influence varies with the type and the source of materials. The MK and BFS have been found to improve the performance of limestone cement, showing that when mineral admixtures react sufficiently quickly, they offer an effective resistance against thaumasite formation. The mixtures prepared with PFA, known to hydrate slowly, remain vulnerable to thaumasite sulfate attack. However the PFA retards sulfate attack, as discussed earlier. It is also important to note that since thaumasite does not contain alumina, sulfate-resistant Portland cement does not give an improvement in resistance against its formation [55,242,243].

Finally, the question is, how widespread is the damage due to sulfate attack? Mehta puts it in appropriate words [244]: "… the threat of structural failures due to sulfate attack…seems to be even less of a threat than that caused by alkali–silica reaction." Similar views have been expressed by other researchers also [60,189,245].

7.4.5 Internal Sulfate Attack or Delayed Ettringite Formation

The delayed ettringite formation (DEF), an ISA, may be defined as the formation of expansive ettringite in the hydrated cementitious material in concrete by a process that begins after the hardening is substantially complete. The DEF occurs when ettringite, which normally forms during hydration, is decomposed and subsequently formed again in the hardened concrete. A characteristic feature of this type of damage is the conspicuous formation of ettringite in voids, cracks and the contact zone between the aggregate and the hardened cement paste, without any external sulfate attack having taken place. In concrete, DEF is evidenced as cracks and the loss of strength. Taylor et al. gave an excellent review on the subject [246]. The DEF has been mostly observed in precast concrete products, cured at elevated temperature. In some cases, it has also been observed in mass concrete, wherein the temperature rose excessively on account of the heat of hydration. There is a difference of opinion as to whether, under certain conditions, the DEF can occur in concrete not subjected to elevated temperature. There are two necessary but not the only conditions that promote DEF in concrete, first, the internal temperature must reach above 70°C, for a sufficiently long period of time, to decompose the ettringite formed during the initial hydration, and, second, after its return to the normal temperature, concrete must be exposed to moist or wet surroundings intermittently or permanently [246–250]. When exposed to moist surroundings, the minimum level of exposure for DEF expansion to occur has been reported to be between 90% and 92% RH, for the heat-cured mortars [251]. It is worthwhile making a mention of a typical case reported by Sahu and Thaulow [252], wherein a petrographic examination of cracked Swedish concrete railroad ties, cured below 60°C, identified DEF as a damaging mechanism. The authors mention that DEF can occur, curing temperature below 60°C notwithstanding, when an unfavorable combination of parameters related to cement composition (alkalis, C_3S, C_3A, SO_3, and MgO) and cement fineness occurs.

The DEF is distinct from the secondary ettringite formation (SEF), in that SEF can occur in any concrete member subject to severe drying, wherein first calcium sulfate is formed upon decomposition of ettringite or monosulfate, quickly followed by its dissolution upon re-wetting and reaction with Al-bearing compounds and subsequent precipitation of products in the cracks [253].

The main reactants in DEF are the C-S-H, monosulfate, and pore solution. The reaction takes place through dissolution and precipitation processes. The ettringite ($3CaO \cdot Al_2O_3 \cdot 3CaSO_4 \cdot 32H_2O$) is formed mainly around the monosulfate phase ($3CaO \cdot Al_2O_3 \cdot CaSO_4 \cdot nH_2O$, $n = 12$–18), which is the principal supplier of aluminum constituent required for the ettringite formation. The SO_4^{2-} ions, mainly desorbed from the C-S-H, participate in the reaction. There are two views on the ettringite and crack formation: (a) through formation of relatively large crystals of ettringite at the aggregate–paste interface and (b) through the formation and growth of much smaller ettringite crystals within the paste [247,254]. Both the views are supported by the experimental and analytical evidence. Therefore it is reasonable to assume that under the field conditions, the cracks are likely to form at both around the aggregate–paste interface as well as within the paste [246].

The national standards on precast concrete products recommend curing temperature below 60°C–70°C. A test to determine if the precast concrete product shall suffer DEF, when exposed to water or moisture, has not been developed as yet. The test should be such as to establish the acceptability of concrete under standard conditions of curing and wet/moist surroundings; the earlier reported Duggan test did not perform well on that score [255–258]. Fu [258] and Pavoine [259,260] proposed new test methods to predict the susceptibility of concrete to DEF.

The incidence of DEF in air-entrained concrete has been found relatively less [250]. The voids created in concrete with the bubbles formed by air entraining agent (AEA) may provide additional space for the substances, such as ettringite growing in concrete. However, when the voids get filled with such substances, they may become ineffective toward providing resistance against freeze–thaw conditions.

The addition of mineral admixtures to concrete reduces the DEF, slows the rate of expansion and delays the onset of expansion. However, the efficacy of a particular mineral admixture in controlling expansion is found to depend on its SO_3/Al_2O_3 molar ratio; the potential of expansion is increased with the increasing molar ratio (typically above 0.8). The MK, which contains high amount of reactive Al_2O_3, was found the most effective at controlling expansion at relatively low cement replacement levels (8% or more). The BFS (25% or more) and PFA (Type F: 15%–25% and Type C: 25%–35%), which are sources of Al_2O_3, are also effective at suppressing expansion at higher replacement levels. The SF was found less effective at controlling expansion at conventional replacement levels and even at higher replacement levels the expansion may only be delayed [258,261–263].

7.4.6 Decalcification or Leaching

The decalcification or leaching process is usually described by the dissolution of CH (portlandite) and C-S-H in hydrated cement systems exposed to water. Water reaching the surface may be in various forms: dew, fog, or rain or some combination. Under atmospheric conditions, water is saturated with CO_2 and CO_3^{2-} and other ions are formed. The concentration gradient at the cement paste-water contact drives Ca^{2+} and OH^- ions in the pore solution toward the surface. The Ca^{2+} from cement combines with CO_3^{2-} to precipitate $CaCO_3$ on the surface. These surface deposits of $CaCO_3$ are termed efflorescence. The loss of calcium leads to the dissolution of CH and secondary precipitations of monosulfate, ettringite, and calcite. The precipitation of these minerals takes place in the innermost part of the degraded concrete. The leaching increases with the water-to-cement ratio, as it affects the permeability and pore volume of concrete. The efflorescence or surface deposits develop on new constructions with Portland cement concrete and on masonry units, including bricks and tiles, bonded with Portland cement. The efflorescence is not normally damaging but aesthetically undesirable [55,264–266]. The decalcification can also occur due to the attack of acidic water, carbonation, or sulfates [66].

The decalcification changes the bulk density, pore structure, and the pore volume of the hydrated cement paste [56,267–269]. It has a negative effect on the compressive strength of concrete [269,270].

The use of mineral admixtures, combined with adequate curing, decreases the permeability of concrete and reduces leaching and decalcification. The beneficial influence is due to the reduction in the initial CH content, as a result of cementitious or pozzolanic action [270,271]. The modification of ITZ, in terms of reduction in CH content, makes it offer better resistance toward leaching [272].

As the rate of leaching is usually very low, the degradation of cement paste due to decalcification rarely affects common concrete structures. However it is relevant to hydraulic structures, such as dams and radioactive disposal facilities, wherein long-term stability must be guaranteed [273–275]. The long-term and safe disposal of nuclear wastes is a matter of intense research, all over the world. One of the ways to dispose such wastes is underground concrete tunnels. These are built with HPC with a safe design life of 1000 years. The consequences of calcium leaching, in terms of mechanical behavior has to be taken into account under such situation. Sellier et al. [276] presented the progress in that regard of a research project undertaken by the French National Radioactive Waste Management Agency (ANDRA).

7.4.7 Frost or Freeze–Thaw Action

The frost or freeze–thaw action is the distress and deterioration of critically saturated concrete due to freezing of water. Water, upon freezing, occupies about 9% more volume. Thus when 91% or more of the pore volume of concrete is occupied by water, it is susceptible to distress and deterioration by freeze–thaw action. However, further investigations revealed that distress can occur even at lower saturation levels [277]. The typical signs of freeze–thaw deterioration that could be observed on the surface are spalling and scaling, large chunks of concrete (cm size) breaking off, exposed but mostly uncracked aggregate, and surface parallel cracking.

Powers first proposed the "hydraulic pressure theory" to explain this phenomenon [277,278], which was followed by the "diffusion and growth of capillary ice theory" by Powers and Helmuth [279], "dual mechanism theory" by Larson and Cady [280], and the "desorption theory" by Litvan [281]. While these theories disagree as to whether water moves toward or away from the point of ice formation, they agree that the amount of water in the pores and the resistance to the movement of that water play a role in the frost resistance of concrete. It is generally accepted that the pore system in concrete is susceptible to damage from freeze–thaw action. The efforts to produce frost-resistance remain primarily focused on providing a proper system of entrained air voids within concrete. The pore system in some types of aggregates show susceptibility to damage from freeze–thaw; hence, efforts are also focused on identifying damage-resistant aggregates [282].

The most commonly used procedure to test the freeze–thaw resistance of concrete is ASTM C666 or AASHTO T161, titled "Standard method of test for resistance of concrete to rapid freezing and thawing." The method covers the determination of the resistance of concrete specimens to rapidly repeated cycles of freezing and thawing in the laboratory by two different procedures: procedure A, rapid freezing and thawing in water, and procedure B, rapid freezing in air and thawing in water. Another test associated with the freeze–thaw resistance of concrete is the "Scaling resistance of concrete surfaces exposed to deicing chemicals (ASTM C672)." There is a difference of opinion over the limitations of the procedures A and B and on the appropriateness of using the test to predict field durability of concrete. The limitations of procedure A are (a) the rigid physical confinement of specimens could cause damage and (b) the problem of maintaining the correct thickness of water surrounding the specimens. The limitation of procedure B is that the specimens are allowed to dry during freezing. Janssen and Snyder [282] report the development of a new procedure that is claimed to remove the limitation of procedure B. The new procedure C consists of wrapping the specimens with absorbent cloth to keep the specimens wet during freezing. It should be noted that the procedures are not intended to provide a quantitative measure of the length of service that may be expected from a specific type of concrete. The specifications relating to the frost resistance of concrete are generally based on laboratory tests of laboratory-produced concrete. The quality control for laboratory-produced concrete is usually significantly better than what is achieved in the field and the laboratory freeze–thaw tests generally produce conditions that are not close to any real field exposure conditions. The field exposure conditions are not the same from one location to the next and may not even be the same from 1 year to the next at the same location. Janssen [283] provided guidance to interpret the results of field tests of concrete exposed to natural freezing and thawing conditions. The types of frost damage as well as types of field test sites were discussed and recommendations provided for the use of field test results to modify frost-resistance specifications.

In order to protect concrete from freeze–thaw damage, it should be air-entrained by adding an organic surface active agent to the concrete mixture. The addition creates a large number of closely spaced, small air bubbles in the hardened concrete. The air bubbles relieve the pressure buildup caused by ice formation by acting as expansion chambers [284,285]. ASTM C260 [286] specifies requirements for air entraining admixtures. About 4% air entrainment by concrete volume is needed and the air bubbles should be well distributed and have a distance between each other of less than 0.25 mm in the cement paste. In general, concrete with high water content and high water-to-cement ratio is less frost resistant than concrete with lower water content. The freeze–thaw resistance of concrete containing 30%–50% PFA replacing cement, showed excellent results [287].

The cracking of concrete, mostly in pavements, caused by the freeze–thaw deterioration of the coarse aggregate within concrete is called D-cracking [288]. The D-cracks are a series of cracks in concrete observed near and roughly parallel to joints, edges and structural cracks. The problem can be reduced either by selecting aggregate that perform better in freeze–thaw cycles or where marginal aggregate are required to be used, by reducing the maximum particle size (<13.2 mm). Some aggregate, such as high quality dolomite and rocks of igneous origin are not so prone to such cracking, while aggregate with high proportions of chert and certain types of limestone are risky for use. The experiments have shown that susceptible aggregate get saturated just by contact with damp ground. The presence of a small 15% of a poor material can cause deterioration of the concrete. Therefore, the basic preventive measure against the D-cracking is the selection of durable materials. The installation of effective drainage systems to carry free water out from under the pavement and sealing of joints may also be helpful but cannot prevent deterioration of susceptible aggregate [289]. Kaneuji et al. [290] reported that a correlation exists between the pore size distribution of a coarse aggregate and its durability toward freeze–thaw damage in concrete. The expected durability factor (EDF), calculated from the pore size distribution curve, can be used to distinguish between aggregate that are durable or nondurable with respect to D-cracking in concrete pavements. Janssen and Snyder took an extensive review of D-cracking, which also included different tests available to identify freeze–thaw durable aggregate [282].

7.5 Durability Indices for Performance-Based Design of Structures

The durability and service life of RC structures depend, besides the strength, on the quality and thickness of concrete cover and its ability to protect the reinforcing steel against the attack of deleterious agents. The durability indices or indicators are the measurable parameters that indicate the quality of concrete cover. The service life of concrete can be predicted linking the durability indices to the transport mechanisms (gaseous, liquid and ionic) that control the deterioration. Thus the durability index (DI) approach helps performance-based design of concrete structures.

The DI could be classified as "general" (relevant to many degradation processes), which include CH (portlandite) content, porosity or air, water permeability, and "specific," which are transport properties related to specific degradation processes, such as oxygen or chloride diffusion coefficient (carbonation or corrosion) and electrical conductivity (corrosion). The DI could be determined on test samples in laboratory under controlled conditions or by nondestructive tests carried in situ on the existing structure. The nondestructive tests are simple, less time consuming, and relatively cheaper.

The DI could be used as input parameters for a performance model to design concrete mixture and to decide cover thickness, to protect structures against degradation, for the target lifetime and environmental exposure conditions, or conversely to predict the service life of a new structure at the design stage or the "residual" lifetime of an existing and possibly deteriorated structure [291,292].

The performance-based design of structures, using durability indices, has been evolved in South Africa and practiced under aggressive marine conditions in Durban (subtropical, high temperature and RH, strong salt-laden onshore winds). The three indices used are oxygen permeability index (OPI), water sorptivity index (WSI), and chloride conductivity (CC). The information on the design methodology has been published in series of papers by Ballim, Alexander and others [293–296]. The step-by-step approach to the design is summarized as follows:

Step I: Define exposure class related to the mechanism of deterioration, as per the relevant national standard.

Step II: Derive a quantitative design methodology, accounting for the required service life. That can be done on the basis of predictive service life models.

Step III: Develop test methods that relate to the input parameters of the design method. The DI are used as input parameters to the service life models. The specification contains reference to DI values to meet the criteria in Step II.

Step IV: Produce provisional conformity criteria and calibrate against traditional solutions. The conformity criteria should contain limiting DI values to assure desired service life. Also, define "deemed-to-satisfy" values for both the material supplier and the constructor, subject to inherent variability.

Step V: Establish limitations of test applicability—exclude very high strength (>60 MPa) and other special application concretes.

Step VI: Ensure production control and acceptance testing—distinguish between "as supplied" and "as built" concrete with a stricter requirement for the "as supplied" concrete.

Step VII: Conduct full-scale trials and long-term monitoring to confirm conformity requirements.

Andrade and Martinez [297] developed different methodologies that use indices to assess the structural performance, with regard to corrosion of reinforcement, at different stages of the service life of structure. These indices are (a) "Simplified Index of Structural Damage (SISD)" to assess the deteriorated structure, (b) "Repair Index Method (RIM)" to select the best repair option, and (c) "Repair Performance Indicator (RPI)" to monitor the repair work.

It is observed that incorporation of mineral admixtures in concrete improves DI values, making the structure survive longer against the aggressive environment [293,295].

7.6 Sustainable Cement and Concrete

It is appropriate to culminate the discussion on strength and durability, refocusing on the sustainability of cement and concrete as building materials. It is important because concrete is second only to water in total volume produced and consumed by the society [298]. The conservation of materials and energy and the minimization of emission of greenhouse gases and pollutants at all stages, starting from quarrying of raw materials for cement manufacture, aggregate, through transportation, materials handling and unit operations of size reduction, blending, pyro-processing, packaging, and distribution, till it reaches the user site, are the key to ensure the sustainability. Firstly there is a need to create general awareness on the subject and next is to develop tools for the evaluation of performance at every stage and finally to develop methodology and technology that take the cement and concrete industry forward in the direction of sustainability.

The global average gross CO_2 emission per ton of cementitious products in 2006 was 679 kg. Out of the cement industry's direct CO_2 emission, nearly 60% is produced by the decomposition of limestone in the raw materials and the remaining comes directly and indirectly from the fuel combustion and electricity consumption, respectively. According to the study reported by the World Business Council for Sustainable Development (WBCSD), on 844 cement installations worldwide for the period 1990–2006, the industry achieved significant decoupling of economic growth and CO_2 emissions. During that period, the cement production increased by 53%, whereas the net CO_2 emissions increased by only 35% [298]. In fact, the same could be said about the concrete industry also. The reduction in CO_2 emission has been effectively achieved through the following measures [299–315]:

a. *Improving process technology*: The CO_2 emission from cement production is continuously reduced by (i) using alternative raw materials including utility bottom ash, utility boiler slag, foundry sands, iron mill scale, and limestone fines; (ii) increasing the use of industrial wastes as fuel and the CO_2 emission factor (ton CO_2/ton clinker produced) has been reduced from 0.9 to 0.8, as a result; and (iii) reducing and optimizing the heat and electrical energy consumption, with the application of new technology [299].

b. *Substituting clinker or cement with mineral admixtures*: The mineral admixtures, which are in fact industrial and agricultural wastes, such as PFA, BFS, RHA substitute clinker or cement on large scale, in cement and concrete manufacturing, respectively. The appropriate admixtures are limited in their regional availability. New materials may also play a role as cement constituents in the future. The resulting product shows high strength, low thermal and drying shrinkage, high resistance to cracking, high durability, and, consequently, excellent potential for use as a sustainable structural material, for general construction. As a result of these practices, the average clinker factor for cement (ton clinker/ton cement) has been reduced from 0.83 to 0.60, which is expected to bring the clinker requirement down from 2.30×10^9 ton/annum in 2010 to 1.18×10^9 ton/annum in 2030, despite increase in the production. The development of gap-graded blended cement (see Chapter 1), HVFAC [300–302,304] are the further steps in that direction.

c. *Consuming less cement in concrete mixtures*: The concrete mixtures are developed with proper selection of the cement type and the type and dosage of appropriate mineral admixture to substitute cement and the concrete quality to best suit the use in question. The development of self-compacting concrete, making binary, ternary, and quaternary blends of different mineral admixtures with Portland cement, with improved strength and durability characteristics, is one such example [307]. The cement consumption in concrete can also be brought down by (a) specifying concrete on the basis of 56 or 90-day compressive strength, wherever possible; (b) using water-reducing admixture to obtain the desired consistency and to reduce water and cement content as a consequence; and (c) improving the packing factor, optimizing aggregate size and grading.

d. *Enhancing service life of structures*: The enhancement of service life means designing concrete structures based on performance criteria. The addition of mineral admixtures to concrete affects reduction in cement consumption, energy consumption, and cost. The gain is maximized, when HPC is utilized, with the durability criteria predominating over strength. The HPC, with a well-designed mix proportion, produces structures with adequate durability for the desired service life and reduces cost and energy demand for the construction-repair-demolition-recycling-reconstruction cycle, making the construction process self-sustainable, contributing toward the conservation of natural resources. The HPC designed in this manner, satisfies all the requirements of eco-efficiency [308].

e. *Recycling concrete*: The construction and demolition waste (C&D waste), a major portion of all generated solid waste, is used as recycled concrete aggregate (RCA) substituting natural aggregate.

The recycled aggregate is crushed and stored in open for considerable period, before use. Such aggregate can be considered as a carbon dioxide sink, as it undergoes carbonation reaction, absorbing carbon dioxide from the atmosphere. As per the Environmental Protection Agency (EPA) estimate, in 2003, the United States produced 164×10^9 ton building-related waste, of which 9% was construction waste, 38% was renovation waste and 53% was demolition debris [309]. One major success story in the United States is the recycling of 6.5×10^6 ton of C&D waste from Denver's former Stapleton International Airport [310]. Some American states estimate a saving of 50%–60% from the use of recycled aggregate versus fresh aggregate, taking into consideration the savings from disposal costs and potential road damage from the transport of fresh or waste material [308].

f. *Designing energy efficient buildings*: The high heat capacity of concrete and the high degree of air-tightness of concrete buildings is used to design and construct buildings with reduced energy consumption for heating and cooling as well as with improved thermal comfort. This can be done, fruitfully utilizing the free heat gains such as solar radiation, heat from occupants and equipment within the building. The project Eco-Culture Project, Denmark, demonstrated how modern technology could be used to construct environmentally friendly buildings [303,311]. At the international level, United Nations Environment Programme's Sustainable Buildings and Climate Initiative (UNEP-SBCI) promotes sustainable building practices worldwide. This is a joint effort undertaken by the sector's key stakeholders: industry, business, governments, local authorities, research institutions, academia, experts, and NGO [312].

g. *Educating and training manpower*: The sustainable cement production and utilization rely on well educated and well trained manpower at all levels in cement plants as well as in the construction companies using the product. The challenge of reducing energy and raw material consumption and at the same time complying with quality, performance and cost requirements in the context of the huge demand for cement as a construction materials in the future will only be met with highly efficient manpower education and training [313].

However, in spite of the environmentally friendly profile displayed by cement and concrete, the growing demand for these materials caused by the population growth shall lead to an increased utilization of raw materials and natural resources. In comparison to 2010, the world population shall nearly double by 2050. That means doubling the current consumption of cement to nearly 4.4×10^9 ton/annum and the aggregates consumption to nearly 43×10^9 ton/annum. These figures emphasize the need to redouble our efforts toward the materials and energy conservation. There is a need to develop performance indicators to obtain quantitative measure of how cement and concrete is being used eco-efficiently.

It is possible to define concrete efficiency in terms of the total amount of binder, the total cost of concrete production, or the environmental loads imposed to deliver one unit of functional performance. At present, the cement industry generates nearly 5% of the global anthropogenic CO_2 emission. As explained earlier, the production of cement shall increase with the demand and so will the share of CO_2 emission, in the future. The traditional strategies to mitigate emissions, focused on the production of cement alone, will be inadequate to compensate for such growth. Therefore, additional mitigation strategies are needed, including those on more efficient use of cement. The study reported by Damineli et al. [314] proposes two simple indices to measure the eco-efficiency of cement and concrete. First is the Binder Intensity Index, b_i, which measures the total amount of binder necessary to deliver one unit of performance, which is compressive strength (MPa) in most cases, however other measurable parameter relevant for the specific situation may also be chosen. Second is the CO_2 Intensity Index, c_i, which allows estimating the global warming potential of concrete formulations. The authors, considering compressive strength (MPa) as the performance indicator, established a benchmark based on literature data. On the basis of large number of data points, it was found that, for the compressive strength above 50 MPa, the minimum b_i of the order of $5 \text{ kg/m}^3 \cdot \text{MPa}$ is feasible and has already been achieved in practice. However, the value of minimum b_i increases with the lowering of compressive strength, ranging between 10 and 20 $\text{kg/m}^3 \cdot \text{MPa}$. In the authors' opinion, these values could be a result of the minimum cement content stipulated in many standards and reveal a significant potential to improve the performance. The data points indicated a minimum achievable CO_2 intensity of $1.5 \text{ kg/m}^3 \cdot \text{MPa}$. However, while estimating the ci, the emissions resulting from the production and transportation of raw materials, fuel, mineral admixtures, and aggregate have been neglected; only CO_2 emission from the production of clinker (1 ton CO_2/ton clinker) was considered. In another study reported by Habert et al. [315] the potential of modern technological developments in achieving the CO_2 emission targets envisaged by the Intergovernmental Panel on Climate Change (IPCC) is evaluated. The CO_2 emissions in developed countries in 1990 have been taken as the basis. The IPCC report envisages achieving reduction in CO_2 generation by a factor of 2 (50%) by 2020 and by a factor of 4 (25%) by 2050, in comparison to what it was in 1990. The authors are of the view that it would be possible to achieve a "factor 2" reduction, with the current technological development but achieving a "factor 4" reduction shall require a technological turnaround in cement production and construction technology.

7.7 Summary

The strength and durability of concrete structure must go hand in hand. Durability is the ability of a structure to resist weathering action, chemical attack, and abrasion, while maintaining minimum strength and other desired engineering properties. In today's context, designing for strength and durability is synonymous to design for sustainability. The objective of the national structural design codes should be to enable the design decision based on the life cycle cost. In order to do that, the mechanisms and the mitigation of structural deterioration due to the attack of deleterious agents need to be understood. It requires a multidisciplinary approach. The three factors common to the concrete structural design, with the prescriptive or the performance-based approach, are the general acceptance of "permeability" as a criterion for durability, maintaining the quality of construction work, and the continuous maintenance of structures during use. The national codes of practice, which mostly follow prescriptive approach, do not explicitly specify the minimum service life that can be achieved by following the durability provisions specified. The hallmark of the performance-based approach is structural design based on durability-based service life. The structure's reliability is the probability of a structure to fulfill the given functions during its service life, that is, to keep the performance characteristics: safety, durability, and serviceability, within the given limits.

When mineral admixtures are used, the strength of concrete can be considered as a result of three principal factors, first accounting for the reduction in the quantity of cement (dilution), second heterogeneous nucleation (physical) and third pozzolanic reaction (chemical). The net result is higher long-term strength, in most cases.

The structures satisfying the requirement of cost, service life, strength and durability require the use HPC. The HPC is often of high strength, but high strength concrete may not necessarily be of high performance. The judicious choice of chemical and mineral admixtures reduces the cement content and that results in economical HPC.

The curing is a process of preventing loss of moisture from concrete, while maintaining a satisfactory temperature regime. The early and long term curing is beneficial for the development of concrete properties, including strength.

The important factors that control concrete deterioration, besides strength, are the near-surface quality of the finished concrete and the aggressiveness of the environment. The durability specifications therefore increasingly rely on the measurement of the fluid flow properties of the surface or cover zone of concrete, usually 28 days after casting. The commonly observed processes that are responsible, individually or together, for the deterioration of concrete are carbonation, AAR, chloride initiated corrosion of reinforcement, external sulfate attack, ISA or DEF, decalcification or leaching and frost or freeze–thaw action. It is generally accepted that under the optimum conditions of effective blending of components, transportation, placing, and curing, the addition of mineral admixtures to concrete improves its resistance toward the deteriorating agents.

The carbonation refers to the precipitation of calcite ($CaCO_3$) as well as other CO_2-based solid phases, through the reaction of penetrating atmospheric CO_2 with the calcium ions in the pore solution. The main consequence of carbonation is the drop in the pH of the pore solution of concrete so that the passive layer that usually covers and protects the reinforcing steel against corrosion becomes unstable. The continuous diffusion of CO_2 inside concrete may also lead to decomposition of C-S-H. The consequences are loss of strength, shrinkage, cracking and increase in the porosity of concrete. In concrete with mineral admixture, where the amount of CH is reduced due to pozzolanic or cementitious reaction, the carbonation is dependent on permeability and the resultant lower permeability hinders the ingress of CO_2. The main factors affecting concrete carbonation are the type and the content of binder, water-to-binder ratio, degree of hydration, concentration of CO_2 and relative humidity of the surrounding. Under normal atmospheric conditions, the relative humidity in the range of 50%–80% is optimum for carbonation to progress. Although the accelerated carbonation and the phenolphthalein colorimetric methods, to measure carbonation depth, are widely used, they are still very diversified in terms of time and type of curing, preconditioning of specimen, surrounding temperature, and specimen dimensions.

The aggregate containing certain dolomitic or siliceous minerals react with soluble alkalies in concrete (termed as AAR) and sometimes result in detrimental expansion, cracking, and the premature loss of serviceability of concrete structures affected. All kinds of concrete structures may be affected, although structures in direct contact with water, such as dams and bridges, are particularly susceptible to AAR. The conditions that initiate and propagate AAR in concrete are as follows: (a) critical quantity of reactive aggregate, (b) sufficient alkali in concrete, and (c) sufficient moisture. It should be noted that the attack and the damage due to AAR can occur only when all the three conditions prevail in concrete. The AAR is time dependent and its occurrence is unpredictable, as there is wide variability in the conditions causing it. The two types of AAR are (a) ACR and (b) ASR. They differ in the type of aggregate mineral phases and reaction mechanisms. The alkali content of concrete (expressed as Na_2O equivalent) is determined both by the alkali content of cement as well as the total cement content. Usually an alkali content less than $1.8\,kg/m^3$ of concrete is innocuous. The structural behavior of concrete affected by the AAR is difficult to model due to the number of variable parameters that govern the chemical reactions. There is no universally accepted standard testing method for all cases of AAR.

The national standards, world over, have several test methods to identify potential reactivity of aggregate and to test the concrete durability against AAR. The mineral admixtures replacing cement, such as BFS, PFA, and natural pozzolans, mitigate or eliminate AAR in concrete. Once the structure gets affected by AAR, it is impossible to interrupt in most cases. The only way to lessen its harmful effects is by taking appropriate remedial measures, whose effectiveness depends strongly on an adequate prediction of the stress and strain field development in the built structure.

Under marine conditions, chloride ions penetrate through porous concrete and build up around the reinforcement and the alkalinity (pH) of the surrounding pore solution falls substantially. At that stage, the protective iron oxide film around reinforcing bars depassivates and cracks, exposing the steel. The exposed steel gets corroded in the presence of water and oxygen, resulting in the formation of expansive corrosion products (rust) that occupy several times the volume of the original steel consumed. The expansive corrosion products create tensile stresses on the concrete surrounding the corroding steel reinforcing bar, leading to cracking and spalling of concrete cover. The CTL is the content of chloride that is necessary to sustain the breakdown of passive oxide film and initiate the corrosion of exposed steel. The time needed to reach the CTL corresponds to the corrosion initiation period. The onset and propagation of corrosion takes place under the following conditions: (a) availability of iron in metallic (Fe) state at the surface of reinforcing steel, (b) continuous availability of oxygen and moisture, (c) low electrical resistivity of concrete surrounding the reinforcement, and (d) substantially increased permeability of concrete. The comparison of the chloride resistance of cementitious materials in terms of diffusion coefficients is useful, particularly in the design of concrete structures for durability against aggressive environment. The appearance of the first corrosion crack is usually used to define the end of functional service life, when rehabilitation of a corroding structural element is required. The mathematical models are available, which predict the time from corrosion initiation to corrosion cracking. The national standards are available to measure chloride permeability on long-term (in years), short-term (90 days), and nondestructive, rapid basis. Out of these, the last one, the RCPT has gained more acceptance in the construction industry. Measuring the half-cell potential of steel reinforcement in RCC structures for SCE (ASTM C876) gives qualitative assessment on the state of corrosion. The addition of mineral admixtures to concrete inhibits corrosion of reinforcement, improving the resistance toward chloride penetration, and reducing the quantity of free (soluble) chloride in concrete. Besides delaying the initiation, the corrosion propagation period is also extended.

The deterioration of concrete due to external sulfate attack is a commonly observed phenomenon, when structures are exposed to sulfate solutions or built in sulfate bearing soil and/or ground water. All commonly obtained water soluble sulfates are deleterious (Mg > Na > Ca) to concrete, but the effect is severe when it is associated with Mg cations. The cement-based materials subjected to sulfate attack suffer from two types of damages: loss of strength due to decalcification of cement hydrates, mainly C-S-H, and cracking due to the formation of expansive compounds, mainly gypsum and ettringite. As the solubility of calcium sulfate in water is low, the higher concentration of sulfate ions is generally due to the presence of magnesium sulfate and sodium sulfate. Whereas the action of Na_2SO_4 is seen in the expansion, that of $MgSO_4$ is seen in the loss of strength of concrete, in addition. When calcareous aggregate is used, at low temperature (below 15°C), a nonbinding compound (thaumasite) is formed from C-S-H, under the conditions of destructive sulfate attack. It should be noted that the understanding on sulfate attack is mostly developed on the basis of the experiments on concrete conducted in laboratory, the situation is more complicated in actual practice, in the field. The modeling and parametric studies revealed that the sulfate diffusivity plays a dominant role, in comparison with other parameters, during sulfate attack. The national standards to test concrete durability against sulfate attack have limitations in terms of applicability and utility, simulation of field conditions and accounting for concentration and temperature effects. The accelerated test methods need to give reliable results within a relatively short time (4 weeks or less) and should be applicable to both Portland and blended cement, for practical benefit to the cement manufacturers and the users. The concrete with mineral admixtures, exposed to Na_2SO_4 environment, in general, shows lower expansion. It is attributed to the lower content of CH and the formation of secondary C-S-H due to pozzolanic/cementitious reactions. The lower availability of CH in hardened concrete is believed to create a negative effect, during magnesium sulfate attack. However, this is often offset by the reduced permeability and densification caused by the use of mineral admixtures.

The DEF, an ISA, may be defined as the formation of expansive ettringite in the hydrated cementitious material in concrete by a process that begins after the hardening is complete. The DEF occurs when ettringite, which normally forms during hydration, is decomposed and subsequently formed again in the hardened concrete. A characteristic feature of this type of damage is the conspicuous formation of ettringite in voids, cracks, and contact zone between the aggregate and the hardened cement paste, without any external sulfate attack having taken place. In concrete, DEF is evidenced as cracks and the loss of strength. The DEF has been mostly observed in precast concrete products, cured at elevated temperature. In some cases, it has also been observed in mass concrete, wherein the temperature rose excessively on account of the heat of hydration. There are two necessary but not the only conditions that promote DEF in concrete, firstly the internal temperature must reach above 70°C, for sufficiently long period of time, to decompose the ettringite formed during the initial hydration and secondly, after its return to the normal temperature, concrete must be exposed to moist or wet surroundings intermittently or permanently. The national standards on precast concrete products recommend curing temperature below 60°C–70°C. The incidence of DEF in air-entrained concrete has been found relatively less. The addition of mineral admixtures to concrete reduces the DEF, slows the rate of expansion and delays the onset of expansion. However, the efficacy of a particular mineral admixture in controlling expansion is found to depend on its SO_3/Al_2O_3 ratio.

The decalcification or leaching process is usually described by the dissolution of CH (portlandite) and C-S-H in hydrated cement systems exposed to water. It results in surface deposits of $CaCO_3$, termed efflorescence and secondary precipitations of monosulfate, ettringite and calcite, deep within concrete. The efflorescence or surface deposits develop on new constructions with Portland cement concrete and on masonry units, including bricks and tiles, bonded with Portland cement. It is normally not damaging but aesthetically undesirable. It has negative effect on the compressive strength of concrete. The use of mineral admixtures, combined with adequate curing, decreases the permeability of concrete and reduces leaching and decalcification. It is relevant to hydraulic structures, such as dams and radioactive disposal facilities, wherein long-term stability must be guaranteed.

The frost or freeze–thaw action is the distress and deterioration of critically saturated concrete due to freezing of water. However distress can occur even at lower saturation levels. The typical signs of freeze–thaw deterioration that could be observed on the surface are spalling and scaling, large chunks of concrete (cm size) breaking off, exposed but mostly uncracked aggregate, surface parallel cracking. The amount of water in the pores and the resistance to the movement of that water play a role in the frost resistance of concrete. The efforts to produce frost-resistant concrete remain primarily focused on providing a proper system of entrained air voids. The pore system in some types of aggregate shows susceptibility to damage from freeze–thaw; hence, using damage-resistant aggregate is also important. The specifications relating to frost resistance of concrete are generally based on laboratory tests of laboratory-produced concrete. The field tests are required to be carried out and interpreted to suitably modify the specifications. In order to protect concrete from freeze–thaw damage, it should be air-entrained by adding an organic surface active agent to the concrete mixture. The cracking of concrete, mostly in pavements, caused by the freeze–thaw deterioration of the coarse aggregate within concrete is called D-cracking. It can be controlled either by selecting aggregate that perform better under freeze–thaw conditions or where marginal aggregate are required to be used, by reducing the maximum particle size (<13.2 mm).

The DI could be used as input parameters for a performance model to design concrete mixture and to decide cover thickness, to protect structures against degradation, for the target lifetime and environmental exposure conditions, or conversely to predict the service life of a new structure at the design stage or the "residual" lifetime of an existing and possibly deteriorated structure.

The cement and concrete, to survive as building materials on the long term, must satisfy the sustainability criteria. The conservation of materials and energy and the minimization of emission of greenhouse gases and pollutants at all stages, starting from quarrying of raw materials for cement manufacture, aggregate, through transportation, materials handling and unit operations of size reduction, blending, pyro-processing, packaging, and distribution, till it reaches the user site, are the key to ensure the sustainability.

8

New Mineral Admixtures

8.1 Introduction

In the preceding chapters, the characteristics and applications of commonly used mineral admixtures (Chapter 1 through Chapter 5) and their contributions toward the hydration (Chapter 6), strength and durability (Chapter 7) of cement and concrete were discussed. In this chapter, some new, lesser known but potentially useful, mineral admixtures will be introduced. It should be noted that most of these admixtures are still under investigation at various stages. The purpose of this discussion, besides providing an introduction, is to generate, among the researchers, more interest in the subject.

Section 8.2 discusses the applications of ash produced from biomass, such as corn cob ash (CCA), palm oil residue ash (PORA), sugarcane bagasse ash (SBA), wheat straw ash (WSA), and wood waste ash (WWA). Section 8.3 briefs us on how the calcined wastepaper sludge (CWS) can be utilized in concrete. Section 8.4 gives some details on the utilization of dust collected from electric-arc furnace (EAF) in concrete. Finally, two sections deal with the utilization of sewage sludge ash (SSA, in Section 8.5) and municipal solid waste ash (MSWA, in Section 8.6). Section 8.7 summarizes this chapter.

8.2 Biomass Combustion Ash

In the bio-based economy, renewable herbaceous biomass such as straw and perennial grasses (like Miscanthus, switchgrass) will become important cellulosic feedstock for conversion to biofuels, chemicals, electricity, and heat. A significant fraction—up to one-fifth—of the herbaceous biomass consists of inorganic constituents, commonly referred as ash, that cannot be converted to energy [1]. The minerals and silicates from earth are stored in the plants during the natural growth process. The inorganic materials, especially silicates, are found in higher proportions in seasonally grown plants, such as rice, wheat, and sugarcane in comparison to the long-lived trees [2]. The ASTM C618 [3] specifies that the sum of $SiO_2 + Al_2O_3 + Fe_2O_3$ in the

TABLE 8.1

Chemical Composition of Ash Produced from Biomass vis-à-vis ASTM Type F and Type C PFA

Sl No	Particulars (in %)	FA ASTM Type F (Range)[a]	FA ASTM Type C (Typical)[a]	Ash from Biomass (Range)[b]				
				Corn Cob	Palm Oil Residue	Sugarcane Bagasse	Wheat Straw	Wood
i	SiO$_2$	48–60	37	65–67	44–65	64–78	51–74	5–68
ii	Al$_2$O$_3$	19–32	20	6–9	2–11	6–9	1–5	1–15
iii	Fe$_2$O$_3$	2–16	6	4–6	1–8	4–6	1–5	1–10
iv	CaO	1–5	24	10–13	5–8	2–11	3–13	6–70
v	MgO	0.5–2	5	2	3–4	2–3	2–3	1–15
vi	Na$_2$O	0.3–0.8	2	0.4–0.5	0.1–5	1–3	1–2	1–5
vii	K$_2$O	0.1–4	0.4	4–6	4–8	1–7	6–17	2–25
viii	SO$_3$	0.2–1	2	1	0.2–3	0.4–2	2	1–5
ix	LOI	1–12	1	1–3	10–18	1–5	5–15	5–10

[a] See Chapter 1.

[b] The total of all percentages does not add to 100, as minor constituents have not been mentioned. The values or the range of values are rounded off and indicative. The actual values may be lower or higher in the individual cases.

chemical composition should be minimum of 70% for all kinds of PFA, to be suitable as pozzolan. It is found that, ash produced from some biomass possesses pozzolanic properties and can be used as partial cement replacement material in cement and concrete making. Table 8.1 gives the comparison of chemical composition of ash produced from biomass vis-à-vis that of the ASTM type F and Type C PFA. As seen in Table 8.1, the ash from neat biomass combustion typically has more alkali (Na and K) and less alumina (Al$_2$O$_3$) in comparison to coal fly ash. The biomass fuels exhibit more variation in both composition and amount of inorganic materials in comparison to that of coal. The ash composition seems to be related to the combustion conditions. Therefore, the composition of biomass fly ash, obtained from co-combustion with coal, varies more than that of coal fly ash (PFA). Even for the same type of biomass, the properties of ash depend on some growth and production factors such as weather, season, storage, and geographic origin, besides combustion conditions [2,4–11]. The following sections review the useful properties of some biomass ash varieties and their applications in manufacturing structural grade concrete.

8.2.1 Corn Cob Ash

The corn cob is the central core of a maize ear (Plate 8.1). It is a waste product. The estimated world corn production during 2009–2010 was over 800 × 10^6 ton [12]. It is an important cereal crop in sub-Saharan Africa, and the United States is the largest producer. On a dry matter basis, cob yield

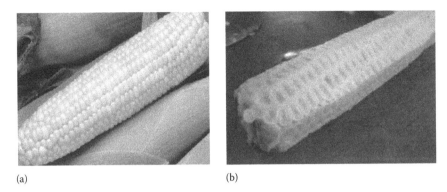

(a)   (b)

PLATE 8.1
(a) Corn and (b) corn cob.

averages about 14% of the grain yield and accounts for about 13%–23% of the corn stover (leaves and stalks of maize) biomass. Upon complete burning, corn cob yields about 1.5% ash.

Adesanya et al. [10] extensively studied the pozzolanic properties of CCA. The sum of SiO_2 + Al_2O_3 + Fe_2O_3 in the chemical composition of ash (Table 8.1) satisfies the requirement of ASTM C618 (minimum sum 70%), suggesting its pozzolanic nature. It is reported that ground CCA with fineness comparable with PC may be used as replacement material for cement, up to 15%, in structural grade concrete [10,13,14].

8.2.2 Palm Oil Residue Ash

The palm oil industry is one of the most important agro industries in Malaysia, Indonesia, and Thailand. It produces a large quantity of solid waste, besides the crude palm oil. On average, processing of 1 ton fresh palm fruit bunches in oil mill produces 0.21 ton palm oil and residues consisting of 0.06–0.07 ton kernels, 0.06–0.07 ton shells, 0.14–0.15 ton fibers, 0.23 ton empty fruit bunches (Plate 8.2), and 0.65 ton effluent [15,16]. The solid palm oil residue is mostly burned as fuel in power plants and generates ash, about 5% of its mass [17]. With the increase in palm oil production, the quantity of PORA continues to increase. However, its utilization remains minimal and most of it is disposed of in landfills, causing environmental hazard.

The PORA is characterized by a spongy and porous structure. The particles have an angular and irregular form and a typical specific gravity of 2.33 (Plate 8.3) [18,19]. The chemical composition of PORA (Table 8.1) shows that the sum of SiO_2 + Al_2O_3 + Fe_2O_3 may not or just marginally satisfy the requirement of ASTM C618. Although it is not a natural pozzolan, when ground finely, it can be classified as a Class N (natural) pozzolan, based on the chemical composition according to ASTM C618 [16]. The pozzolanic reactivity improves with the size reduction [21]. Foo and Hameed [16] summarized

(a)

(b)

(c)

PLATE 8.2
(a) Fresh oil palm fruit bunch, (b) empty fruit bunch, and (c) empty fruit bunch fiber.

PLATE 8.3
Porous PORA particles.

the results of earlier studies, where PORA was used as a substitute for cement to make structural grade concrete. It is observed that finely ground PORA (typically 1%–3% retained on 45 μm or sieve 325), as a mineral admixture, enhances the strength and durability characteristics of concrete, for cement replacement on the order of 20%–30% [19–22].

(a)

(b)

PLATE 8.4
(a) Sugarcane and (b) sugarcane bagasse waste.

8.2.3 Sugarcane Bagasse Ash

It is a waste product of sugar and alcohol production (Plate 8.4). The sugar factory produces nearly 0.3 ton of wet bagasse for each ton of sugarcane crushed. It contains 1%–4% ash (dry mass basis). As seen in Table 8.1, the chemical composition of SBA is comparable with ASTM Type F PFA, except the silica content that is high; a part of silica is present in the amorphous form, depending upon the burning conditions. The ash generally has higher surface area, depending upon grinding condition and lower specific gravity in comparison with cement. The pozzolanic activity is found proportional to the Blaine surface area [23]. The cement replacement with SBA, up to 10%–20%, can produce concrete without any adverse effect on strength, with reduced heat evolution and low water permeability [4,5,24].

8.2.4 Wheat Straw Ash

The world wheat production is likely to grow to 841×10^6 ton/annum by 2020 [25]. It is one of the major agricultural products of developing countries. The yield of wheat straw (Plate 8.5) is about 0.7–0.75 ton, per ton of wheat crop produced. The ash content is found to vary between 8% and 11% (dry mass basis) [26]. The WSA, obtained by burning wheat straw under controlled conditions and subsequent grinding, possesses more surface area and smaller average particle size (4800–5500 cm²/g), in comparison to PC and a specific gravity in the range of 1.98–2.41 [2,11]. The particles have irregular shape (Plate 8.6). The rice husk ash (RHA) is the mineral admixture nearest to WSA. The comparison of the morphology and chemical composition (Table 8.1 and Chapter 4) of both the admixtures (Figure 8.1) reveals that both contain amorphous silica (hump in the x-ray diffractogram) but the silica content of WSA is relatively less. The sum of $SiO_2 + Al_2O_3 + Fe_2O_3$ percentage in the chemical composition of WSA satisfies the requirement of ASTM C618 and suggests pozzolanic character, in most cases. The WSA can be used as a cement replacement material in manufacturing structural grade concrete,

PLATE 8.5
Wheat straw.

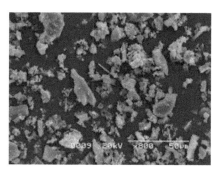

PLATE 8.6
SEM micrograph of WSA.

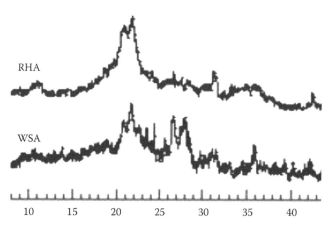

FIGURE 8.1
X-ray diffractograms of RHA and WSA.

up to the replacement level of 10%; however, relatively high content of alkali should be noted. The ternary blends could be useful from that point of view. Such blends of cement with one less reactive and another more reactive pozzolan have been found to produce synergistic effect in terms of strength and higher replacement levels are possible [11,27–29].

8.2.5 Wood Waste Ash

In the current trend of power generation, emergence of biomass-fuelled (forestry and agricultural waste) power plants is a significant development, as they use continuously renewable, zero net-GHG fuel with low operating cost. In Portugal, two units of pilot biomass-fuelled power plants have been constructed to supplement the power demand of the national electric gridlines [30]. Wiltsee lists summary information on 20 biomass power plants—18 in the United States, 1 in Canada, and 1 in Finland [31]. The ash generated through the incineration of wood in these power plants is mostly disposed of through the conventional method, that is, land filling. Typically between 0.4% and 1.8% of the mass of burned wood (dry mass basis) results in ash. WWA contains fine particles that easily get airborne and may cause respiratory problems for the residents living in the surroundings. The possible contamination of ground water by leaching of heavy metals from such landfills is also a cause of concern [32]. The disposal of WWA therefore requires greater attention. The chemical compositions of ash and leachates are important criteria for landfill disposal.

WWA is a fine powder. It is a heterogeneous mixture of variable size particles, highly porous in nature (Plate 8.7). The particle porosity contributes to the surface area [33]. The particles are unburned or partially burned wood or bark. The bulk density and specific gravity of wood waste FA are in the range of $490\,kg/m^3$ and 2.48, respectively. The relatively lower value of specific gravity indicates the possibility of reduction in the mass of concrete per unit volume, when cement is partially substituted with WWA. The physical properties of WWA vary significantly with the type of wood species and also have a significant effect on its pozzolanic and hydraulic activity.

The proportion of essential oxide compounds that governs the suitability of WWA as a cement replacement material, as given in the chemical composition (Table 8.1), such as silica (SiO_2), alumina (Al_2O_3), iron oxide (Fe_2O_3), and lime (CaO), varies significantly with the species of wood. It is observed that it

PLATE 8.7
SEM micrograph of WWA.

×2000   20 nm   5 kV

has low content of Al_2O_3 and Fe_2O_3. WWA with higher sum of $SiO_2 + Al_2O_3 + Fe_2O_3$ in the chemical composition (>70%) are found to possess better pozzolanic activity. The industrial wood-fired boilers mostly operate above 1000°C. The ash yield decreases, when the combustion temperature goes above 500°C. The potassium, sodium, zinc, and carbonate content decreases, whereas other metal ions remain constant or increase with temperature. The ash composition also changes during storage and under varying environmental conditions, as carbon dioxide and moisture react with ash to form carbonates, bicarbonates, and hydroxides [9,34,35]. In view of these findings, it is necessary that WWA is properly characterized before using as a substitute for cement.

The research reported in the literature suggests that WWA can replace cement, while making structural grade concrete. The properties of fresh concrete—standard consistency and other rheological properties, setting time, heat evolution, as well as those of hardened concrete, namely, soundness and mechanical strength (compressive, flexural, and splitting tensile strength)— are significantly affected. The requirement of air-entraining agent increases. In general, there is a reduction in the mechanical strength to varying extents, depending upon the characteristics of ash. On the other hand, the improvement in durability characteristics has been observed against certain specific deleterious agents [9,32,33,36,37]. A significant amount of ash is generated from burning wood with supplementary fuels such as coal, oil, natural gas, and coke by pulp and paper mills and wood-products manufacturers. The ash generated from such facilities is a mixture of wood ash and other ashes generated from the supplemental fuels. The controlled low-strength materials (CLSM) developed [38] using such ash and tested for flow, bleed water, settlement, shrinkage and cracking, setting characteristics, density, compressive strength, and permeability satisfied the requirements of ACI 229 [39].

It may be noted that the blending of biomass feedstock and coal does not necessarily result in an additive effect. The prediction of these effects is based on the understanding of the manner in which the inorganic constituents of the fuel interact during combustion and their effect on the chemical and physical properties of the ash and gas phases in the system. The biofuels having a higher concentration of alkali metals and alkaline earth metals have higher risk of ash deposition and slagging in combustion systems, during co-firing. It is reported that substitution of coal up to 20%, in general, does not lead to an increase in the risk of ash deposition or slagging, that would compromise a combustion system [40].

The concentration of environmentally relevant heavy metals (especially Cd, Hg, Pb, and Zn) and organic contaminants (polychlorinated dibenzo-dioxins/furans [PCDD/F] and polycyclic aromatic hydrocarbons [PAH]) in biomass ash, irrespective of the type of biomass, as well as that of alkalies, chlorides and sulfates, increases with decreasing ash precipitation temperature and particle size, although in most cases, the concentration of metals in the ash is lower than the EPA maximum limits for safe disposal. The biomass affects the radiological characteristics of ash produced after combustion, through the addition of ^{40}K or ^{137}Cs radio nuclides [30,41,42]. The effect of biomass co-combustion on the speciation of metals and other toxic or deleterious agents and radioactive nuclides should be examined carefully.

The foregoing discussion reveals that it is possible to produce PFA with pozzolanic properties, burning and processing biomass under controlled conditions, either through co-firing with mineral coal or separately. However, it is necessary to control the carbon, alkali, chloride, and sulfate content. In addition, leaching and toxicity properties also need to be tested before its use in concrete [30,42–44]. It is possible to produce structural grade concrete, partially replacing cement with PFA obtained from co-combustion of coal and biomass, to the extent of 20%–25%. In general, it is observed that the water demand, requirements of air-entraining agent, and setting time increase with the application of such PFA. The durability parameters also need to be tested on a case-to-case basis [7,9,45].

The biomass ash produced in thermal power plants is currently disposed of mostly in landfills or recycled to agricultural fields or forests and most of the times it goes on without any form of control. The strategies for sustainable management of biomass ash will have to be worked out, considering the disposal cost and the volume, increasing worldwide. The European List of Wastes [46] classifies biomass ash from thermal power plants as an industrial waste, with Code 100101 or 100103, and it should be managed accordingly. The ASTM C618 [3] prohibits the use of biomass fly ash in concrete. The European Standard [47] allows co-firing of biomass with coal, up to 25% on mass basis. However, on account of the wide range of biomass resources and combustion conditions, an upper limit has been specified for the content of alkali (5%), chloride (0.1%), and unburned carbon (5%).

8.3 Calcined Wastepaper Sludge

The production of metakaolin (MK) by calcination of wastepaper sludge has already been discussed in Chapter 5 on metakaolin. This section may be read along with that.

Industrial paper making impacts the environment, both upstream (raw materials procurement and processing) and downstream (waste disposal). The paper production worldwide today accounts for about 35% of all trees harvested. Many of the trees used for paper come from tree farms that are planted and replenished for that purpose [48]. There is a worldwide trend toward recovery and recycling of wastepaper. The production of a ton of paper from the recycled paper saves up to 17 trees and requires 50% less water than that from virgin pulp [49]. The recycling process results in significant quantities of sludge as a by-product of the ink removal process. Typically 20%–35% of the wastepaper feedstock is lost as sludge, which is mostly disposed of as a landfill, raising environmental concerns [50,51].

PLATE 8.8
Wastepaper sludge under a binocular microscope (40×) showing cluster of particles with
different sizes and shapes. (From Sabador, E.
et al., *Mater. Construct.*, 57(285), 45, 2007.)

The wastepaper sludge is an agglomeration of organic materials (cellulose), calcium carbonate, and clayey minerals (mostly kaolin clay used for
paper coating), depending upon the process. The sludge is a cluster of particles with different sizes and shapes (Plate 8.8) [52]. It has grayish color (due
to de-inking process) with a typical specific gravity of 2.67 [53]. It is found
that calcination of sludge, under controlled condition, produces ash with
pozzolanic characteristics (Plate 8.9) [52]. The calcination temperature and

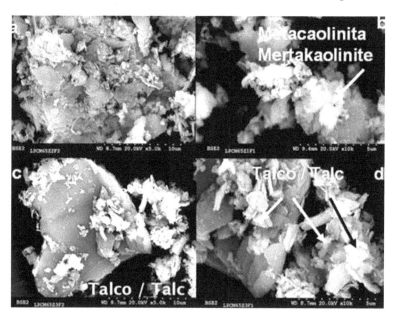

PLATE 8.9
SEM micrographs of wastepaper sludge incinerated at 600°C: (a) highly porous sample surface,
(b) metakaolinite crystal, (c) large talc crystal, and (d) bundled talc clusters. *Note*: The point
analyses of the material incinerated at 600°C showed a large amount of calcium on the porous
surfaces (a). The metakaolinite clusters with very small grain size were observed (b) alongside
fibrous structures with a higher magnesium content than that found in talc (c), with a tendency
to bundle in some cases (d). (From Sabador, E. et al., *Mater. Construct.*, 57(285), 45, 2007.)

retention time affect the characteristics of wastepaper sludge ash. In general, calcination in the range of 600°C–750°C and the retention time of about 2 h produces CWS with good pozzolanic characteristics. At higher calcination temperature, the pozzolanic activity decreases on account of (a) decarbonation of calcite, (b) formation of new mineral phases, and (c) decrease in the reactive surface as a consequence of the formation of new crystalline phases and big-size aggregates [51,54,55].

Table 8.2 illustrates the mineralogical and chemical composition of wastepaper sludge and CWS, vis-à-vis ASTM Type C PFA. The ground CWS generally has finer particle size and it can be used as pozzolanic material, partially substituting cement in concrete up to 20%. Sabador et al. [52], after studying the whiteness index, recommended that CWS manufactured at incineration temperature above 600°C could be used as an addition to white cement. It is reported that MK and calcite in CWS affect properties of concrete in fresh and hardened states. The rate of heat evolution, water demand, workability

TABLE 8.2

Mineralogical and Chemical Composition of Wastepaper Sludge and CWS vis-à-vis ASTM Type C PFA

Sl No	Particulars (in %)	PFA ASTM Type C (Typical)[a,c]	Wastepaper Sludge[c]	Calcined Wastepaper Sludge[c]
Mineralogical composition[c]				
i	Kaolinite		12–36	
i	Calcite		12–45	
iii	Phyllosilicates[b]		7–15	
iv	Organic matter		29–40	
Chemical composition[c]				
v	SiO_2	37	11–22	21–30
vi	Al_2O_3	20	7–16	14–19
vii	Fe_2O_3	6	0.4–0.8	0.5–0.9
viii	CaO	24	7–25	31–44
ix	MgO	5	0.9–4	2–5
x	Na_2O	2	0.1–0.3	0.1–0.2
xi	K_2O	0.4	0.2–0.7	0.3–1
xii	SO_3	2	0.3	0.3–1
xiii	LOI	1	46–56	15–23

[a] See Chapter 1.

[b] The phyllosilicates or sheet silicates are an important group of minerals that includes muscovite or micas, chlorite, serpentine, talc, and the clay minerals.

[c] The total of all percentages does not add to 100, as minor constituents have not been mentioned. The values or the range of values are rounded off and indicative. The actual values may be lower or higher in the individual cases.

characteristics, and drying shrinkage need to be observed [50,51,53,55–58]. The development of low-grade cementitious material using a mixture of CWS and GGBS has been reported, where the CWS was manufactured by flash calcination of sludge in the fluidized bed at 850°C–1200°C and subsequent rapid cooling, down to 200°C [59,60].

8.4 Electric-Arc Furnace Dust

The production of steel by EAF technology is increasing in importance, in comparison to the traditional open hearth and basic oxygen converter technology. It reached nearly one-third of the global share in 1999 and the share continues to rise [61]. The EAF produces carbon steel and alloy steel from scrap metal along with variable quantities of direct reduced iron, hot briquetted iron, and cold pig iron. The furnace produces dust containing particulate matter and gas. The particulate matter removed in dry form, in the gas cleaning system, is referred as electric-arc furnace dust (EAFD); nearly 15–20 kg dust is generated per ton of steel produced. It is listed by the U.S. Environmental Protection Agency as a hazardous waste under the Resource Conservation and Recovery Act, as it contains zinc, lead, cadmium, and hexavalent chromium, mostly beyond the acceptable limits [61]. Thus, all the EAFD need to be safely disposed of. The dust also generally contains significant amount of chlorides [62]. In a typical EAF operation, approximately 1%–2% of the charge is converted into dust [63]. According to the International Iron and Steel Institute (IISI) estimate [64], in 2000, world production of crude steel by EAF was about 286×10^6 ton. Thus, the production of EAFD was about $2.9–5.7 \times 10^6$ ton, in that year. Wherever economically feasible, heavy metals are recovered from the dust (especially zinc). However, in other cases, it needs to be disposed of. Worldwide approximately 70% of EAFD is sent to landfill [65]. The prevention of pollution of surroundings due to leaching of toxic metals is the principal issue that is required to be addressed in such cases.

The EAFD, as received from the plant, has a specific gravity of 4–4.2 (higher than cement) and fine grained with the surface area of 4800–7300 cm²/g Blaine [66,67]. The substitution of cement with EAFD to the extent of 7.5%–15% is reported to improve the strength and durability, whereas higher additions may have adverse effect on the strength. The setting time and drying shrinkage and along with that workability and slump retention also increase [65,66,68]. Maslehuddin et al. [65] recommend that the compatibility of EAFD with the blended cement (containing mineral admixture) should be evaluated before addition. However, from the studies reported, it appears that the maximum amount of EAFD that could be added to cement should be decided on the basis of the possible environmental impact, that is, the leachability of the heavy metals. The potentially toxic elements such as cadmium and lead present in EAFD get stabilized in concrete [67]. The cement blended with GGBS is found more effective in incorporating heavy metals leached by EAFD [69,70].

8.5 Sewage Sludge Ash

The treatment of wastewater generates sludge and its incineration produces SSA. About 300–400 kg ash is produced per ton of dried sludge. Since SSA is a waste material, attention must be paid to its environmental impact when reused.

Cyr et al. [71] characterized and studied the application of SSA as an admixture in cement mortar, including its environmental impact. The authors also took a review of the application of SSA in cement based materials, as reported in the literature. The SSA is polyphasic material consisting of several crystalline and amorphous (40%–74%) phases. It is composed of particles with irregular shape and porous structure. The particle shape is related to the method of incineration. The chemical (oxide) composition is characterized by wide variation. It mainly comprises of SiO_2, Al_2O_3, Fe_2O_3, CaO, and P_2O_5. The sum of SiO_2 + Al_2O_3 + Fe_2O_3 varies around the mean value of 59% (weak pozzolan). The CaO and P_2O_5 as well as sulfate (gypsum) and alkali content are also high, in comparison with PFA. The trace element analysis shows the presence of heavy metals such as zinc, chromium, and copper; the leachability tests, therefore, become important.

When used in concrete, reduced workability and increased setting time has been reported. The negative effect on workability can be countered using superplasticizer. The results on compressive strength are not uniform; reduced, comparable, and better performance have been reported [72–74]. In view of that, 10%–15% cement replacement may be tried, subject to the satisfactory leachability tests.

8.6 Municipal Solid Waste Ash

The generation of solid waste is rapidly increasing due to the growth in population, living standards, and industry. Thousands of million tons of municipal solid waste (MSW) is produced, globally. The ever growing MSW load has a great impact on the ambient environment and public health, such as malodors from landfill sites, possible explosions resulting from combustible gases and contamination of groundwater and soil by hazardous organics and heavy metals in MSW. The increased environmental awareness, continuously increasing landfill costs, scarcity of landfill sites, and slow compost process underline the need to take into account alternative disposal methods. The municipal waste management and utilization strategies, therefore, are the main concern in many countries. The incineration is a common technique for treatment as it reduces solid waste by mass and volume, provides for energy recovery and leads to complete disinfection. It has technically proven as an effective waste treatment approach [75,76]. In view of its hazardous nature, guidelines on the treatment of emissions and disposal of incinerator waste have been issued by different countries [77].

The incineration process produces two types of residues. First are those collected by the air pollution control (APC) equipment: (a) semidry scrubber residue called the reaction product (RP) and (b) dry residue from the electrostatic precipitator or the bag filter, the so-called fly ash. Second are solid residues like bottom ash, grate siftings, and heat recovery ash [78]. The gas emissions from modern MSW incineration plants are cleaned thoroughly and made practically pollutant free. The hazardous fractions in MSW get concentrated in the solid ash residue. The total amount of municipal solid waste incineration ash (MSWA) generation ranges from 4% to 10% by volume and 15%–20% by mass of the original quantity of waste and the fly ash ((b) above) amounts to about 10%–20% of the total ash [79].

The physical and chemical characteristics of solid residue depend on many factors, such as the composition of feed MSW, type of incinerator, APC devices, operating conditions, and so on. The majority (>50%) of MSW consists of paper, food, and garden waste, followed by the textiles, plastics, glass, and metal waste [80]. The ash contains valuable earth elements, such as silicon, aluminum, iron, and calcium as well as soluble salts of sodium, potassium, and calcium—chlorides or sulfates. The main oxides indicated in the chemical analysis are SiO_2, CaO, and Al_2O_3. It also contains heavy metals, around 2% on mass basis, principally lead and zinc but also cadmium, chromium, copper, mercury, and nickel. The LOI* at 975°C is high. The chemical analysis shows that in the solid residues, less volatile elements with high boiling temperature remain in the bottom ash and grate siftings, while more volatile elements with low boiling temperature are captured in the fly ash. The leachability of heavy metals from the fly ash residue has been found high, hence it could be classified as hazardous waste. In comparison, the concentration of heavy metals in the leachates of the grate siftings and the bottom ash is relatively lower; hence, these residues can be considered for land filling, after conducting the required tests. In general, it is necessary to make a detailed analysis for better management, control, and utilization of the solid residue obtained from the MSW incinerators [81–85]. Amongst the solid residues, MSW incinerator fly ash is the most

---

* The LOI, that is, loss on ignition at 975°C ± 25°C, suggested by many specifications is an arbitrary value. The sample is ignited in a furnace under controlled temperature conditions, for 1–2h. It is an indication of chemically combined water and CO_2 emission.

hazardous; the following discussion mainly focuses on the characteristics and utilization of such ash in the untreated form.

The study reported by Remond et al. [86] on the MSW incineration fly ash obtained from a incineration plant in France showed that the particle size ranged between that of cement and sand (1–600 μm). Plate 8.10 is a back-scattered electron image of the MSW incineration fly ash. As could be seen, unlike coal fly ash that is composed of spherical particles, a relatively large fraction of MSW fly ash particles appear to have a vitreous form, although some particles are spherical (solid or hollow). One can also observe elon-gated, angular, highly porous particles and clusters of sintered particles. The x-ray powder diffraction (XRD) results corroborate the results of chemical analysis. However, it may be noted that below a mass percentage of 5%, com-pounds are difficult to be detected by XRD and for that reason the heavy metals of concern are not seen in the XRD diagram [86]. Similar observations are also reported in the literature [81,87].

The Portland cement, blended with PFA and BFS, containing up to 20% MSWA, is found good in mechanical properties as well as for the immobiliza-tion of lead and sulfate and somewhat lesser for zinc and chloride, into stable, less soluble compounds. The capacity increases with the hardening age [88].

The study reported by Huang and Chu [78] revealed that both the fly ash and the RP collected from the APC equipment have cement-like properties and can be easily immobilized by cement. The study reported by Gougar et al. on the waste ion immobilization [89] shows that ettringite formed during the cement hydration incorporates number of heavy metal ions like lead, zinc, cadmium, nickel, cobalt, and so on. C-S-H, which is also formed during cement hydration, immobilizes ions and salts of the waste species. However, high concentration of water soluble salts may not get effectively immobi-lized. Since APC equipment residues contain high level of soluble salts, the leaching of these salts from the mortar or concrete matrix, over time, may

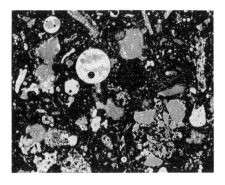

PLATE 8.10
Backscattered electron image of the MSW incinerator fly ash. (From Remond, S. et al., *Cem. Concr. Res.*, 32(2), 303, 2002; Garboczi, E.J. et al., An electronic monograph: Modeling and mea-suring the structure and properties of cement-based materials, National Institute of Standards and Technology, Gaithersburg, MD, February 23, 2011.)

result in poor performance. The subsequent loss of physical strength and durability may lead to an increase in metal leaching. In fact, loss of strength has been observed, even in nonaggressive environment. In general, irrespective of the solidification or stabilization method used to immobilize the hazardous constituents, the treatment is carried out in order to comply with landfill criteria [86,90]. In Taiwan, therefore, the solidification operation usually consists of a maximum of 25% cement (mass basis) and heavy metal chelator, with ash waste. The solidified ash wastes are required to comply with the landfill regulatory directive, with 10 kg/cm² for unconfined single axial compressive strength and 5 ppm for lead, for toxicity characteristic leaching procedure (TCLP) test. The addition of sulfide base additive, however, is found to reduce leachable lead levels, apparently due to the formation of insoluble PbS. It could be said that the technologies for the immobilization of toxic wastes are still under development. Although MSWA can be added to concrete as partial cement replacement, there is still some scope for the development, before it could be effectively used as a mineral admixture [91–93].

8.7 Summary

In this chapter, some new, less known but potentially useful, mineral admixtures have been introduced. Most of these admixtures are still under investigation at various stages. The purpose of this discussion, besides providing an introduction, is to generate, among the researchers, more interest in the subject. The new mineral admixtures covered in this chapter are (a) ash produced from biomass, such as CCA, PORA, SBA, WSA, WWA, (b) CWS, (c) EAFD (d) SSA, and (e) MSWA. A significant fraction, up to one-fifth, of the herbaceous biomass consists of inorganic constituents, commonly referred to as ash, which cannot be converted to energy through combustion. It is possible to produce PFA with pozzolanic properties, burning and processing biomass under controlled conditions, either through co-firing with mineral coal or separately. However, it is necessary to control the carbon, alkali, chloride and sulfate content. In addition, leaching and toxicity properties also need to be tested before its use in concrete. It is possible to produce structural grade concrete, partially replacing cement with PFA obtained from co-combustion of coal and biomass, to the extent of 20%–25%. In general, it is observed that the water demand, requirement of air-entraining agent and setting time increase with the application of such PFA. The durability parameters also need to be tested, on a case-to-case basis. The strategies for sustainable management of biomass ash will have to be worked out, considering the disposal cost and the volume, increasing worldwide. The ASTM C618 prohibits use of biomass fly ash in concrete. The European Standard allows co-firing of biomass with coal, up to 25% on mass basis. However, on account of the wide range of biomass resources and combustion conditions, upper limit has been specified for the content of alkali (5%), chloride (0.1%), and unburned carbon (5%).

There is a worldwide trend toward recovery and recycling of wastepaper. The recycling process results in significant quantities of sludge as a byproduct of the ink removal process. Typically 20%–35% of the wastepaper feedstock is lost as sludge, which is mostly disposed of as a landfill, raising environmental concerns. It is found that calcination of sludge, under controlled condition, produces ash with pozzolanic characteristics. The ground CWS generally has finer particle size and it can be used as pozzolanic material, partially substituting cement in concrete up to 20%.

The EAF produces carbon steel and alloy steel from scrap metal along with variable quantities of direct reduced iron, hot briquetted iron, and cold pig iron. The furnace produces dust containing particulate matter and gas. The particulate matter removed in dry form, in the gas cleaning system, is referred as EAF dust (EAFD); nearly 15–20 kg dust is generated per tonne of steel produced. It is hazardous, as it contains heavy metals like zinc, lead, cadmium, and hexavalent chromium, mostly beyond the acceptable limits. Wherever economically feasible, heavy metals are recovered from the dust (especially zinc); however, in other cases, it needs to be disposed of. Worldwide, approximately 70% of EAFD is sent to landfill. The prevention of pollution of surroundings due to leaching of toxic metals is the principal issue that is required to be addressed in such cases. The substitution of cement with EAFD to the extent of 7.5%–15% is reported to improve the strength and durability, whereas higher additions may have an adverse effect on the strength. The maximum amount of EAFD that could be added to cement should be decided on the basis of the possible environmental impact, that is, the leachability of the heavy metals.

The treatment of wastewater generates sludge and its incineration produces SSA. About 300–400 kg ash is produced per ton of dried sludge. Since SSA is a waste material, attention must be paid to its environmental impact, when reused. When used in concrete, reduced workability and increased setting time has been reported. The negative effect on workability can be countered using superplasticizer. The results on compressive strength are not uniform; reduced, comparable, and better performances have been reported. In view of that, 10%–15% cement replacement may be tried, subject to the satisfactory leachability tests.

Thousands of million tons of MSW is produced globally. The increased environmental awareness, continuously increasing landfill costs, scarcity of landfill sites, and slow compost process underline the need to take into account alternative disposal methods. Incineration is a common technique for treatment as it reduces solid waste by mass and volume, provides for energy recovery, and leads to complete disinfection. The total amount of municipal solid waste incineration ash (MSWA) generation ranges from 4% to 10% by volume and 15%–20% by weight of the original quantity of waste and the fly ash (obtained from the dust collection equipment) amounts to about 10%–20% of the total ash. Portland cement, blended with PFA and BFS, containing up to 20% MSWA, is found to be good in mechanical properties as well as for the immobilization of lead and sulfate and somewhat lesser for zinc and chloride, into stable, less soluble compounds. The capacity increases with the hardening age. Although MSWA can be added to concrete as a partial cement replacement, there is still some scope for the development, before it could be effectively used as a mineral admixture.

References

Chapter 1

1. Humphreys K., Mahasenan M., *Towards Sustainable Cement Industry—Substudy 8: Climate Change*, with contributions from M. Placet, K. Fowler, Battelle and World Business Council for Sustainable Development, Geneva, Switzerland, March 2002.
2. Business Solutions for Sustainable World, http://www.wbcsd.org
3. Bapat J. D., Performance of cement concrete with mineral admixtures, *J. Adv. Cem. Res.*, 13(4), 2001, 139–155.
4. Bapat J. D., Manufacturing blended cements for better performance, *Zement Kalk Gips Int.*, 51(12), 1998, 702–706.
5. ASTM C618, Standard specification for coal fly ash and raw or calcined natural pozzolan for use in concrete, ASTM International, West Conshohocken, PA, 2008, doi: 10.1520/C0618-08, http://www.astm.org
6. Pietersen H. S., Reactivity of fly ash and slag in cement, Thesis, Delft University of Technology, Delft, the Netherlands, 1993.
7. Cabrera J. G., Hopkins C. J., Woolley G. R., Lee R. E., Shaw J., Plowman C., Fox H., Evaluation of the properties of British pulverized fuel ashes and their influence on the strength of concrete. In *Proceedings of the Second International Conference on the Use of Fly Ash, Silica Fume, Slag and other Mineral Byproducts in Concrete*, Ed. V. M. Malhotra, Madrid, Spain, April 21–25, 1986, Vol. 2, SP 91-5, pp. 115–144.
8. ASTM C188-95, Standard test method for density of hydraulic cement, American Society for Testing and Materials, West Conshohocken, PA, 2003.
9. ASTM C204-05, Standard test method for fineness of hydraulic cement by air permeability apparatus, American Society for Testing and Materials, West Conshohocken, PA.
10. IS 3812: Part I: 2003 (second revision), Pulverised fuel ash—Specification—Part I: For use as pozzolana in cement, cement mortar and concrete, Bureau of Indian Standards, New Delhi, India.
11. ASTM 430-96, Standard test method for fineness of hydraulic cement by the 45 μm (No. 325) sieve, American Society for Testing and Materials, West Conshohocken, PA, 2003.
12. Bapat J. D., Higher qualities from modern finish grinding process, *Int. Cem. Rev.*, January 1998, p. 30.
13. Lee S. H., Sakai E., Daimon M., Bang W. K., Characterization of fly ash directly collected from electrostatic precipitator, *J. Cem. Concr. Res.*, 29, 1999, 1791–1797.
14. Monk M., Portland-PFA cement: A comparison between intergrinding and blending, *Mag. Concr. Res.*, 35(124), 1983, 131–141.
15. Ha T., Muralidharan S., Bae J., Ha Y., Lee H., Park K. W., Kim D., Effect of unburnt carbon on the corrosion performance of fly ash cement mortar, *J. Constr. Build. Mater.*, 19(7), 2005, 509–515.

16. Das S. K., Yudhbir, A simplified model for prediction of pozzolanic character-istics of fly ash, based on chemical composition, *J. Cem. Concr. Res.*, 36, 2006, 1827–1832.
17. Malhotra V. M., Ramezaniapour A. A., *Fly Ash in Concrete*, 2nd edn., CANMET, Natural Resources of Canada, Ottawa, Ontario, Canada, 1994.
18. Carette G. G., Malhotra V. M., Characterization of Canadian fly ashes and their relative performance in concrete, *Can. J. Civil Eng.*, 14, 1987, 667–680.
19. Narang K. C., Portland and blended cement. In *Ninth International Congress on the Chemistry of Cement*, National Council for Cement and Building Materials, New Delhi, India, 1992, Vol. 1, pp. 213–257.
20. Turkdogan E. T., *Physicochemical Properties of Molten Slags and Glasses*, The Metals Society, London, U.K., 1983.
21. Mehta P. K., Pozzolanic and cementitious byproducts and mineral admixtures for concrete—A critical review. In Ed. V. M. Malhotra, ACI Publications, SP-79-1, American Concrete Institute, Farmington Hills, MI, 1983, Vol. 1, pp. 1–46.
22. Ghosh S. N. Ed., *Progress in Cement and Concrete—Mineral Admixtures in Cement and Concrete*, Vol. 4, Academia Books International, New Delhi, India, 1993.
23. Young L., Environmental assessment of fluidized bed Combustion waste. In *Fifth International Conference on Circulating Fluidized Beds*, Beijing, China, 1996.
24. Lu X., Amano R. S., Feasible experimental study on the utilization of a 300 MW CFB boiler desulfurising bottom ash for construction applications, *J. Energy Resour. Technol.*, 128(4), 2006, 311–318.
25. Conn R. E., Wu S., Sellakumar K. M., Environmental assessment and utilization of CFB ash. In *Pittsburgh Coal Conference*, Taiyuan, China, 1997.
26. Sellakumar K. M., Conn R. E., Bland A. E., A comparison study of ACFB and PCFB ash characteristics. In *Sixth International Conference on Circulating Fluidized Beds*, Wurzburg, Germany, August 22–27, 1999.
27. Fukudome K., Shintani N., Saitoh T., Kita T., Sasaki H., Utilisation of coal ash produced from pressurized fluidized bed combustion power plant as concrete mineral admixture. In *Sixth CANMET/ACI International Conference on Fly Ash, Silica Fume, Slag and Natural Pozzolans in Concrete*, Ed. V. M. Malhotra, Bangkok, Thailand, May 31–June 5, 1998, Vol. II, SP-178-29, pp. 528–544.
28. Li F., Zhai J., Fu X., Sheng G., Characterization of fly ashes from circulating fluidized bed combustion (CFBC) boilers cofiring coal and petroleum coke, *J. Energy Fuels*, 20, 2006, 1411–1417.
29. Kaplan E., Nedder N., Petroleum coke utilization for cement kiln firing. In *Cement Industry Technical Conference*, IEEE-IAS/PCA 2001, Vancouver, British Columbia, Canada, April 29–May 3, 2001, pp. 251–263.
30. Scott, A. N., Thomas, M. D. A., Evaluation of fly ash from co-combustion of coal and petroleum coke for use in concrete, *ACI Mater. J.*, 104(1), 2007, 62–69.
31. Yu J., Kulaots J., Sabanegh N., Gao Y., Hurt R., Suuberg E., Mehta A., Adsorptive and optical properties of fly ash from coal and petroleum coke co-firing, *Energy Fuels*, 14(3), 2000, 591–596.
32. Lamers F., Vissers J., van den Berg J., Effects of co-combustion of secondary fuels on fly ash quality. In *Seventh CANMET/ACI International Conference on Fly Ash, Silica Fume, Slag and Natural Pozzolans in Concrete*, Ed. V. M. Malhotra, SP-199, American Concrete Institute, Farmington Hills, MI, 2001, pp. 433–455.
33. Jia L., Anthony E., Charland J., Investigation of vanadium compounds in ashes from a CFBC firing 100% petroleum coke, *Energy Fuels*, 16(2), 2002, 397–403.

34. Andrade C., Maringolo V., Kihara T., Incorporation of V, Zn and Pb into the crystal phases of Portland clinker, *Cem. Concr. Res.*, 33, 2003, 63–71.
35. Sear L. K. A., Weatherley A. J., Dawson A., The environmental impacts of using fly ash—The UK producers' perspective. In *International Ash Utilization Symposium*, Kentucky, U.K., 2003.
36. Brown J., Ray N. J., Ball M., The disposal of pulverised fuel ash in water supply catchment areas, *Water Res.*, 10, 1976, 1115–1121.
37. Scott P. E., Baldwin G., Sopp C., James L., Harissa, Leaching trials on materials proposed as infill for Combe Down Stone Mines, AEA Technology Report AEA/CS/18303036/19, 1994.
38. Young P. J., Wilson D. C., Testing of hazardous waste to assess their suitability for landfill disposal, AERE Report R10737, 1987.
39. Baldwin G., Addis R., Clark J., Rosvear A., Use of industrial by-products in road construction—Water quality effects, CIRIA Report 167, 1997.
40. DIN 38414-S4, German Standard methods for the examination of water, waste water and sludge. Sludge and sediments group (Group S). Determination of leachability by water, Deutsches Institut für Normung e. V., Berlin, Germany.
41. Baba A., Gurdal G., Sengunalp F., Leaching characteristics of fly ash from fluidized bed combustion thermal power plant: Case study: Can (Canakkale-Turkey), *Fuel Process. Technol.*, 91, 2010, 1073–1080.
42. The Energy Resources Institute (TERI), New Delhi, India, http://www.teriin.org
43. Arnold G. K., Dawson A. R., Muller M., Determining the extent of ground and surface water contamination adjacent to embankments comprising pulverised fuel ash (PFA), Project report by the University of Nottingham, School of Civil Engineering, Nottingham Centre for Pavement Engineering, December 2003.
44. Born P. J. A., Toxicity and occupational health hazards of coal fly ash (CFA). A review of data and comparison to coal mine dust, *Ann. Occup. Hyg.*, 41(6), 1997, 659–676.
45. Meij R., Nagengast S., Winkel H. T., The occurrence of quartz in coal fly ash particles, *J. Inhal. Toxicol.*, 12(10 Suppl. 1), 2000, 109–116.
46. Meij R., Status report on the health issues associated with pulverised fuel ash and fly dust, 50131022-KPS/MEC 01-6032, KEMA Nederland B.V., Arnhem, the Netherlands, 2003.
47. GLOBE-Net, Report by the GLOBE Foundation of Canada, May 16, 2007.
48. Kumar V., Mathur M., Sharma P., Fly ash management: Indian endeavor. In *Proceedings of the National Seminar on Utilisation of Flyash in Water Resources Sector, Central Soil and Materials Research Station*, New Delhi, India, April 11–12, 2001.
49. Agarwal V. K., Technological options for handling and transportation of fly ash. In *Proceedings of the National Seminar on Utilisation of Flyash in Water Resources Sector, Central Soil and Materials Research Station*, New Delhi, India, April 11–12, 2001.
50. Bapat J. D., Application of ESP for gas cleaning in cement industry—With reference to India, *J. Hazard. Mater.*, B81(3), 2001, 285–308.
51. Harder J., Dry ash handling systems for advanced coal-fired boilers, *Bulk Solids Handling Int. J.*, 17(1), 1997, 65–70.
52. Mills D., Agarwal V. K., *Pneumatic Conveying Systems, Design, Selection and Troubleshooting with Particular Reference to Pulverised Fuel Ash*, Trans Tech Publications, Germany, 2001.

53. Barry E. E., *Beneficiated Fly Ash: Hydration, Microstructure, and Strength Development in Portland Cement Systems*, American Concrete Institute Special Publication, Farmington Hills, MI, Vol. 114, 1989, pp. 241–274.
54. Florida Department of Transportation (FDOT), USA, Specs and Estimates, July 2008 Workbook (Rev 2-5-07), FA 3-14-07, 7-08.
55. Chen Y., Shah N., Huggins F. E., Huffman G. P., Dozier A., Characterisation of ultrafine fly ash fly ash particles by energy filtered TEM, *J. Microsc.*, 217(Pt. 3), 2005, 225–234.
56. Kandie B. K. T., Byars E. A., Ultra fine fly ash concrete. In *Proceedings of the International Conference Sustainable Construction Materials and Technologies*, June 11–13, 2007, Coventry, Taylor & Francis, London, U.K., pp. 121–129.
57. Yijin L., Shiqiong Z., Jian Y., Yingli G., The effect of fly ash on the fluidity of cement paste, mortar and concrete. In *International Workshop on Sustainable Development of Concrete Technology*, Beijing, China, May 20–21, 2004.
58. Baoju L., Youjun X., Shiqiong Z., Qianlian Y., Influence of ultrafine fly ash composites on the fluidity and compressive strength of concrete, *Cem. Concr. Res.*, 30(9), 2000, 1489–1493.
59. Yingli G., Shiqiong Z., Influence of ultra-fine fly ash on hydration shrinkage of cement paste, *J. Cent. South Univ. Technol.*, 12(5), 2005, 596–600.
60. Subramaniam K. V., Gromotka R., Shah S. P., Obla K., Hill R., Influence of ultra-fine fly ash on the early age response and the shrinkage cracking potential of concrete, *ASCE J. Mater. Civil Eng.*, 17(1), 2005, 45–53.
61. Rathbone R. F., Mahboub K. C., Superpozzolanic concrete for sustainable construction and CO_2 emission reduction, Publication of the Centre for Applied Energy Research (CAER), University of Kentucky, Kentucky, U.K., 2006.
62. Blanco F., Garcia M. P., Ayala J., Mayoral G., Garcia M. A., The effect of mechanically and chemically activated fly ashes on mortar properties, *J. Fuel*, 85(16), 2006, 2345–2351.
63. Poon C. S., Qiao X. C., Lin Z. S., Pozzolanic properties of reject fly ash in blended cement pastes, *J. Fuel*, 85(14–15), 2006, 2018–2026.
64. Sable S. P., Mercury speciation in pulverized fuel co-combustion and gasification, PhD thesis, Technische Universiteit Delft, Delft, the Netherlands, 2007.
65. Chang R., Beneficiation of fly ash containing mercury and carbon, 1004267, EPRI Technical Update, Palo Alto, CA, March 2005.
66. Duijn G. S., DUOS coal fly ash upgrading technology. In *Sixth ACI International Conference on Fly Ash, Silica Fume, Slag and Natural Pozzolans in Concrete*, Bangkok, Thailand, 1998.
67. Moret J. B. M., Maasvlakte fly ash plant, In *The 3rd International Conference on the Environmental and Technical Implications of Construction with Alternative Materials: Putting Theory into Practice*, The International Society ISCOWA, Houthem St. Gerlach, Limburg, the Netherlands, June 4–6, 1997.
68. Moret J. B. M., Process, marketing and sales of upgraded fly ash in the Dutch market. In *Twelfth International Symposium on Management and Use of Coal Combustion By-Products*, American Coal Ash Association, Alexandria, VA, January 1997.
69. Sheu T. C., Quo L. W., Kuo S. T., *Class F Fine Fly Ash*, Materials Research Society, Fall Meeting, Boston, MA, November 1989.
70. Bittner J. D., Gasiorowski S. A., Triboelectrostatic fly ash beneficiation: An update on separation technologies' international operations. In *2005 Conference on Unburned Carbon on Utility Fly Ash*, Lexington, KY, April 13, 2005.

71. Gasiorowski S. A., Bittner J. D., Mackay B., Whitlock D., Applications of carbon concentrates derived from fly ash. In *15th International American Coal Ash Association Symposium on Management and Use of Coal Combustion Products (CCPs)*, St. Petersburg, FL, January 2003.

72. Bittner J. D., Gasiorowski S. A., Hrach F. J., Carbon separation and ammonia removal at Jacksonville electric St. Johns River Power Park. In *2003 International Ash Utilization Symposium*, Lexington, KY, October 20–22, 2003.

73. Shilling M., Carolina Power & Light Co. carbon/ash separation. In *1999 International Ash Utilization Symposium*, Lexington, KY, October, 1999.

74. Bittner J. D., Gasiorowski S. A., Five years of commercial fly ash beneficiation by Separation Technologies, Inc. In *1999 International Ash Utilization Symposium*, Lexington, KY, October, 1999.

75. Jeffcoat J., Vasiliauskas A., Boyer J. P., Fly Ash beneficiation at Baltimore Gas & Electric Brandon Shores Plant. In *1999 International Ash Utilization Symposium*, Lexington, KY, October, 1999.

76. Lockert C., Lister R., Stencel J. M., Commercialization status of a pneumatic transport, triboelectrostatic system for carbon/ash separation. In *2001 Conference on Unburned Carbon (UBC) on Utility Fly Ash*, Pittsburgh, PA, May 15–16, 2001.

77. Lockert C. A., Dunn K. T., Stencel J. M., Pneumatic transport triboelectric carbon/ash separation, full scale installation. In *2003 International Ash Utilization Symposium*, Lexington, KY, October 20–22, 2003.

78. Stencel J. M., Li T. X., Gurupira T., Jones C., Neathery J. K., Ban H., Altman R., Technology development for carbon-ash beneficiation by pneumatic transport, triboelectric processing. In *1999 International Ash Utilization Symposium*, Lexington, KY, October, 1999.

79. Frady W. T., Keppeler J. G., Knowles J. C., South Carolina electric & gas successful application of carbon burn-out at the Wateree Station. In *1999 International Ash Utilization Symposium*, Lexington, KY, October, 1999.

80. Giampa V. M., The fate of ammonia and mercury in the carbon burn-out (CBO) process. In *2003 International Ash Utilization Symposium*, Lexington, KY, October 20–22, 2003.

81. ASTM C311, Standard methods of sampling and testing fly ash or natural pozzolans for use as a mineral admixture in Portland cement concrete, American Society for Testing and Materials, West Conshohocken, PA.

82. BS EN 450, Fly ash for concrete—Definitions, requirements and quality control, British Standards Institution, London, U.K.

83. Ishikawa Y., Research on the quality distribution of JIS type II fly ash in Japan. In *2007 World of Coal Ash (WOCA) Conference*, Covington, KY, May 7–10, 2007.

84. ASTM C595-06, Standard specification for blended hydraulic cements, American Society for Testing and Materials, West Conshohocken, PA.

85. ACI 211.1: Standard practice for selecting proportions for normal, heavyweight and mass concrete, ACI Manual of Concrete Practice, Part 1. American Concrete Institute, Detroit, MI, 1996.

86. BS EN 206-1, Concrete. Specification, performance, production and conformity, British Standards Institution, London, U.K., 2000.

87. BS 3892: Part 1, Specification for pulverised-fuel ash for use with Portland cement, British Standards Institution, London, U.K., 1997.

88. JIS R 5213: Portland fly-ash cement, Japan Cement Association, Tokyo, Japan.

89. JIS A 6201, Fly ash for use in concrete, Japan Cement Association, Tokyo, Japan.

90. IS 1489: Part 1, Specification for Portland pozzolana cement fly ash based, Bureau of Indian Standards, New Delhi, India.
91. IS 456-2000: Indian standard code of practice for plain and reinforced concrete, Bureau of Indian Standards, New Delhi, India.
92. Bouzoubaa N., Zhang M. H., Bilodeau A., Malhotra V. M., Mechanical properties and durability of concrete made with high volume fly ash blended cements, SP 178.31. In *Sixth CANMET/ACI/JCI Conference on Fly Ash, Silica Fume, Slag and Natural Pozzolans in Concrete*, Ed. V. M. Malhotra, Tokushima, Japan, Vol. I, 1998.
93. Zhang T., Yu Q., Wei J., Zhang P., A new gap-graded particle size distribution and resulting consequences on properties of blended cement, *Cem. Concr. Compos.*, 33, 2011, 543–550.
94. Sonebi M., Medium strength self-compacting concrete containing fly ash: Modelling using factorial experimental plans, *Cem. Concr. Res.*, 34, 2004, 1199–1208.

Chapter 2

1. ASTM C989-99, Standard specification for ground granulated blast-furnace slag for use in concrete and mortars, ASTM International, West Conshohocken, PA, 2003, doi: 10.1520/C0989-99, http://www.astm.org
2. Banerjee A. K., Blast-furnace slag granulation process, for making better quality slag cement, *Iron & Steel Rev.*, December 1994, 79–81.
3. Toshio M., Jun-Ichiro Y., Tomohiro A., Granulation of molten slag for heat recovery. In *37th International Energy Conversion Engineering Conference*, Washington, DC, July 29, 2002, pp. 641–646.
4. Piplal K. C., Ramesh H. S., Ganesan K. V., Competitiveness of Portland slag cement Tata Steel experience. In *Proceedings of the Sixth NCB International Seminar on Cement and Building Materials*, New Delhi, India, 1994, pp. X-50–X-55.
5. Regourd M., Structure and behaviour of slag Portland cement hydrates. In *Proceedings of Seventh International Congress on the Chemistry of Cement*, Vol. I, Principal Reports, Paris, France, 1980, pp. III-2/10–III-2/26.
6. Douglas E., Wilson H., Malhotra V. M., Production and evaluation of a new source of granulated blast furnace slag, *J Cem. Concr. Aggregate*, 10(2), 1988, 75–87.
7. ASTM E11, Standard specification for woven wire test sieve cloth and test sieves, 2009, doi: 10.1520/E0011-09E01, http://www.astm.org
8. BS 6699, Specification for ground granulated blast furnace slag for use with Portland cement, 1992, British Standards Institution, London, U.K.
9. Asim M. E., Blast furnace slag processing to blended cements, *Zement kalk Gips*, 45(10), 1992, 519–528.
10. Lang E., Influence of the consolidation of stored granulated blastfurnace slag on its properties, *Cem. Int.*, 5(3), 2007, 85–94.
11. Petzelt N., Modern systems for grinding granulated blast furnace slag, *Zement Kalk Gips*, 45(9), 1992, E 239–E 244.
12. Bapat J. D., Manufacturing blended cements for better performance, *Zement Kalk Gips Int.*, 51(12), 1998, 702–706.
13. Wiegmann D., Muller-Pfeiffer M., Production of slag cements at the Schwelgern cement works using different plant systems, *Zement Kalk Gips Int.*, 50(3), 1997, 154–160.

14. Xianfan S., The relationship of water requirement and cementitious property of granulated alkaline blast furnace slag with its grinding fineness. In *Third Beijing International Symposium on Cement and Concrete*, Ed. W. Zhaogi, Beijing, China, Vol. 2, 1993, pp. 542–547.
15. Osbaeck B., Ground granulated blast furnace slags grinding methods, particle size distribution and properties, *Am. Conc. Inst. Pub.*, SP-114-60, 2, 1989, 1239–1263.
16. Sumner M. S., Hepher N. M., Moir G. K., The influence of narrow particle size distribution on cement paste and concrete water demand. In *Proceedings of the Eighth International Congress on the Chemistry of Cement*, Vol. II, 1986, pp. 310–315.
17. Kuhlmann K., Ellerbrock H. G., Sprung S., Particle size distribution and properties of cement Part I: Strength of Portland cement, *Zement Kalk Gips Int.*, 38(4), 1985, 169–178.
18. Aiqin W., Chengzhi Z., Ningsheng Z., The theoretic analysis of the influence of particle size distribution of cement system on the property of cement, *Cem. Concr. Res.*, 29, 1999, 1721–1726.
19. DIN66145, Graphical representation of particle size distributions; RRSB-grid, Deutsches Institut für Normung E. V., Berlin, Germany, 1976.
20. Schnatz R., Ellerbrock H. G., Sprung S., Description and reproducibility of measured particle size distribution of finely ground substances, Part 1, *Zement Kalk Gips Int.*, 52(2), 1999, 57–67.
21. Wan H., Shui Z., Lin Z., Analysis of geometric characteristics of GGBS particles and their influences on cement properties, *Cem. Concr. Res.*, 34(1), 2004, 133–137.
22. Rojak S. M., Structure et Activite Hydraulique des Laitiers de Haut Fourneau, *Cement*, 8, 1978, 4–5.
23. Smolczyk H. G., Structure et Caracterisation des Laitiers. In *Seventh International Congress on the Chemistry of Cement*, Paris, France, Vol. I—Rapports principaux, 1980, pp. III 1/3–1/16.
24. Demoulian E., Gourdin P. et Autres, Influence de la Composition Chimique et de la Texture des Laitiers sur leur Hydraulicite. In *Seventh International Congress on the Chemistry of Cement*, Paris, France, Vol. II—Communications, 1980, pp. III 89–III 94.
25. Lang E., Blast furnace cements. In *Structure and Performance of Cements*, Eds. J. Bensted, P. Barnes, 2nd edn., Spon Press, New York, 2002, pp. 310–325, Chapter 12.
26. Galibert R., Glass content influence upon hydraulic potential of blast-furnace slag, National Slag Association, Wayne, PA, NSA 184-2, 1984.
27. Lea F. M., *The Chemistry of Cement and Concrete*, 3rd edn., Chemical Publishing Co. Inc., New York, 1971.
28. Gribko W. F., Satarin V. I., Knolodnyi A. G., Stschotkina T. J., Hydraulic activity of glasses of the melilite series, *Silikattechnik*, 25, 1974, 222.
29. Zachariassen W. J., The atomic arrangement in glass, *J. Am. Ceram. Soc.*, 54, 1932, 3841–3951.
30. Pietersen H. S., Reactivity of fly ash and slag in cement, thesis, Delft University of Technology, Delft, the Netherlands, 1993.
31. Satarin V. I., Slag Portland cement. In *The Sixth International Congress on the Chemistry of Cement (VI ICCC)*, Moscow, Russia, 1974, Principal Paper p. 1/51.
32. Ghosh S. N. Ed., *Progress in Cement and Concrete—Mineral Admixtures in Cement and Concrete*, Vol. 4, Academia Books International, New Delhi, India, 1993.

33. ASTM C595, Standard specification for blended cements, American Society for Testing and Materials, ASTM International, West Conshohocken, Pennsylvania, PA, 2006, http://www.astm.org
34. United States Department of Transportation—Federal Highway Administration, http://www.fhwa.dot.gov
35. ACI 318-11: Building code requirements for structural concrete and commentary, ACI Committee 318, American Concrete Institute, Farmington Hills, MI, 2011.
36. BS EN 197-1: Cement—Part 1: Composition, specifications and conformity criteria for common cements, British Standards Institution, London, U.K., 2000.
37. BS EN 206-1, Concrete—Part 1: Specification, performance, production and conformity, British Standards Institution, London, U.K., 2000.
38. BS 8500,: Concrete—Complementary British Standard to BS EN 206-1. Part 1: 2002 Method of specifying and guidance for the specifier. Part 2: 2002 Specification for constituent materials and concrete, British Standards Institution, London, U.K., 2002.
39. JIS R 5211, Portland blast-furnace slag cement, Japan Cement Association, Tokyo, Japan, 2003.
40. IS 455-1989, Specification for Portland slag cement, Bureau of Indian Standards, New Delhi, India, 1989.
41. IS 456-2000, Indian Standard Code of practice for plain and reinforced concrete, Bureau of Indian Standards, New Delhi, India, 2000.
42. Tomisawa T., Fuji M., Effects of high fineness and large amounts of ground granulated blast furnace slag on properties and microstructure of slag cements, In *Proceedings of the Fifth International Conference on Fly Ash, Slag and Natural Pozzolans in Concrete*, Milwaukee, WI, ACI SP 153, Vol. 2, 1995, pp. 951–973.
43. Skripkiunas G., Dauksys M., Stuopys A., Levinskas R., The Influence of cement particles shape and concentration on the rheological properties of cement slurry, *Mater. Sci. (Medziagotyra)*, ISSN 1392–1320, 11(2), 2005, 150–158.
44. Oner M., A study of intergrinding and separate grinding of blast furnace slag cement, *J. Cem. Concr. Res.*, 30, 2000, 473–480.
45. Opoczky L., Grinding technical questions of producing composite cement, *Int. J. Miner. Process.*, 44–45, 1996, 395–404.
46. Zhang X., Han J., The effect of ultra-fine admixture on the rheological property of cement paste, *J. Cem. Concr. Res.*, 30, 2000, 827–830.
47. Escalante-Garcia J. I., Espinoza-Perez L. J., Gorokhovsky A., Gomez-Zamorano L. Y., Coarse blast furnace slag as a cementitious material, comparative study as a partial replacement of Portland cement and as an alkali activated cement, *J. Constr. Build. Mater.*, 23(7), 2009, 2511–2517, doi: 10.1016/j.conbuildmat.2009.02.002.
48. Tan K., Pu X., Strengthening effects of finely ground fly ash, granulated blast furnace slag, and their combination, *J. Cem. Concr. Res.*, 28(12), 1998, 1819–1825.
49. Binici H., Temiz H., Kose M. M., The effect of fineness on the properties of the blended cements incorporating ground granulated blast furnace slag and ground basaltic pumice, *J. Constr. Build. Mater.*, 21, 2007, 1122–1128.
50. Zhang X., Wu K., Yan A., Carbonation property of hardened binder pastes containing super-pulverized blast-furnace slag, *J. Cem. Concr. Compos.*, 26, 2004, 371–374.
51. Bougara A., Lynsdale C., Milestone N. B., Reactivity and performance of blast-furnace slags of differing origin, *Cem. Concr. Compos.*, 32, 2010, 319–324.

52. Oner A., Akyuz S., An experimental study on optimum usage of GGBS for the compressive strength of concrete, *J. Cem. Concr. Compos.*, 29, 2007, 505–514.
53. Papadakis V. G., Tsimas S., Supplementary cementing materials in concrete, part I: Efficiency and design, *J. Cem. Concr. Res.*, 32, 2002, 1525–1532.
54. Papadakis V. G., Antiohos S., Tsimas S., Supplementary cementing materials in concrete, part II: A fundamental estimation of the efficiency factor, *J. Cem. Concr. Res.*, 32, 2002, 1533–1538.
55. Swamy R. N., Design for durability and strength through the use of fly ash and slag in concrete. In *Fifth CANMET/ACI International Conference on Superplasticizers and other Chemical Admixtures in Concrete*, Rome, Italy, October 1997.
56. Cantharin P., The quantitative determination of blastfurnace slag in cements, *Zement Kalk Gips*, 29(2), 1976, 71–77.

Chapter 3

1. ACI 116R: Cement and concrete terminology, American Concrete Institute, Farmington Hills, MI, 1990, 58pp.
2. Kuennen T., Silica fume resurges, *Concr. Prod.*, Technology Series, March 1996, 5pp.
3. Holland T. C., Silica fume user's manual, U.S. Federal Highway Administration and Silica Fume Association, Washington, DC, 2005, 193pp.
4. Helland S., Acker P., Gram H. E., Sellevold E. J., Condensed silica fume in concrete, FIP State of the Art Report, Thomas Telford, London, U.K., 1988.
5. Malhotra V. M., Caretta G. G., Sivasundaram V., Role of silica fume in concrete: A review. In *Advances in Concrete Technology*, Ed. V. M. Malhotra, 2nd edn., CANMET, Ottawa, Ontario, Canada, 1992, pp. 915–990.
6. St. John D. A., Dispersion of silica fume. In *Mechanism of Chemical Degradation of Cement-based Systems*, Eds. K. L. Scrivener, J. F. Young, Taylor & Francis, Boca Raton, FL, 1997.
7. Wolsiefer J., The measurement and analysis of silica fume particle size distribution and de-agglomeration of different silica fume product forms. In *Ninth CANMET/ACI International Conference on Fly Ash, Silica Fume, Slag and Natural Pozzolans in Concrete*, Warsaw, Poland, May 20–25, 2007, Vol. 242, pp. 111–132.
8. Diamond S., Sahu S., Densified silica fume: Particle sizes and dispersion in concrete, *J. Mater. Struct.*, 39(9), 2006, 849–859.
9. Baweja D., Cao T., Bucea L., *Investigation of Dispersion Levels of Silica Fume in Pastes, Mortars, and Concrete*, American Concrete Institute Special Publication, Farmington Hills, MI, 2003, Vol. 212, pp. 1019–1034.
10. Diamond S., Sahu S., Densified silica fume—Is it what you think it is. In *Conference on Advances in Cement and Concrete IX: Volume Changes, Cracking and Durability*, Copper Mountain, CO, August 10–14, 2003, pp. 233–248.
11. Guide for the use of silica fume in concrete. In *Manual of Concrete Practice*, American Concrete Institute, Farmington Hills, MI, 2003.
12. Diamond S., Sahu S., Thaulow N., Reaction products of densified silica fume agglomerates in concrete, *J. Cem. Concr. Res.*, 34(9), 2004, 1625–1632.
13. Rangaraju P. R., Olek J., Evaluation of the potential of densified silica fume to cause alkali-silica reaction in cementitious matrices using a modified ASTM C1260 test procedure, *Cem. Conc. Aggr. J.*, 22(2), 2000, 150–160.

14. Maas A. J., Ideker J. H., Juenger M. C. G., Alkali silica reactivity of agglomerated silica fume, *J. Cem. Concr. Res.*, 37(2), 2007, 166–174.

15. Justnes H., Condensed silica fume as a cement extender. In *Structure and Performance of Cements*, Eds. J. Bensted, P. Barnes, 2nd edn., Spon Press, London, U.K., 2002.

16. Gupta S. M., Sehgal V. K., Kaushik S. K., Shrinkage of high strength concrete, *Proc. World Acad. Sci. Eng. Technol.*, 38, 2009, 268–271.

17. Wallewick O., Practical description of the rheology of fresh concrete. In *Proceedings of Borregaard Symposium on Workability and Workability Retention*, Hanko, Norway, 1998.

18. Davraz M., Gunduz L., Reduction of alkali silica reaction risk in concrete by natural (micronised) amorphous silica, *Constr. Build. Mater.*, 22, 2008, 1093–1099.

19. Yuan Q., Shi C., Schutter G. D., Audenaert K., Deng D., Chloride binding of cement-based materials subjected to external chloride environment—A review, *Constr. Build. Mater.*, 23, 2009, 1–13.

20. National Institute for Occupational Safety and Health (NIOSH), Centers for Disease Control and Prevention (CDC), The U.S. Department of Health and Human Services, Washington, DC.

21. United States Department of Labor, Occupational Safety and Health Administration (OSHA), Regulations (Standards—29 CFR), Ventilation—1910.94.

22. Cunningham E. A., Todd J. J., Jablonski W., Was there sufficient justification for the 10-fold increase in the TLV for silica fume? A critical review, *Am. J. Ind. Med.*, 33(3), 1998, 212–223.

23. Investigation of crystalline phases in silica fume, Safety in Mines Research Advisory Committee (SIMRAC), Department of Science and Technology, Pretoria, South Africa, 2002.

24. ASTM C1240, Standard specification for silica fume used in cementitious mixtures, American Society for Testing and Materials, ASTM International, West Conshohocken, Pennsylvania, PA, 2011, DOI: 10.1520/C1240-11, http://www.astm.org

25. EN 13263-1, Silica fume for concrete. Definitions, requirements and conformity criteria, 2005 and EN 13263-2: Silica fume for concrete. Conformity evaluation, European Committee for Standardization (CEN), Brussels, Belgium, 2006.

26. CAN/CSA-A23.5, Supplementary cementing materials, Canadian Standards Association, Ontario, Canada, 1998.

27. JIS A 6207, Silica fume for use in concrete, Japan Cement Association, Tokyo, Japan, 2000.

28. IS 15388, Silica fume—Specification, Bureau of Indian Standards, New Delhi, India, 2003.

29. Aitcin P. C., *High-Performance Concrete*, Taylor & Francis, London, U.K., 1998.

30. Coleman S. E., Hwu S. J., Vogt W. L., Determination of silica fume in unhydrated, blended, dry-packaged mixture, and hydrated mortar, *Cem. Concr. Aggregates*, 17(1), 1995, 61–68.

31. Kumar A., Roy D. M., A study of silica-fume-modified cements of varied fineness, *J. Am. Ceram. Soc.*, 67(1), 1984, 61–64.

32. Thomas M., Hopkins D. S., Perreault, M., Cail K., Ternary cement in Canada: Factory blends allow widespread application, *Concr. Int.*, 29(7), 2007, 59–64.

33. Using coal ash to mitigate alkali silica reactivity, Resource Paper, Combustion and By-product Use Program, EPRI, November 2003.
34. Thomas M. D. A., Shehata M. H., Shashiprakash S. G., Hopkins D. S., Cail K., Use of ternary cementitious systems containing silica fume and fly ash in concrete, *Cem. Concr. Res.*, 29, 1999, 1207–1214.
35. Yajun J., Cahyadi J. H., Simulation of silica fume blended cement hydration, *Mater. Struct.*, 37(270), 2004, 397–404.
36. Holland T. C., Working with silica-fume concrete, *Concr. Constr.*, 32(3), 1987, 261–267.

Chapter 4

1. Houston D. F., *Rice Hulls: Rice Chemistry and Technology*, American Association of Cereal Chemists, Inc., St. Paul, MN, 1972, pp. 301–352.
2. Mehta P. K., Siliceous ashes and hydraulic cements prepared therefrom, Belgium Patent 802909, July 1973 and U.S. Patent 4105459, August 1978.
3. Mehta P. K., Pitt N., Energy and industrial materials from crop residues, *J. Resour. Recovery Conserv.*, 2, 1976, 23–38.
4. Pitt N., Process for preparation of siliceous ashes, U.S. Patent 3959007, May 1976.
5. Mehta P. K., Properties of blended cements made from RHA, *ACI J. Proc.*, 74(9), 1977, 440–442.
6. Nehdi M., Duquette J., Damatty A. E., Performance of rice husk ash produced using new technology as a mineral admixture in concrete, *Cem. Concr. Res.*, 33, 2003, 1203–1210.
7. Mehta P. K., Pozzolanic and cementitious byproducts as mineral admixtures for concrete—A critical review. In *Proceedings of CANMET/ACI First International Conference on the use of Fly Ash, Silica Fume, Slag and other Mineral By-products in Concrete*, Ed. V. M. Malhotra, American Concrete Institute Publication, Detroit, MI, SP-79, Vol. 1, 1983, pp. 1–46.
8. Zhang W. J., Liu G. D., Bai C. J., A forecast analysis on global production of staple crops. In *Fourth International Conference on Agriculture Statistics (ICAS-4)—Advancing Statistical Integration and Analysis (ASIA)*, Beijing, China, October 22–24, 2007.
9. U.N. Food and Agriculture Organization, Rice market monitor, XII(1), February 2009.
10. Demonstration of rice husks-fired power plant in An Giang province—A pre-feasibility study report, PREGA National Technical Experts from Institute of Energy, Hanoi, Vietnam, May 2004.
11. United Nations Framework Convention on Climate Change (UNFCCC), Clean Development Mechanism (CDM), http://cdm.unfccc.int
12. Intergovernmental Panel on Climate Change. Set up by World Meteorological Organization (WMO) and the United Nations Environment Programme (UNEP), Report of Working Group II, 1995.
13. Govindarao V. M. H., Utilization of rice husk—A preliminary analysis, *J. Sci. Ind. Res.*, 39, 1980, 495–515.
14. Kamiya K., Oka A., Nasu H., Hashimoto T., Comparative study of structure of silica gels from different sources, *J. Sol-Gel Sci. Technol.*, 19, 2000, 495–499.
15. Jauberthie R., Rendell U. F., Tamba S., Cisse I., Origin of the pozzolanic effect of rice husks, *Constr. Build. Mater.*, 14, 2000, 419–423.

16. Salas A., Delvasto S., Gutierrez R. M., Lange D., Comparison of two processes for treating rice husk ash for use in high performance concrete, *Cem. Concr. Res.*, 39(9), 2009, 773–778.
17. Kapur P. C., Production of reactive bio-silica from the combustion of rice husk in a tube-in-basket (TiB) burner, *Powder Technol.*, 44, 1985, 63–57.
18. James J., Chemistry of silica from rice husk ash and rice husk ash cement, PhD thesis, Indian Institute of Science, Bangalore, India, 1986.
19. Kumar A., Rice husk ash based cements. In *Mineral Admixtures in Cement and Concrete*, Ed. S. N. Ghosh, ABI Books Private Ltd., New Delhi, India, Vol. 4, 1993.
20. James J., Rao M. S., Silica from rice husk through thermal-decomposition, *Thermochim. Acta*, 97, 1986, 329–336.
21. Nair D. G., Fraaij A., Klaassen A. A. K., Kentgens A. P. M., A structural investigation relating to the pozzolanic activity of rice husk ashes, *Cem. Concr. Res.*, 38, 2008, 861–869.
22. Asavapisit S., Ruengrit N., The role of RHA-blended cement in stabilizing metal-containing wastes, *Cem. Concr. Compos.*, 27, 2005, 782–787.
23. James J., Rao M. S., Characterisation of silica in rice husk ash, *Am. Ceram. Soc. Bull.*, 65(8), 1986, 1177–1180.
24. James J., Rao M. S., Rice-husk-ash cement—A review, *J. Sci. Ind. Res.*, 51, 1992, 383–393.
25. Chopra S. K., Ahluwalia S. C., Laxmi S., Technology and manufacture of rice husk masonry cement. In *ESCAP/RCTT Third Workshop on Rice Husk Ash Cements*, New Delhi, India, 1981.
26. Rozainee M., Ngo S. P., Salema A. A., Tan K. G., Fluidized bed combustion of rice husk to produce amorphous siliceous ash, *J. Energy Sust. Dev.*, XII(1), 2008, 1–10.
27. Nair D. G., Jagadish K. S., Fraaij A., Reactive pozzolanas from rice husk ash: An alternative to cement for rural housing, *Cem. Concr. Res.*, 36, 2006, 1062–1071.
28. Ichiro W., Toshio K., Makoto K., Process technology of the production of rice husk ash for a concrete admixture, *JCA Proc. Cem. Concr.*, 53, 1999, 347–353.
29. Sugita S., Shoya M., Tokuda H., Evaluation of pozzolanic activity of rice husk ash. In *Proceedings of the Fourth International Conference on Fly Ash, Silica Fume, Slag, and Natural Pozzolans in Concrete*, Istanbul, Turkey, May 1992, pp. 495–512.
30. Sugita S., Method of producing active rice husk ash, U.S. Patent 5329867, 1994.
31. Beagle E. C., Basic and applied research needs for optimizing utilization of rice husk. In *Proceedings of the Rice Husk By-Products Utilization International Conference*, Valencia, Spain, 1974, pp. 1–43.
32. Allen M. L., The manufacture of a cement extender from rice-husks using a basket-burner, UNESCO-sponsored Technology Networks of Southeast Asia, http://journeytoforever.org/farm_library/RiceHusks.pdf
33. Yogananda M. R., Jagadish K. S., Pozzolanic properties of rice husk ash, burnt clay and red mud, *Build. Environ.*, 23(4), 1988, 303–308.
34. Cordeiro G. C., Filho R. D. T., Fairbairn E. M. R., Use of ultrafine rice husk ash with high-carbon content as pozzolan in high performance concrete, *Mater. Struct.*, 42, 2009, 983–992.
35. Bouzoubaa N., Fournier B., Concrete incorporating rice husk ash: Compressive strength and chloride-ion permeability, Research report MTL 2001-5 (TR), Materials Technology Laboratory, CANMET, Natural Resources Canada, Ottawa, Ontario, Canada, July 2001.

36. Ismail M. S., Waliuddin A. M., Effect of rice husk ash on high strength concrete, *Constr. Build. Mater.*, 10(7), 1996, 521–526.
37. Agarwal S. K., Pozzolanic activity of various siliceous materials, *Cem. Concr. Res.*, 36, 2006, 1735–1739.
38. Mehta P. K., The chemistry and technology of cement made from rice husk ash. In *Proceedings of the UNIDO/ESCAP/RCTT Workshop on Rice Husk Ash Cements*, Peshawar, Pakistan, Regional Centre for Technology Transfer, Bangalore, India, January 1979, pp. 113–22.
39. Sensale G. R., Ribeiro A. B., Gonçalves A., Effects of RHA on autogenous shrinkage of Portland cement pastes, *Cem. Concr. Compos.*, 30, 2008, 892–897.
40. Ganesan K., Rajagopal K., Thangave K., Rice husk ash blended cement: Assessment of optimal level of replacement for strength and permeability properties of concrete, *Constr. Build. Mater.*, 22, 2008, 1675–1683.
41. Rukzon S., Chindaprasirt P., Mahachai R., Effect of grinding on chemical and physical properties of rice husk ash, *Int. J. Miner. Metall. Mater.*, 16(2), 2009, 242–247.
42. Laskar A. I., Talukdar S., Rheological behavior of high performance concrete with mineral admixtures and their blending, *Constr. Build. Mater.*, 22, 2008, 2345–2354.
43. Hiemstra T., Riemsdijk W. H., Multiple activated complex dissolution of metal (hydr)oxides: A thermodynamic approach applied to quartz, *J. Colloid Interface Sci.*, 136, 1990, 132–150.
44. Mehta P. K., Monteiro P. J. M., *Concrete: Microstructure, Properties and Materials*, 3rd edn., McGraw-Hill, New York, 2006.
45. Chandra S., *Waste Materials Used in Concrete Manufacturing*, 1st edn., Noyes Publications, Park Ridge, NJ, 1997.
46. Zhang M. H., Malhotra V. M., High-performance concrete incorporating rice husk ash as a supplementary cementing material, *ACI Mater. J.*, 93(6), 1996, 629–636.
47. Habeeb G. A., Fayyadh M. M., Rice husk ash concrete: The effect of RHA average particle size on mechanical properties and drying shrinkage, *Aust. J. Basic Appl. Sci.*, 3(3), 2009, 1616–1622.
48. Neville A. M., *Properties of Concrete*, 3rd edn., Longman Scientific & Technical, Essex, U.K., 1981.
49. Omatola K. M., Onojah A. D., Elemental analysis of rice husk ash using X-ray fluorescence technique, *Int. J. Phys. Sci.*, 4(4), 2009, 189–193.
50. Gastaldini A. L. G., Isaia G. C., Gomes N. S., Sperb J. E. K., Chloride penetration and carbonation in concrete with rice husk ash and chemical activators, *Cem. Concr. Compos.*, 29, 2007, 176–180.
51. ASTM C618, Standard specification for coal fly ash and raw or calcined natural pozzolan for use in concrete, American Society for Testing and Materials, ASTM International, West Conshohocken, PA, 2012, DOI: 10.1520/C0618-12, http://www.astm.org
52. Saraswathy V., Song H. W., Corrosion performance of rice husk ash blended concrete, *Constr. Build. Mater.*, 21, 2007, 1779–1784.
53. Bui D. D., Hu J., Stroeven P., Particle size effect on the strength of rice husk ash blended gap-graded Portland cement concrete, *Cem. Concr. Compos.*, 27, 2005, 357–366.
54. Chindaprasirt P., Rukzon S., Sirivivatnanon V., Resistance to chloride penetration of blended Portland cement mortar containing palm oil fuel ash, rice husk ash and fly ash, *Constr. Build. Mater.*, 22, 2008, 932–938.

Chapter 5

1. ACI 116R, Cement and concrete terminology, American Concrete Institute, Farmington Hills, MI, March 16, 2000.
2. ASTM C618, Standard specification for coal fly ash and raw or calcined natural pozzolan for use in concrete, American Society for Testing and Materials, ASTM International, West Conshohocken, PA, 2012, DOI: 10.1520/C0618-12, http://www.astm.org
3. Mehta P. K., Natural pozzolans: Supplementary cementing materials for concrete, CANMET-SP-86-8E, Canadian Government Publishing Center, Supply and Services, Ottawa, Canada, K1A0S9, 1987.
4. Stanley C. C., *Highlights in the History of Concrete*, Cement and Concrete Association, London, U.K., 1979.
5. Saad M. N. A., de Andrade W. P., Paulton V. A., Properties of mass concrete containing active pozzolan made from clay, *Concr. Int.*, July 1982, 59–65.
6. Gruber K. A., Sarkar S. L., Exploring the pozzolanic activity of high reactivity metakaolin, *World Cem.*, 27(2), 1996, 78–80.
7. De Silva P. S., Glasser F. P., Pozzolanic activation of metakaolin, *Adv. Cem. Res.*, 4(16), 1992, 167–178.
8. ACI 232.1R-00, Use of raw or processed natural pozzolans in concrete (reapproved 2006), American Concrete Institute, Farmington Hills, MI, 2000.
9. Sabir B. B., Wild S., Bai J., Metakaolin and calcined clays as pozzolans for concrete: A review, *Cem. Concr. Compos.*, 23, 2001, 441–454.
10. Siddique R., Klaus J., Influence of metakaolin on the properties of mortar and concrete: A review, *Appl. Clay Sci.*, 43, 2009, 392–400.
11. Gamiz E., Melgosa M., Sanchez-Maranon M., Martın-Garcia J. M., Delgado M. R., Relationships between chemico-mineralogical composition and color properties in selected natural and calcined Spanish kaolins, *Appl. Clay Sci.*, 28, 2005, 269–282.
12. Miyazaki M., Kamitani M., Nagai T., Kano J., Saito F., Amorphization of kaolinite and media motion in grinding by a double rotating cylinders mill—A comparison with a tumbling ball mill, *Adv. Powder Technol.*, 11, 2000, 235–244.
13. Cassagnabere F., Escadeillas G., Mouret M., Study of the reactivity of cement/metakaolin binders at early age for specific use in steam cured precast concrete, *Constr. Build. Mater.*, 23, 2009, 775–784.
14. Changling H., Osbaeck B., Makovicky E., Pozzolanic reaction of six principal clay minerals: Activation reactivity assessments and technological effects, *Cem. Concr. Res.*, 25(8), 1995, 1691–1702.
15. Zhang M. H., Malhotra V. M., Characteristics of a thermally activated aluminosilicate pozzolanic material and its use in concrete, *Cem. Concr. Res.*, 25(8), 1995, 1713–1725.
16. Wild S., Khatib J. M., Jones A., Relative strength, pozzolanic activity and cement hydration in superplasticised metakaolin concrete, *Cem. Concr. Res.*, 26(10), 1996, 1537–1544.
17. Badogiannis E., Kakali G., Dimopoulou G., Chaniotakis E., Tsivilis S., Metakaolin as a main cement constituent. Exploitation of poor Greek kaolins, *Cem. Concr. Compos.*, 27, 2005, 197–203.

18. Moulin E., Blanc P., Sorrentino D., Influence of key cement chemical parameters on the properties of metakaolin blended cements, *Cem. Concr. Compos.*, 23(6), 2001, 463–469.
19. Brinkley G. W., *Ceramic Fabrication Processes*, Technology Press, Cambridge, MA, 1958.
20. Kingery W. D., Uhlmann D. R., Bowen H. K., *Introduction to Ceramics*, 2nd edn., John Wiley & Sons, New York, 1976.
21. Bensted J., Barnes P., *Structure and Performance of Cements*, 2nd edn., Spon Press, New York, 2002.
22. Ambroise J., Murat M., Pera J., Hydration reaction and hardening of calcined clays and related minerals. V: Extension of the research and general conclusions, *Cem. Concr. Res.*, 15, 1985, 261–268.
23. Murat M., Comel C., Hydration reaction and hardening of calcined clays and related minerals. III. Influence of calcination process of kaolinite on mechanical strengths of hardened metakaolinite, *Cem. Concr. Res.*, 13, 1983, 631–637.
24. Salvador S., Pozzolanic properties of flash-calcined kaolinite: A comparative study with soak-calcined products, *Cem. Concr. Res.*, 25(1), 1995, 102–112.
25. Kakali G., Perraki T., Tsivilis S., Badogiannis E., Thermal treatment of kaolin: The effect of mineralogy on the pozzolanic activity, *Appl. Clay Sci.*, 20, 2001, 73–80.
26. Ambroise J., Martin-Calle S., Pera J., Pozzolanic behavior of thermally active kaolin. In *Proceedings of the Fourth International Conference on Fly Ash, SF, Slag and Natural Pozzolans in Concrete*, Ed. V. M. Malhotra, Istanbul, Turkey, Vol. 1, 1992, pp. 731–741.
27. Walters G. V., Jones T. R., Effect of metakaolin on alkali-silica reaction (ASR) in concrete manufactured with reactive aggregate. In *Second International Conference on Durability of Concrete*, Ed. V. M. Malhotra, Montreal, Quebec, Canada, American Concrete Institute, Detroit, MI, ACI SP-126, Vol. 2, 1991, pp. 941–953.
28. Murat M., Hydration reaction and hardening of calcined clays and related minerals: II. Influence of mineralogical properties of the raw-kaolinite on the reactivity of metakaolinite, *Cem. Concr. Res.*, 13(4), 1983, 511–518.
29. Bich C., Ambroise J., Pera J., Influence of degree of dehydroxylation on the pozzolanic activity of metakaolin, *Appl. Clay Sci.*, 44, 2009, 194–200.
30. Shvarzman A., Kovler K., Grader G. S., Shter G. E., The effect of dehydroxylation/amorphization degree on pozzolanic activity of kaolinite, *Cem. Concr. Res.*, 33, 2003, 405–416.
31. Barger G. S., Hansen E. R., Wood M. R., Neary T., Beech D. J., Jacquier D., Production and use of calcined natural pozzolans in concrete, *ASTM J. Cem. Concr. Aggregates*, 23(2), 2001, 73–80.
32. Aglietti E. F., Porto Lopez J. M., Pereira E., Mechanochemical effects in kaolinite grinding. I. Textural and physicochemical aspects, *Int. J. Miner. Process.*, 16, 1986, 125–133.
33. Aglietti E. F., Porto Lopez J. M., Pereira E., Mechanochemical effects in kaolinite grinding. II. Structural aspects, *Int. J. Miner. Process.*, 16, 1986, 135–146.
34. Mako E., Frost R. L., Kristog J., Horvath E., The effect of quartz content on the mechanochemical activation of kaolinite, *J. Colloid Interface Sci.*, 244, 2001, 359–364.

35. Sekulic Z., Petrov M., Zivanovic D., Mechanical activation of various cements, *Int. J. Miner. Process.*, 74S, 2004, S355–S363.

36. Vizcayno C., Gutierrez R. M., Castello R., Rodriguez E., Guerrero C. E., Pozzolan obtained by mechanochemical and thermal treatments of kaolin, *Appl. Clay Sci.*, 49(4), 2009, 405–413, doi: 10.1016/j.clay.2009.09.008.

37. Pera J., Amrouz A., Development of highly reactive metakaolin from paper sludge, *Adv. Cem. Based Mater.*, 7, 1998, 49–56.

38. Frias M., Garcia R., Vigil R., Ferreiro S., Calcination of art paper sludge waste for the use as a supplementary cementing material, *Appl. Clay Sci.*, 42, 2008, 189–193.

39. Villa R. V., Frias M., Rojas M. I. S., Vegas I., Garcia R., Mineralogical and morphological changes of calcined paper sludge at different temperatures and retention in furnace, *Appl. Clay Sci.*, 36, 2007, 279–286.

40. Banfill P., Frias M., Rheology and conduction calorimetry of cement modified with calcined paper sludge, *Cem. Concr. Res.*, 37, 2007, 184–190.

41. Mozaffari E., Kinuthia J. M., Bai J., Wild S., An investigation into the strength development of wastepaper sludge ash blended with ground granulated blast-furnace slag, *Cem. Concr. Res.*, 39, 2009, 942–949.

42. Vegas I., Urreta J., Frias M., Garcia R., Freeze-thaw resistance of blended cements containing calcined paper sludge, *Constr. Build. Mater.*, 23, 2009, 2862–2868.

43. Garcia R., Villa R. V., Vegas I., Frias M., Rojas M. I. S., The pozzolanic properties of paper sludge waste, *Constr. Build. Mater.*, 22, 2008, 1484–1490.

44. Curcio F., DeAngelis B. A., Pagliolico S., Metakaolin as a pozzolanic microfiller for high-performance mortars, *Cem. Concr. Res.*, 28(6), 1998, 803–809.

45. Wild S., Khatib J. M., Portlandite consumption in metakaolin cement pastes and mortars, *Cem. Concr. Res.*, 27(1), 1997, 137–146.

46. Li Z., Ding Z., Property improvement of Portland cement by incorporating with metakaolin and slag, *Cem. Concr. Res.*, 33, 2003, 579–584.

47. DMS 4635, Texas Department of Transportation, Angleton, TX, 2004, http://www.dot.state.tx.us/

48. Lee S. T., Moon H. Y., Hooton R. D., Kim J. P., Effect of solution concentrations and replacement levels of metakaolin on the resistance of mortars exposed to magnesium sulfate solutions, *Cem. Concr. Res.*, 35, 2005, 1314–1323.

49. Frias M., Rojas M. I. S., Cabrera J., The effect that the pozzolanic reaction of metakaolin has on the heat evolution in metakaolin-cement mortars, *Cem. Concr. Res.*, 30, 2000, 209–216.

50. Ramlochan T., Thomas M., Gruber K. A., The effect of metakaolin on alkali-silica reaction in concrete, *Cem. Concr. Res.*, 30, 2000, 339–344.

51. Sha W., Pareira G. B., Differential scanning calorimetry study of ordinary Portland cement paste containing metakaolin and theoretical approach of metakaolin activity, *Cem. Concr. Compos.*, 23, 2001, 455–461.

52. Ambroise, J., Maximilien, S., Pera, J., Properties of metakaolin blended cements, *Adv. Cem. Based Mater.*, 1, 1994, 161–168.

53. Razak H. A., Wong H. S., Strength estimation model for high-strength concrete incorporating metakaolin and silica fume, *Cem. Concr. Res.*, 35, 2005, 688–695.

54. Gruber K. A., Ramlochan T., Boddy A., Hooton R. D., Thomas M. D. A., Increasing concrete durability with high-reactivity metakaolin, *Cem. Concr. Compos.*, 23, 2001, 479–484.

55. Al-Akhras N. M., Durability of metakaolin concrete to sulfate attack, *Cem. Concr. Res.*, 36, 2006, 1727–1734.
56. Gleize P. J. P., Cyr M., Escadeillas G., Effects of metakaolin on autogenous shrinkage of cement pastes, *Cem. Concr. Compos.*, 29, 2007, 80–87.
57. Boddy A., Hooton R. D., Gruber K. A., Long-term testing of the chloride-penetration resistance of concrete containing high-reactivity metakaolin, *Cem. Concr. Res.*, 31(5), 2001, 759–765.
58. Khatib J. M., Metakaolin concrete at a low water to binder ratio, *Constr. Build. Mater.*, 22, 2008, 1691–1700.
59. Poon C. S., Kou S. C., Lam L., Compressive strength, chloride diffusivity and pore structure of high performance metakaolin and silica fume concrete, *Constr. Build. Mater.*, 20, 2006, 858–865.
60. Brooks J. J., Johari M. A. M., Effect of metakaolin on creep and shrinkage of concrete, *Cem. Concr. Compos.*, 23(6), 2001, 495–502.
61. Vu D. D., Stroeven P., Bui V. B., Strength and durability aspects of calcined kaolin- blended Portland cement mortar and concrete, *Cem. Concr. Compos.*, 23(6), 2001, 471–478.
62. Wong H. S., Razak H. A., Efficiency of calcined kaolin and silica fume as cement replacement material for strength performance, *Cem. Concr. Res.*, 35, 2005, 696–702.
63. Ding J. T., Li Z. J., Effects of metakaolin and silica fume on properties of concrete, *ACI Mater. J.*, 99(4), 2002, 393–398.
64. Shi C., An overview on the activation of reactivity of natural pozzolans, *Can. J. Civil Eng.*, 28, 2002, 778–786.
65. Bredy P., Chabannet M., Pera J., Microstructure and porosity of metakaolin blended cements, *Mater. Res. Soc. Symp. Proc.*, 137, 1989, 431–436.
66. He C., Osbaeck B., Makavicky E., Pozzolanic reactions of six principal clay minerals: Activation, reactivity assessment and technological effects, *Cem. Concr. Res.*, 25(8), 1995, 1691–1702.
67. He C., Makovicky E., Osbaeck B., Thermal stability and pozzolanic activity of calcined kaolin, *Appl. Clay Sci.*, 9, 1994, 165–187.
68. Asbridge A. H., Walters G. V., Jones T. R., Ternary blended concretes—OPC/GGBFS/metakaolin. In *Concrete across Borders International Conference*, Odense, Denmark, June 22–25, 1994.
69. CAN/CSA-A23.5-03, Supplementary cementing materials, Canadian Standards Association, Mississauga, Ontario, Canada, 2003.
70. ASTM C595, Standard specification for blended cements, American Society for Testing and Materials, West Conshohocken, Pennsylvania, PA, 2006, http://www.astm.org
71. Kostuch J. A., Walters G. V., Jones T. R., High performance concrete incorporating metakaolin—A review. In *Concrete 2000*, University of Dundee, Dundee, U.K., 1993, pp. 1799–1811.
72. Oriol M., Pera J., Pozzolanic activity of metakaolin under microwave treatment, *Cem. Concr. Res.*, 25(2), 1995, 265–270.
73. Golaszewski J., Szwabowski J., Bisok B., Interaction between cement and superplasticizer in presence of metakaolin. In *Proceedings of the International Conference on Admixtures—Enhancing Concrete Performance*, Warsaw, Poland, 2005, pp. 47–57.
74. Koehler E. P., Fowler D. W., Summary of concrete workability test methods, Report Number ICAR 105-1, International Center for Aggregates Research, The University of Texas at Austin, Austin, TX, 2003.

75. Guneyisi E., Gesoglu M., Properties of self-compacting mortars with binary and ternary cementitious blends of fly ash and metakaolin, *Mater. Struct.*, 41(9), 2008, 1519–1531.

76. Samet B., Mnif T., Chaabouni M., Use of a kaolinitic clay as a pozzolanic material for cements: Formulation of blended cement, *Cem. Concr. Compos.*, 29, 2007, 741–749.

77. Andriolo F. R., Sgaraboza B. C., The use of pozzolans from calcined clays in preventing excessive expansion due to alkali-silica in some Brazilian dams. In *Proceedings of the Seventh International Conference of AAR*, Grattan Bellow, NJ, 1986, pp. 66–70.

78. Khatib J. M., Sabir S., Wild S., Some properties of MK paste and mortar. In *Concrete for Environmental Enhancement and Protection*, Eds. R. K. Dhir, T.D. Dyer, Spon Press, London, U.K., 1996.

79. Poon C. S., Lam I., Kou S. C., Wong Y. L., Wong R., Rate of pozzolanic reaction of metakaolin in high-performance cement pastes, *Cem. Concr. Res.*, 31, 2001, 1301–1306.

80. Badogiannis E., Tsivilis S., Papadakis V. G., Chaniotakis E., The effects of MK on concrete properties. In *Proceeding of the Dundee Conference*, Dundee, U.K., 2002, pp. 81–89.

81. Courard L., Darimont A., Schouterden M., Ferauche F., Willem X., Degeimbre R., Durability of mortar modified with metakaolin, *Cem. Concr. Res.*, 33(9), 2003, 1473–1479.

82. Qian X., Zhan S., Li Z., Research of the physical and mechanical properties of the high performance concrete with metakaolin, *J. Build. Mater.*, 4(1), 2001, 75–78.

83. Badogiannis E., Papadakis V. G., Chaniotakis E., Tsivilis S., Exploitation of poor Greek kaolins: Strength development of metakaolin concrete and evaluation by means of k-value, *Cem. Concr. Res.*, 34, 2004, 1035–1041.

84. Batis G., Pantazopoulou P., Tsivilis S., Badogiannis E., The effect of metakaolin on the corrosion behavior of cement mortars, *Cem. Concr. Compos.*, 27, 2005, 125–130.

85. Khatib J. M., Hibbert J. J., Selected engineering properties of concrete incorporating slag and metakaolin, *Constr. Build. Mater.*, 19, 2005, 460–472.

86. Bai J., Wild S., Sabir B. B., Kinuthia J., Workability of concrete incorporating pulverised fuel ash and metakaolin, *Mag. Concr. Res.*, 51(3), 1999, 207–216.

87. Bai J., Wild S., Ware J. A., Sabir B. B., Using neural networks to predict workability of concrete incorporating metakaolin and fly ash, *Adv. Eng. Software*, 34, 2003, 663–669.

88. Wild S., Khatib J., Roose I. J., Chemical and autogenous shrinkage of Portland cement-metakaolin pastes, *Adv. Cem. Res.*, 10(3), 1998, 109–119.

89. Pera J., Bonnin E., Inertization of toxic metals in metakaolin-blended cements, environmental issues and waste management technologies in the ceramic and nuclear industries II. *Ceramic Transactions.*, Eds. V. Jain, D. Peeler, American Ceramic Society, Vol. 72, 1996, 365–374.

90. Pera J., Bonnin E., Chabannet M., Immobilization of wastes by metakaolin-blended cements. In *Fly Ash, Silica Fume, Slag, and Natural Pozzolans in Concrete, Proceedings of the Sixth International CANMET/ACI/JCI Conference*, Bangkok, Thailand, 1998, pp. 997–1006.

91. Chusid M., Reed A., Better, brighter, whiter concrete, *Constr. Specifier*, 57(9), 2004, 34–39.

Chapter 6

1. Kondo R., Diamond M., Phase composition of hardened cement paste. In *Sixth International Congress on the Chemistry of Cement*, Moscow, Russia, September 1974, pp. 2–55.
2. Taylor H. F. W., Chemistry of cement hydration. In *Eighth International Congress on the Chemistry of Cement*, Rio de Janeiro, Brazil, September 22–27, 1986, pp. 82–100.
3. Bensted J., The hydration of Portland cement, *World Cem.*, 22(8), 1991, 27–32.
4. Mehta P. K., Monteiro J. M., *Concrete: Microstructure, Properties and Materials*, 1st edn., Indian Concrete Institute, Chennai, India, 1997.
5. Lea F. M., *The Chemistry of Cement and Concrete*, Chemical Publishing Company, Inc., New York, 1971.
6. Stutzman P. E., Leigh S., Compositional analysis of NIST reference material clinker 8486. In *Accuracy in Powder Diffraction III, Proceedings. National Institute of Standards and Technology*, Poster #2, Gaithersburg, MD, April 22–25, 2001.
7. Harada T., Ohta M., Takagi S., Effects of polymorphs of tricalcium silicate on hydration and structural characteristics of hardened paste, *Yogyo Kyokui Shi*, 88(5), 1978, 195–202.
8. Scrivener K. L., Fullmann T., Gallucci E., Walenta G., Bermejo E., Quantitative study of Portland cement hydration by X-ray diffraction/Rietveld analysis and independent methods, *Cem. Concr. Res.*, 34, 2004, 1541–1547.
9. Breval E., Gas-phase and liquid-phase hydration of C_3A, *Cem. Concr. Res.*, 7, 1977, 297–304.
10. Bensted J., Gypsum in cements. In *Structure and Performance of Cements*, Eds. J. Bensted, P. Bernes, 2nd edn., Spon Press, London, U.K., 2002, pp. 253–264.
11. Bensted J., Effects of the clinker-gypsum grinding temperature upon early hydration of Portland cement, *Cem. Concr. Res.*, 12, 1982, 341–348.
12. Odler I., Abdul-Maula S., Investigation on the relationship between porosity structure and strength of hydrated Portland cement pastes III. Effect of clinker composition and gypsum addition, *Cem. Concr. Res.*, 17, 1987, 22–30.
13. Roszczynialski W., Gawlicki M., Nocun-Wczelik W., Production and use of by-product gypsum in the construction industry. In *Waste Materials Used in Concrete Manufacturing*, Ed. S. Chandra, Elsevier, Amsterdam, the Netherlands, 1996, pp. 53–141.
14. Strydom C. A., Potgieter J. H., Dehydration behaviour of a natural gypsum and a phosphogypsum during milling, *Thermochim. Acta*, 332, 1999, 89–96.
15. Graf L. A., Isothermal dehydration of natural and synthetic gypsum and the effect of clinker, PCA R&D Serial No. 2490, Portland Cement Association, Skokie, IL, 2010.
16. Mehta P. K., Pozzolanic and cementitious byproducts and mineral admixtures for concrete—A critical review. In *1st International Conference on the Use of Fly Ash, Silica Fume, Slag and Other Mineral Byproducts in Concrete*, Ed. V. M. Malhotra, Montebello, PQ, July 31–Aug 5, 1983, American Concrete Institute, Detroit, MI, Special Publication SP-79-1, Vol. 1, 1983, pp. 1–46.
17. Malhotra V. M., Mehta P. K., *High Performance, High Volume Fly Ash Concrete*, Marquardt Printing Ltd., Ottawa, Ontario, Canada, August 2002.
18. Tenoutasse N., Marion A. M., Microscopical investigation on the pozzolanic behavior of fly ashes. In *Proceedings of the 10th International Congress on Cement Microscopy*, San Antonio, TX, April 11–14, 1988, pp. 58–70.

19. Pietersen H. S., Reactivity of fly ash and slag in cement, thesis, Delft University of Technology, Delft, the Netherlands, 1993.
20. Ghosh S. N. Ed., *Progress in Cement and Concrete—Mineral Admixtures in Cement and Concrete*, Vol. 4, Academia Books International, New Delhi, India.
21. Petrographic methods of examining hardened concrete: A petrographic manual, FHWA-HRT-04–150, Federal Highway Administration, U.S. Department of Transportation, Washington, DC, 2006.
22. Larbi J. A., The cement paste-aggregate interfacial zone in concrete, Thesis, Materials Science Group, Faculty of Civil Engineering, Technical University of Delft, Delft, the Netherlands, 1991.
23. ACI 116R-00, Cement and concrete terminology, ACI manual of concrete practice, Part 1, American Concrete Institute, Farmington Hills, MI, 2000.
24. Koehler E. P., Fowler D. W., Summary of concrete workability test methods, International Center for Aggregates Research, The University of Texas at Austin, Austin, TX, August 2003.
25. Previte R. W., Concrete slump loss, *J. Am. Concr. Inst.*, 74(8), 1977, 361–367.
26. Soroka I., Ravina D., Hot weather concreting with admixtures, *Cem. Concr. Compos.*, 20, 1998, 129–136.
27. Erdogdu S. S., Effect of retempering with superplasticizer admixtures on slump loss and compressive strength of concrete subjected to prolonged mixing, *Cem. Concr. Res.*, 35, 2005, 907–912.
28. Tatersall G. H., Banfill P. G. F., *The Rheology of Fresh Concrete*, Pitman, London, U.K., 1983.
29. Kovler K., Roussel N., Properties of fresh and hardened concrete, *Cem. Concr. Res.*, 41, 2011, 775–792.
30. Chidiac S. E., Maadani O., Razaqpur A. G., Mailvaganam N. P., Controlling the quality of fresh concrete—A new approach, NRCC-43146, National Research Council, Ottawa, Ontario, Canada, 1999.
31. Roussel N., Rheology of fresh concrete: From measurements to predictions of casting processes, *Mater. Struct.*, 40(10), 2007, 1001–1012.
32. ASTM C232/C232M, Standard test methods for bleeding of concrete, ASTM International, West Conshohocken, PA, 2009, DOI: 10.1520/C0232_C0232M-09, http://www.astm.org
33. ASTM C403/C403M, Standard test method for time of setting of concrete mixtures by penetration resistance, ASTM International, West Conshohocken, PA, 2008, DOI: 10.1520/C0403_C0403M-08, http://www.astm.org
34. Schindler A., Prediction of concrete setting. In *RILEM International Symposium on Advances in Concrete through Science and Engineering*, Evanston, IL, March 22–24, 2004.
35. ACI 305R, Hot weather concreting, American Concrete Institute, Farmington Hills, MI, 2000.
36. ACI 306.1-90, Standard specification for cold weather concreting, American Concrete Institute, Farmington Hills, MI, 1990.
37. Holt E. E., Early age autogenous shrinkage of concrete, Technical Research Centre of Finland (VTT), Vuorimiehentie, Finland, 2001.
38. Bentz D. P., Sant G., Weiss J., Early-age properties of cement-based materials. I: Influence of cement fineness, *ASCE J. Mater. Civil Eng.*, 20(7), 2008, 502–508.
39. Sanchez de Rojas M. I., Frias M., The pozzolanic activity of different materials, its influence on the hydration heat in mortars, *Cem. Concr. Res.*, 26(2), 1996, 203–213.

40. Chandra S., Properties of concrete with mineral and chemical admixtures. In *Structure and Performance of Cements*, Eds. J. Bensted, P. Bernes, 2nd edn., Spon Press, London, U.K., 2002, pp. 140–185.
41. Bernal J. D., The structures of cement hydration compounds. In *Proceedings of the Third International Symposium on the Chemistry of Cement*, Cement and Concrete Association, London, U.K., 1954, pp. 216–236.
42. Gard J. A., Taylor H. F. W., Calcium silicate hydrate (II) ("C-S-.(II)"), *Cem. Concr. Res.*, 6, 1976, 667–678.
43. Taylor H. F. W., Proposed structure for calcium silicate hydrate gel, *J. Am. Ceram. Soc.*, 69(6), 1986, 464–467.
44. Richardson I. G., Groves G. W., Models for the composition and structure of calcium silicate hydrate (C-S-H) gel in hardened tricalcium silicate pastes, *Cem. Concr. Res.*, 22, 1992, 1001–1010.
45. Taylor H. F. W., Nanostructure of C-S-H: Current status, *Adv. Cem. Based Mater.*, 1, 1993, 38–46.
46. Richardson I. G., Groves, G. W., The incorporation of minor and trace elements into calcium silicate hydrate (C-S-H) gel in hardened cement pastes, *Cem. Concr. Res.*, 23, 1993, 131–138.
47. Richardson I. G., The nature of C-S-H in hardened cements, *Cem. Concr. Res.*, 29, 1999, 1131–1147.
48. Nonat A., The structure and stoichiometry of C-S-H, *Cem. Concr. Res.*, 34, 2004, 1521–1528.
49. Renaudin G., Russias J., Leroux F., Frizon F., Cau-dit-Coumes C., Structural characterization of C-S-H and C-A-S-H samples—Part I: Long-range order investigated by Rietveld analyses, *J. Solid State Chem.*, 182, 2009, 3312–3319.
50. Renaudin G., Russias J., Leroux F., Cau-dit-Coumes C., Frizon F., Structural characterization of C-S-H and C-A-S-H samples—Part II: Local environment investigated by spectroscopic analyses, *J. Solid State Chem.*, 182, 2009, 3320–3329.
51. Taylor H. F. W., Hydrated calcium silicate Part 1: Compound formation at ordinary temperature, *J. Chem. Soc.*, 726, 1950, 3682–3690.
52. Ferraris C. F., Stutzman P. E., Snyder K. A., Sulfate resistance of concrete: A new approach, PCA R&D Serial No. 2486, Portland Cement Association, Skokie, IL, 2006.
53. Stark J., Moser B., Eckert A., New approaches to cement hydration, Part 1, *ZKG Int..*, 54(1), 2001, 52–60.
54. Richardson I. G., The nature of the hydration products in hardened cement pastes, *Cem. Concr. Compos.*, 22, 2000, 97–113.
55. RILEM Technical Reports, 73-SBC RILEM Committee, Final report siliceous by-products for use in concrete, *Mater. Struct.*, 21(1), 1988, 69–80.
56. Mehta P. K., Pozzolanic and cementitious by-products in concrete—Another look. In *Proceedings of the Third International Conference on the Use of Fly Ash, Silica Fume, Slag, and Natural Pozzolans in Concrete*, Ed. V. M. Malhotra, American Concrete Institute, Farmington Hills, MI, ACI SP-114, Trondheim, Norway, June 18–23, 1989, pp. 1–43.
57. Papadakis V. G., Effect of fly ash on Portland cement systems part I. Low-calcium fly ash, *Cem. Concr. Res.*, 29, 1999, 1727–1736.
58. Papadakis V. G., Effect of fly ash on Portland cement systems part II. High-calcium fly ash, *Cem. Concr. Res.*, 30, 2000, 1647–1654.

59. Pietersen H. S., Bijen J. M., The hydration chemistry of some blended cements. In *Ninth International Congress on Chemistry of Cement*, New Delhi, India, Vol. 6, November 23–28, 1992, pp. 281–290.

60. Fraay A. L. A., Bijen J. M., de Haan Y. M., The reaction of fly ash in concrete. A critical examination, *Cem. Concr. Res.*, 19, 1989, 235–246.

61. Diamond S., Minimal pozzolanic reaction in one year old fly ash pastes: An SEM evaluation. In *Proceedings of the Eleventh Annual International Conference on Cement Microscopy*, New Orleans, LA, April 10–13, 1989, pp. 263–274.

62. Bensted J., A discussion of the paper, The reaction of fly ash in concrete: A critical examination, *Cem. Concr. Res.*, 20, 1990, 317–318.

63. Sakai E., Miyahara S., Ohsawa S., Lee S., Daimon M., Hydration of fly ash cement, *Cem. Concr. Res.*, 35, 2005, 1135–1140.

64. Berry E. E., Hemmings R. T., Zhang M. H., Cornelius B. J., Golden D. M., Hydration in high-volume fly ash concrete binders, *ACI Mater. J.*, 91(4), 1994, 382–389.

65. Vargas J. A., A designer's view of fly ash concrete, *Concr. Int.*, February 2007, 43–46

66. Mehta P. K., Burrows R. W., Building durable structures in the 21st century, *Concr. Int.*, 23(3), 2001, 57–63.

67. Tikalsky P. J., Carrasquillo P. M., Carrasquillo R. L., Strength and durability considerations affecting mix proportioning of concrete containing fly ash, *ACI Mater. J.*, 85(6), 1988, 505–511.

68. Ravina D., Mehta P. K., Properties of fresh concrete containing large amounts of fly ash, *Cem. Concr. Res.*, 16, 1986, 227–238.

69. Wei S., Handong Y., Binggen Z., Analysis of mechanism on water-reducing effect of fine ground slag, high-calcium fly ash and low-calcium fly ash, *Cem. Concr. Res.*, 33, 2003, 1119–1125.

70. Naik T. R., Singh S. S., Influence of fly ash on setting and hardening characteristics of concrete systems, *ACI Mater. J.*, 94(5), 1997, 355–358.

71. Narmluk M., Nawa T., Effect of fly ash on the kinetics of Portland cement hydration at different curing temperatures, *Cem. Concr. Res.*, 41, 2011, 579–589.

72. Brouwers H., Chen W., Hydration models for alkali-activated slag. In *Proceedings of the 8th CANMET/ACI International Conference on Fly Ash, Silica Fume, Slag and Natural Pozzolans*, Ed. M. V. Malhotra, Las Vegas, NV, American Concrete Institute, Farmington Hills, MI, May 23–29, 2004, pp. 303–318.

73. Meinhard K., Lackner R., Multi-phase hydration model for prediction of hydration-heat release of blended cements, *Cem. Concr. Res.*, 38, 2008, 794–802.

74. Wang P. Z., Trettin R., Rudert V., Spaniol T., Influence of Al_2O_3 content on hydraulic reactivity of granulated blast-furnace slag, and the interaction between Al_2O_3 and CaO, *Adv. Cem. Res.*, 16(1), 2004, 1–7.

75. Bapat J. D., Performance of cement concrete with mineral admixtures, *Adv. Cem. Res.*, 13(4), 2001, 139–155.

76. Papadakis V. G., Experimental investigation and theoretical modeling of silica fume activity in concrete, *Cem. Concr. Res.*, 29, 1999, 79–86.

77. Larbi J. A., Fraay A. L. A., Bijen J. M. J. M., The chemistry of the pore fluid of silica fume-blended cement systems, *Cem. Concr. Res*, 20, 1990, 506–516.

78. James J., Subba Rao M., Rice-husk-ash cement: A review, *J. Sci. Ind. Res.*, National Institute of Science Communication and Information Resources, New Delhi, India, 51, 1992, 383–393.

79. James J., Subba Rao M., Reaction product of lime and silica from rice husk ash, *Cem. Concr. Res.*, 16, 1986, 67–73.
80. Murat M., Hydration reaction and hardening of calcined clays and related minerals, *Cem. Concr. Res.*, 13, 1983, 259–266.
81. Wild S., Khatib J., Roose J. L., Chemical and autogenous shrinkage of Portland cement-metakaolin pastes, *Adv. Cem. Res.*, 10(3), 1998, 109–119.
82. Kinuthia J. M., Wild S., Sabir B. B., Bai J., Self-compensating autogenous shrinkage in Portland cement-metakaolin-fly ash pastes, *Adv. Cem. Res.*, 12(1), 2000, 35–43.
83. Siddique R., Klaus J., Influence of metakaolin on the properties of mortar and concrete: A review, *Appl. Clay Sci.*, 43, 2009, 392–400.
84. Frias M., Sanchez de Rojas M. I., Cabrera J., The effect that the pozzolanic reaction of metakaolin has on the heat evolution in metakaolin-cement mortars, *Cem. Concr. Res.*, 30, 2000, 209–216.
85. Ambroise J., Maximilien S., Pera J., Properties of metakaolin blended cements, *Adv. Cem. Based Mater.*, 1(4), 1994, 161–168.
86. Brooks J. J., Johari M. M. A., Effect of metakaolin on creep and shrinkage of concrete, *Cem. Concr. Compos.*, 23, 2001, 495–502.
87. Li Z., Ding Z., Property improvement of Portland cement by incorporating with metakaolin and slag, *Cem. Concr. Res.*, 33, 2003, 579–584.
88. Badogiannis E., Kakali G., Dimopoulou G., Chaniotakis E., Tsivilis S., Metakaolin as a main cement constituent: Exploitation of poor Greek kaolins, *Cem. Concr. Compos.*, 27, 2005, 197–203.
89. Scrivener K. L., Nonat A., Hydration of cementitious materials, present and future, *Cem. Concr. Res.*, 41, 2011, 651–665.

Chapter 7

1. International Federation for Structural Concrete, *Structural Concrete: Textbook on behaviour, design and performance updated knowledge of the CEB/FIP Model Code 1990*, Vol. 3, Lausanne, Switzerland, December 1999.
2. IS 456–2000, Indian Standard Code of practice for plain and reinforced concrete, Bureau of Indian Standards, New Delhi, India, 2000.
3. Litzner H. U., Becker A., Design of concrete structures for durability and strength to Eurocode 2, *Mater. Struct.*, 32, 1999, 323–330.
4. Mehta P. K., Concrete technology at the crossroads—Problems and opportunities. In *Concrete Technology: Past, Present Future*, American Concrete Institute, Detroit, MI, SP-144, 1994, pp. 1–31.
5. Mehta P. K., Durability—Critical issues for the future, *Concr. Int.*, 19(7), 1997, 27–33.
6. Saetta A. V., Schrefler B. A., Vitaliani R. V., The carbonation of concrete and the mechanism of moisture, heat and carbon dioxide flow through porous materials, *Cem. Concr. Res.*, 23, 1993, 761–772.
7. Sarja A., Vesikari E., Eds., *Durability Design of Concrete Structures*, RILEM Report 14, Spon Press, London, U.K., 1996.
8. Folic R., Durability design of concrete structures—Part 1: Analysis of fundamentals, *Facta Univ., Ser. Archit. Civil Eng.*, 7(1), 2009, 1–18, doi: 10.2298/FUACE0901001F.
9. Ho D. W. S., Chua C. W., Tam C. T., Steam-cured concrete incorporating mineral admixtures, *Cem. Concr. Res.*, 33, 2003, 595–601.

10. RILEM TC116-PCD, Permeability of concrete as a criterion of its durability. Final report: Concrete durability—An approach towards performance testing, *Mater. Struct.*, 32(217), 1999, 163–173.
11. RILEM TC116-PCD, Recommendation of TC 116-PCD: Tests for gas permeability of concrete—Preconditioning of concrete test specimens for the measurement of gas permeability and capillary absorption of water—Measurement of the gas permeability of concrete by the RILEM—CEMBUREAU method—Determination of the capillary absorption of water of hardened concrete, *Mater. Struct.*, 32(217), 1999, 174–179.
12. RTA of NSW, *Concrete for Bridgeworks, Part B 80*, 3rd edn., Roads and Transport Authority, New South Wales, Australia, 1995.
13. BS EN 1992, Eurocode 2: Design of concrete structures, British Standards Institution, London, U.K., 2004.
14. Anoop M. B., Balaji Rao K., Appa Rao T. V. S. R., Gopalakrishnan S., International standards for durability of RC structures: Part 1—A critical review, *Indian Concr. J.*, September 2001, 559–569.
15. International Federation for Structural Concrete (FIB), *Durable Concrete Structures—CEB Design Guide*, Lausanne, Switzerland, 1992.
16. EN 206-1, Concrete—Part 1: Specification, performance, production and conformity, European Committee for Standardization, Brussels, Belgium, December 2000.
17. Masters L. W., Brandt E., Prediction of service life of building materials and components, *Mater. Struct.*, 20(1), 1987, 55–57.
18. Folic R., Reliability and maintenance modeling of civil engineering structures, Bulletin for Applied and Computer Mathematics-BAM-2009/2002, Technical University of Budapest, Budapest, Hungary, 2003, pp. 65–76.
19. Cyr M., Lawrence P., Ringot E., Efficiency of mineral admixtures in mortars: Quantification of the physical and chemical effects of fine admixtures in relation with compressive strength, *Cem. Concr. Res.*, 36, 2006, 264–277.
20. Lawrence P., Cyr M., Ringot E., Mineral admixtures in mortars: Effect of inert materials on short-term hydration, *Cem. Concr. Res.*, 33, 2003, 1939–1947.
21. Lawrence P., Cyr M., Ringot E., Mineral admixtures in mortars: Effect of type, amount and fineness of fine constituents on compressive strength, *Cem. Concr. Res.*, 35, 2005, 1092–1105.
22. Cyr M., Lawrence P., Ringot E., Mineral admixtures in mortars: Quantification of the physical effects of inert materials on short-term hydration, *Cem. Concr. Res.*, 35, 2005, 719–730.
23. Stutzman P. E., Scanning electron microscopy in concrete petrography. In *Materials Science of Concrete Special Volume: Calcium Hydroxide in Concrete (Workshop on the Role of Calcium Hydroxide in Concrete). Proceedings*, Eds. J. Skalny, J. Gebauer, I. Odler, The American Ceramic Society, Anna Maria Island, FL, November 1–3, 2000, pp. 59–72.
24. Bentz D. P., Stutzman P. E., Garboczi E. J., Experimental and simulation studies of the interfacial zone in concrete, *Cem. Concr. Res.*, 22, 1992, 891–902.
25. Stroeven P., Stroeven M., Reconstructions by SPACE of the interfacial transition zone, *Cem. Concr. Compos.*, 23, 2001, 189–200.
26. Jiang W., Roy D. M., Strengthening mechanisms of high-performance concrete. In *High Performance Concrete, Proceedings of the ACI International Conference*, Ed. V. M. Malhotra, Singapore, American Concrete Institute, Detroit, MI, 1994, pp. 135–157.

27. Zhang M. H., Lastra R., Malhotra V. M., Rice-husk ash paste and concrete: Some aspects of hydration and the microstructure of the interfacial zone between the aggregate and paste, *Cem. Concr. Res.*, 26(6), 1996, 963–977.
28. Bentur A., Cohen M. D., Effect of condensed silica fume on the microstructure of the interfacial zone in Portland cement mortars, *J. Am. Ceram. Soc.*, 70, 1987, 738–743.
29. Meeks K. W., Carino N. J., Curing of high-performance concrete: Report of the state-of- the-art, NISTIR 6295, Building and Fire Research Laboratory, National Institute of Standards and Technology (NIST), United States Department of Commerce, Gaithersburg, MD, 1999.
30. Liu S. H., Fang K. H., Li Z., Influence of mineral admixtures on crack resistance of high strength concrete, *Key Eng. Mater.*, 302–303, 2006, 150–154.
31. Caldarone M. A., Gruber K. A., Burg R. G., High reactivity metakaolin (HRM): A new generation mineral admixture for high performance concrete, *Concr. Int.*, 16(11), 1994, 37–41.
32. Hassan K. E., Cabrera J. G., Maliehe R. S, The effect of mineral admixtures on the properties of high-performance concrete, *Cem. Concr. Compos.*, 22(4), 2000, 267–271.
33. Li K. L., Tang X. S., Huang G. H., Xu H., Experimental research of high performance concrete with multi-elements mineral admixtures, *Key Eng. Mater.*, 405–406, 2009, 24–29.
34. Zhang M. H., Malhotra V. M., High-performance concrete incorporating rice husk ash as a supplementary cementing material, *ACI Mater. J.*, 93(6), 1996, 634–636.
35. Bharatkumar B. H., Narayanan R., Raghuprasad B. K., Ramachandramurthy D. S., Mix proportioning of high performance concrete, *Cem. Concr. Compos.*, 23(1), 2001, 71–80.
36. Malhotra V. M., Making concrete greener with fly ash, *Concr. Int.*, 21(5), 1999, 61–66.
37. Swamy R. N., Design for durability and strength through the use of fly ash and slag in concrete, *Adv. Concr. Technol.*, American Concrete Institute Publication, Farmington Hills, MI, SP-171, 1997, 1–72.
38. Aitcin P. C., *High-Performance Concrete*, E&FN Spon, New York, 1998.
39. Smith I. A., The design of fly ash concrete. In *Proceedings of the Institution of Civil Engineers*, ICE, London, U.K., Vol. 36, 1967, pp. 769–790.
40. Berry E. E., Malhotra V. M., Fly ash in concrete, CANMET SP85-3, Ottawa, Canada, 1996.
41. FIP, *State of the Art Report: Condensed Silica Fume in Concrete*, Thomas Telford, London, U.K., 1988.
42. Babu K. G., Rama Kumar V. S., Efficiency of GGBS in concrete, *Cem. Concr. Res.*, 30, 2000, 1031–1036.
43. Papadakis V. G., Tsimas S., Supplementary cementing materials in concrete: Part I. Efficiency and design, *Cem. Concr. Res.*, 32, 2002, 1525–1532.
44. Papadakis V. G., Antiohos S., Tsimas S., Supplementary cementing materials in concrete: Part II. A fundamental estimation of the efficiency factor, *Cem. Concr. Res.*, 32, 2002, 1533–1538.
45. Wong H. S., Razak H. A., Efficiency of calcined kaolin and silica fume as cement replacement material for strength performance, *Cem. Concr. Res.*, 35, 2005, 696–702.

46. ACI Committee 211.1–91, Standard practice for selecting proportions for normal, heavyweight and mass concrete, American Concrete Institute, Farmington Hills, MI, 1991.
47. Thurston S. J., Priestly N., Cooke N., Thermal analysis of thick concrete sections, *ACI J. Proc.*, 77(5), 1980, 347–357.
48. Nili M., Salehi A. M., Assessing the effectiveness of pozzolans in massive high-strength concrete, *Constr. Build. Mater.*, 24, 2010, 2108–2116.
49. Sioulas B., Sanjayan J. G., Hydration temperatures in large high-strength concrete columns incorporating slag, *Cem. Concr. Res.*, 30, 2000, 1791–1799.
50. Liu J. Z., Sun W., Miao C. W., Liu J. P., Research on the hydration heat of paste in ultra high strength concrete at low water-binder ratio, *Jianzhu Cailiao Xuebao/J. Build. Mater.*, 13(2), 2010, 139–142, 168.
51. Yazıcı H., Yardımci M. Y., Yigiter H., Aydin S., Turkel S., Mechanical properties of reactive powder concrete containing high volumes of ground granulated blast furnace slag, *Cem. Concr. Compos.*, 32, 2010, 639–648.
52. Hill J., Sharp J. H., The mineralogy and microstructure of three composite cements with high replacement levels, *Cem. Concr. Compos.*, 24, 2002, 191–199.
53. Bapat J. D., Performance of cement concrete with mineral admixtures, *Adv. Cem. Res.*, 13(4), 2001, 139–155.
54. Al-Amoudi O. S. B., Maslehuddin M., Ibrahim M., Shameem M., Al-Mehthel M. H., Performance of blended cement concretes prepared with constant workability, *Cem. Concr. Compos.*, 33, 2011, 90–102.
55. Samson E., Henocq P., Marchand J., Chemical degradation review, CBP 03, Report by SIMCO Technologies Inc., Quebec City, Canada, December 2008.
56. Glasser F. P., Marchand J., Samson E., Durability of concrete—Degradation phenomena involving detrimental chemical reactions, *Cem. Concr. Res.*, 38, 2008, 226–246.
57. Hewlett P. C. Ed., *Lea's Chemistry of Cement and Concrete*, 4th edn., Arnold, London, U.K., 1998.
58. Taylor H. F. W., *Cement Chemistry*, 2nd edn., Thomas Telford, London, U.K., 1997.
59. Maltais Y., Samsona E., Marchand J., Predicting the durability of Portland cement systems in aggressive environments—Laboratory validation, *Cem. Concr. Res.*, 34, 2004, 1579–1589.
60. Neville A., The confused world of sulfate attack on concrete, *Cem. Concr. Res.*, 34, 2004, 1275–1296.
61. Cultrone G., Sebastian E., Huertas M. O., Forced and natural carbonation of lime-based mortars with and without additives: Mineralogical and textural changes, *Cem. Concr. Res.*, 35, 2005, 2278–2289.
62. Rigo da Silva C. A., Reis R. J. P., Lameiras F. S., Vasconcelos W. L., Carbonation-related microstructural changes in long-term durability concrete, *Mater. Res.*, 5, 2002, 287–293.
63. Papadakis V. G., Vayenas C. G., Fardis M. N., Fundamental modeling and experimental investigation of concrete carbonation, *ACI Mater. J.*, 88, 1991, 363–373.
64. Borges P. H. R., Costa J. O., Milestone N. B., Lynsdale C. J., Streatfield R. E., Carbonation of CH and C-S-H in composite cement pastes containing high amounts of BFS, *Cem. Concr. Res.*, 40, 2010, 284–292.

65. Gervais C., Garrabrants A. C., Sanchez F., Barna R., Oszkowicz P., Kosson D., The effects of carbonation and drying during intermittent leaching on the release of inorganic constituents from a cement-based matrix, *Cem. Concr. Res.*, 34, 2004, 119–131.
66. Chen J. J., Thomas J. J., Jennings H. M., Decalcification shrinkage of cement paste, *Cem. Concr. Res.*, 36, 2006, 801–809.
67. Groves G. W., Brough A., Richardson I. G., Dobson C. M., Progressive changes in the structure of hardened C_3S cement pastes due to carbonation, *J. Am. Ceram. Soc.*, 74(11), 1991, 2891–2896.
68. Bary B., Sellier A., Coupled moisture—Carbon dioxide—Calcium transfer model for carbonation of concrete, *Cem. Concr. Res.*, 34, 2004, 1859–1872.
69. Cahyadi J. H., Uomoto T., Influence of environmental relative humidity on carbonation of concrete (mathematical modeling), *Durability of Building Materials and Components*, Omiya, Japan, E & FN Spon, London, U.K., Vol. 6, October 26–29, 1993, pp. 1142–1151.
70. Saetta A. V., Vitaliani R. V., Experimental investigation and numerical modeling of carbonation process in reinforced concrete structures part I: Theoretical formulation, *Cem. Concr. Res.*, 34, 2004, 571–579.
71. Song H. W., Kwon S. J., Byun K. J., Park C. K., Predicting carbonation in early-aged cracked concrete, *Cem. Concr. Res.*, 36, 2006, 979–989.
72. Smith F. L., Harvey A. H., Avoid common pitfalls when using Henry's Law, *Chem. Eng. Prog.*, 103(9), 2007, 33–39.
73. Liang M. T., Qu W., Liang C. H., Mathematical modeling and prediction method of concrete carbonation and its applications, *J. Mar. Sci. Technol.*, 10(2), 2002, 128–135.
74. Carslaw H. S., Jaeger J. C., *Conduction of Heat in Solids*, Oxford University Press, Oxford, U.K., 1959.
75. Ann K. Y., Pack S. W., Hwang J. P., Song H. W., Kim S. H., Service life prediction of a concrete bridge structure subjected to carbonation, *Constr. Build. Mater.*, 24, 2010, 1494–1501.
76. Bertos M. F., Simons S. J. R., Hills C. D., Carey P. J., A review of accelerated carbonation technology in the treatment of cement-based materials and sequestration of CO_2, *J. Hazard. Mater.*, 112(3), 2004, 193–205.
77. Jiang L., Lin B., Cai Y., A model for predicting carbonation of high-volume fly ash concrete, *Cem. Concr. Res.*, 30, 2000, 699–702.
78. EN 13295:2004, Products and systems for the protection and repair of concrete structures. Test methods. Determination of resistance to carbonation, European Committee for Standardisation (CEN), Brussels, Belgium.
79. EN 14630:2006, Products and systems for the protection and repair of concrete structures. Test methods. Determination of carbonation depth in hardened concrete by the phenolphthalein method, European Committee for Standardisation (CEN), Brussels, Belgium.
80. AFNOR. French Standard NF EN 206-1, Concrete, specification, performance, production and conformity, 2004 (in French).
81. CPC 18:1988, Measurement of hardened concrete carbonation depth, International Union of Laboratories and Experts in Construction Materials, Systems, and Structures (RILEM), Bagneux, France.
82. Sideris K. K., Sawa A. E., Papayianni J., Sulfate resistance and carbonation of plain and blended cements, *Cem. Concr. Compos.*, 28(1), 2006, 47–56.

83. Shi H. S., Xu B. W., Zhou X. C., Influence of mineral admixtures on compressive strength, gas permeability and carbonation of high performance concrete, *Constr. Build. Mater.*, 23(5), 2009, 1980–1985.

84. Massazza F., Pozzolanic cements, *Cem. Concr. Compos.*, 15, 1993, 185–214.

85. Bijen J., Wegen G. V., Selst R. V., Carbonation of Portland blast furnace slag cement concrete with fly ash. In *Proceedings of Third International Conference on Fly Ash Silica Fume, Slag and Pozzolans in Concrete*, American Concrete Institute, Detroit, MI, SP 114-31, Vol. I, 1989.

86. Bijen J., Blast furnace slag cement for durable marine structures, VNC/Beton Prisma, 's-Hertogenbosch, the Netherlands, 1996.

87. Lo Y., Lee H. M., Curing effects on carbonation of concrete using a phenolphthalein indicator and Fourier-transform infrared spectroscopy, *Build. Environ.*, 37(5), 2002, 507–514.

88. Balayssac J. P., Detriche C. H., Grandet J., Effects of curing upon carbonation of concrete, *Constr. Build. Mater.*, 9(2), 1995, 91–95.

89. Sisomphon K., Franke L., Carbonation rates of concrete containing high volume of pozzolanic materials, *Cem. Concr. Res.*, 37(12), 2007, 1647–1653.

90. Ati C. D., Accelerated carbonation and testing of concrete made with fly ash, *Constr. Build. Mater.*, 17(3), 2003, 147–152.

91. Wilding C. R., The performance of cement based systems, *Cem. Concr. Res.*, 22, 1992, 299–310.

92. Diamond S., ASR—Another look at mechanisms. In *Proceedings of the Eighth International Conference on Alkali-Aggregate Reaction*, Eds. K. Okada, S. Nishibayashi, M. Kawamura, Kyoto, Japan, 1989, pp. 83–94.

93. Fournier B., Berube M., Alkali-aggregate reactions in concrete: A review of basic concepts and engineering applications, *Can. J. Civil Eng.*, 127(2), 2000, 167–191.

94. Walker H. N., Lane D. S., Stutzman P. E., *Petrographic Methods of Examining Hardened Concrete: A Petrographic Manual*, FHWA-HRT-04-150, Virginia Department of Transportation, Richmond, VA, 2006.

95. ACI 221.1R-98: State-of-the-art report on alkali aggregate reactivity, ACI Committee 221, American Concrete Institute, Farmington Hills, MI, 1998.

96. Swenson E. G., A Canadian reactive aggregate undetected by ASTM tests, *ASTM Bull.*, 226, 1957, 48–51.

97. Gillott J. E., Petrology of dolomitic limestones. Kingston, Ontario, Canada, *Geol. Soc. Am. Bull.*, 74, 1963, 759–778.

98. Hadley D. W., Alkali reactivity of dolomitic carbonate rocks. In *Proceedings, Symposium on Alkali-Carbonate Rock Reactions*, Highway Research Board, Washington, DC, Record No. 45, 1964, pp. 1–20.

99. Qian G., Deng M., Lan X., Xu Z., Tang M., Alkali carbonate reaction expansion of dolomitic limestone aggregates with porphyrotopic texture, *Eng. Geol.*, 63(1–2), 2002, 17–29.

100. Swenson E. G., Gillott J. E., Alkali-carbonate rock reaction, Highway Research Record, Record No. 45, 1964, pp. 21–40.

101. Gillot G. E., Swenson E. G., Mechanism of alkali-carbonate rock reaction, *J. Eng. Geol.*, 2, 1969, 7–23.

102. Special issue on alkali aggregate reaction in Canada, *Can. J. Civil Eng.*, 27(2), 2000, National Research Council, Canada.

103. Spencer T. E., Blaylock A. J., Alkali silica reaction in marine piles, *Concr. Int.*, 19(1), 1997, 59–62.

104. Malvar L. J., Cline G. D., Burke D. F., Rollings R., Sherman T. W., Greene J. L., Alkali-silica reaction mitigation: State-of-the-art, TR-2195-SHR, U.S. Naval Facilities Engineering Service Center, Port Hueneme, CA, 2001.

105. Ahlstrom G., FHWA alkali-silica reactivity development and deployment program, *HPC Bridge Views*, (51), September–October 2008.

106. Swamy R. N., Alkali-aggregate reaction—The bogeyman of concrete. In *Proceedings of the V. M. Malhotra Symposium on Concrete Technology: Past, Present, and Future*, Ed. P. K. Mehta, American Concrete Institute, Farmington Hills, MI, ACI-SP-144–6, 1994, pp. 105–139.

107. Powers T. C., Steinour H. H., An interpretation of some published researches on the alkali-aggregate reaction. Part 2—A hypothesis concerning safe and unsafe reactions with reactive silica in concrete, *J. Am. Concr. Inst.*, 26(8), 1955, 785–811.

108. Oberholster R. E., Alkali reactivity of silicious rock aggregates: Diagnosis of the reaction, testing of cement and aggregate and prescription of preventive measures. In *Proceeding of the Sixth International Conference on Alkalies in Concrete— Research and Practice*, Technical University of Denmark, Copenhagen, Denmark, 1983, pp. 419–433.

109. IS 415, Guide specification for concrete subject to alkali-silica reactions, Durability Subcommittee, PCA R&D Serial No. 2001b, Portland Cement Association, Skokie, IL, 2007.

110. Wang A. Q., Zhang C. Z., Alkali-aggregate reactivity of concrete hydraulic structures, *J. Hydraul. Eng.*, (2), 2003, 117–121.

111. Glasser L. S. D., Osmotic pressure and the swelling of gels, *Cem. Concr. Res.*, 9, 1979, 515–517.

112. Glasser L. S. D., Kataoka N., The chemistry of 'alkali-aggregate' reaction, *Cem. Concr. Res.*, 11, 1981, 1–9.

113. Glasser L. S. D., Kataoka N., On the role of calcium in the alkali-aggregate reaction, *Cem. Concr. Res.*, 12, 1982, 321–331.

114. Bazant Z. P., Steffens A., Mathematical model for kinetics of alkali-silica reaction in concrete, *Cem. Concr. Res.*, 30, 2000, 419–428.

115. Dron R., Brivot F., Chaussadent T., Mechanism of the alkali-silica reaction. In *10th International Conference on the Chemistry of Cement*, Gothenburg, Sweden, Vol. 4, June 2–6, 1997.

116. Folliard K. J., Thomas M. D. A., Fournier B., Kurtis K. E., Ideker J. H., FHWA-HRT-06-073: Interim recommendations for the use of lithium to mitigate or prevent alkali-silica reaction (ASR), Office of Infrastructure Research and Development, Federal Highway Administration, Lakewood, CO, 2006.

117. Groves G. W., Zhang X., A dilatation model for the expansion of silica glass/ OPC mortars, *Cem. Concr. Res.*, 20, 1990, 453–460.

118. Prezzi M., Monteiro P. J. M., Sposito G., The alkali-silica reaction: Part I. Use of double- layer theory to explain the behavior of reaction product gels, *ACI Mater. J.*, 123, 1997, 10–11.

119. Farage M. C. R., Alves J. L. D., Fairbairn E. M. R., Macroscopic model of concrete subjected to alkali-aggregate reaction, *Cem. Concr. Res.*, 34, 2004, 495–505.

120. Ichikawa T., Miura M., Modified model of alkali-silica reaction, *Cem. Concr. Res.*, 37, 2007, 1291–1297.

121. Multon S., Sellier A., Cyr M., Chemo-mechanical modeling for prediction of alkali silica reaction (ASR) expansion, *Cem. Concr. Res.*, 39, 2009, 490–500.

122. Martin L. C., Engineering Technical Letter (ETL) 06-2: Alkali-aggregate reaction in Portland cement concrete (PCC) airfield pavements, U.S. Department of Air Force, Washington, DC, February 9, 2006.

123. Grimal E., Sellier A., Multon S., Pape Y. L., Bourdarot E., Concrete modelling for expertise of structures affected by alkali aggregate reaction, *Cem. Concr. Res.*, 40, 2010, 502–507.

124. Fournier B., Berube M. A., Rogers C. A., Canadian Standards Association (CSA) standard practice to evaluate potential alkali-aggregate reactivity of aggregates and to select preventive measures against AAR in new concrete structures. In *Proceedings of the 11th International AAR Conference*, Quebec City, Canada, June 2000, pp. 633–642.

125. Feng X., Thomas M. D. A., Bremner T. W., Balcom B. J., Folliard K. J., Studies on lithium salts to mitigate ASR-induced expansion in new concrete: A critical review, *Cem. Concr. Res.*, 35, 2005, 1789–1796.

126. Stark D., Morgan B., Okamoto P., Diamond S., Eliminating or minimizing alkali-silica reactivity, Strategic Highway Research Program, SHRP-C-343, National Research Council, Washington, DC, 1993, pp. 75–106.

127. Du C., Dealing with alkali-aggregate reaction in hydraulic structures, HydroWorld.com

128. Glasser F. P., Chemistry of alkali-aggregate reaction. In *The Alkali-Silica Reaction in Concrete*, Ed. R. N. Swamy, Van Nostrand Reinhold, New York, 1992.

129. Monteiro P. J. M., Wang K., Sposito G., Santos M. C., Andrade W. P., Influence of mineral admixtures on the alkali-aggregate reaction, *Cem. Concr. Res.*, 27(12), 1997, 1899–1909.

130. Duchesne J., Berube M. A., Long-term effectiveness of supplementary cementing materials against alkali-silica reaction, *Cem. Concr. Res.*, 31, 2001, 1057–1063.

131. Malvar L. J., Cline G. D., Burke D. F., Rollings R., Sherman T. W., Greene J. L., Alkali-silica reaction mitigation: State of the art and recommendations, *ACI Mater. J.*, 99(5), 2002, 480–489.

132. Kobayashi S., Hozumi Y., Nakano T., Yanagida, T., Study of the effect of the quality of fly ash for controlling alkali-aggregate reaction. In *Fly Ash, Silica Fume, Slag and Natural Pozzolans in Concrete*, Ed. V. M. Malhotra, ACI SP-114, 1989, pp. 403–415.

133. Bremner T. W., http://www.unb.ca/civil/bremner/CIRCA/Images/Circa_8_Use_in_b_v.jpg

134. Thomas M., Fournier B., Folliard K., Ideker J., Shehata M., ICAR 302-1: Test methods for evaluating preventive measures for controlling expansion due to alkali-silica reaction in concrete, International Center for Aggregates Research, The University of Texas at Austin, Austin, TX, 2006.

135. Alasali M. M., Malhotra V. M., Role of concrete incorporating high volume fly ash in controlling expansion due to alkali-aggregate reaction, *ACI Mater. J.*, 88, 1991, 159–163.

136. Hobbs D. W., Deleterious expansion of concrete due to alkali-silica reaction: Influence of PFA and slag, *Mag. Concr. Res.*, 38, 1986, 191–205.

137. Alasali M. M., Alkali-aggregate reaction in concrete. In *Investigations of Concrete Expansion from Alkali Contributed by Pozzolans in Concrete*, Ed. V. M. Malhotra, American Concrete Institute, Farmington Hills, MI, ACI SP-114, Vol. I, 1989, pp. 431–451.

138. Ramachandran V. S., Alkali-aggregate expansion inhibiting admixtures, *Cem. Concr. Compos.*, 20, 1998, 149–161.
139. Oberholster R. E., Alkali-aggregate reactions in South Africa—Some recent developments in research. In *Proceedings of the Eighth International Conference on Alkali Aggregate Reactions in Concrete*, Kyoto, Japan, The Society of Materials Science, Japan, 1989, p. 77.
140. Berube M. A., Duchesne J., Does silica fume merely postpone expansion due to alkali-aggregate reactivity, *Constr. Build. Mater.*, 7, 1993, 137–143.
141. Fournier B., Berube M. A., Thomas M. D. A., Smaoui N., Folliard K. J., Evaluation and management of concrete structures affected by alkali-silica reaction—A review, CANMET Materials Technology Laboratory, Report No. MTL 2004-11 (OP), 2004.
142. Idorn G., *Concrete Progress—From Antiquity to the Third Millennium*, Thomas Telford, London, U.K., 1997.
143. Mehta P. K., Gerwick B. C., Cracking corrosion interaction in concrete exposed to marine environment, *Concr. Int.*, 4(10), 1982, 45–51.
144. Pourbaix M., *Atlas of Electrochemical Equilibria in Aqueous Solutions*, 2nd English edn., National Association of Corrosion Engineers, Houston, TX, 1974.
145. Pilling N., Bedworth R., Oxidation of metals at high temperature, *J. Inst. Metals*, 29, 1923, 529–591.
146. Schiessl P., Raupach M., Influence of concrete composition and microclimate on the critical chloride content in concrete. In *Corrosion of Reinforcement in Concrete*, Eds. C. L. Page, K. W. J. Treadaway, P. B. Bamforth, Elsevier Applied Science, London, U.K., 1990, pp. 49–58.
147. Ann K. Y., Song H. W., Chloride threshold level for corrosion of steel in concrete, *Corros. Sci.*, 49, 2007, 4113–4133.
148. Tuutti K., Corrosion of steel in concrete, Swedish Cement and Concrete Research Institute, Report No. 4:82, CBI, Stockholm, Sweden, 1982.
149. Suryavanshi A. K., Swamy R. N., Cardew G. E., Estimation of diffusion coefficients for chloride ion penetration into structural concrete, *ACI Mater. J.*, 99(5), 2002, 441–449.
150. Hooton R. D., Geiker M. R., Bentz E. C., Effects of curing on chloride ingress and implications on service life, *ACI Mater. J.*, 99, 2002, 201–206.
151. Funahashi M., Predicting corrosion free service life of a concrete structure in a chloride environment, *ACI Mater. J.*, 87, 1990, 581–587.
152. Bamforth P. B., Price W. F., Factors influencing chloride ingress into marine structures. In *Proceedings of the International Conference, Concrete 2000, Economic and Durable Construction through Excellence*, Eds. R. K. Dhir, M. R. Jones, Vol. 2, E & F N Spon, London, U.K., September, 1993, pp. 1105–1118.
153. Bentz E. C., Evans C. M., Thomas M. D. A., Chloride diffusion modeling for marine exposed concretes. In *Corrosion of Reinforcement in Concrete Construction*, Eds. C. L. Page, P. B. Bamforth, J. W. Figg, The Royal Society of Chemistry Publication, Cambridge, U.K., July 1–4, 1996, pp. 136–145.
154. Service-life prediction: State of the art report, ACI Committee 365, Manual of concrete practice, Part 5, American Concrete Institute, Farmington Hills, MI, 2000.
155. Thomas M. D. A., Matthews J. D., Performance of PFA concrete in a marine environment—10-year results, *Cem. Concr. Compos.*, 26, 2004, 5–20.
156. Weyers R. E., Service life model for concrete structures in chloride laden environments, *ACI Mater. J.*, 95(4), 1998, 445–453.

157. Maaddawy T. E., Soudki K., A model for prediction of time from corrosion initiation to corrosion cracking, *Cem. Concr. Compos.*, 29, 2007, 168–175.

158. Lin G., Liu Y., Xiang Z., Numerical modeling for predicting service life of reinforced concrete structures exposed to chloride environments, *Cem. Concr. Compos.*, 32, 2010, 571–579.

159. AASHTO T 259: Standard method of test for resistance of concrete to chloride ion penetration, American Association of State Highway and Transportation Officials (AASHTO), Washington, DC, 2002.

160. Yang C. C., Wang L. C., The diffusion characteristic of concrete with mineral admixtures between salt ponding test and accelerated chloride migration test, *Mater. Chem. Phys.*, 85, 2004, 266–272.

161. FHWA/RD-81/119: Rapid determination of the chloride permeability of concrete, Federal Highway Administration (FHWA), U.S. Department of Transportation, Washington, DC.

162. Graybeal B. A., FHWA-HRT-06-103: Material property characterization of ultra-high performance concrete, Office of Infrastructure Research and Development, Federal Highway Administration (FHWA), U.S. Department of Transportation, Washington, DC, August 2006.

163. McGrath P. F., Hooton R. D., Re-evaluation of the AASHTO T259 90-day salt ponding test, *Cem. Concr. Res.*, 29, 1999, 1239–1248. (iii) Andrade C., Whiting D., A comparison of chloride ion diffusion coefficients derived from concentration gradients and non-steady state accelerated ionic migration, *Mater. Struct.*, 29(8), 1996, 476–484.

164. Andrade C., Whiting D., A comparison of chloride ion diffusion coefficients derived from concentration gradients and non-steady state accelerated ionic migration, *Mater. Struct.*, 29(8), 1996, 476–484.

165. Chini A., Determination of acceptance permeability characteristics for performance-related specifications for Portland cement concrete, BC 354-41, School of Building Construction, University of Florida, Gainesville, FL, 2003.

166. ASTM C876, Standard test method for half-cell potentials of uncoated reinforcing steel in concrete, ASTM International, West Conshohocken, PA, 2009, doi: 10.1520/C0876-09, http://www.astm.org

167. Elsener B., Bohni H., Potential mapping and corrosion of steel in concrete. In *Corrosion Rates of Steel in Concrete*, ASTM STP 1065, Eds. N. S. Berke, V. Chaker, D. Whiting, American Society for Testing and Materials, Philadelphia, PA, 1990, pp. 143–156.

168. Qian S. Y., Reinforcing decay, NRCC-44737, National Research Council Canada, http://www.nrc.ca/irc/ircpubs

169. Bushman J. B., *Calculation of Corrosion Rate from Corrosion Current (Faraday's Law)*, Bushman & Associates, Inc., Medina, OH, 1996.

170. Trejo D., Halmen C., Reinschmidt K., Corrosion performance tests for reinforcing steel in concrete: Technical report, Report No. FHWA/TX-09/0-4825-1, Research and Technology Implementation Office, Texas Department of Transportation, Federal Highway Administration, Washington, DC, 2009.

171. ASTM G109-07: Standard test method for determining effects of chemical admixtures on corrosion of embedded steel reinforcement in concrete exposed to chloride environments, ASTM International, West Conshohocken, PA, doi: 10.1520/G0109-07, http://www.astm.org

172. American Coal Ash Association, FHWA-IF-03-019: Fly ash facts for highway engineers, Federal Highway Administration, Washington DC, 2003.
173. Kouloumbi N., Batis G., Chloride corrosion of steel rebars in mortars with fly ash admixtures, *Cem. Concr. Compos.*, 14, 1992, 199–207.
174. Dhir R. K., Jones M. R., McCarthy M. J., PFA concrete: Chloride-induced reinforcement corrosion, *Mag. Concr. Res.*, 46, 1994, 269–278.
175. Alhozaimy A., Soroushian P., Mirza F., Effects of curing conditions and age on chloride permeability of fly ash mortar, *ACI Mater. J.*, 93(1), 1996, 87–95.
176. Pradham B., Bhattacharjee B., Role of steel and cement type on chloride-induced corrosion in concrete, *ACI Mater. J.*, 104(6), 2007, 612–619.
177. Anwar M., Miyagawa T., Gaweesh M., Using rice husk ash as a cement replacement material in concrete. In *Waste Materials in Construction, Wascon 2000 Proceedings of the International Conference on the Science and Engineering of Recycling for Environmental Protection*, Harrogate, England, May 31, June 1–2, 2000, Eds. G. R. Woolley, J. J. J. M. Goumans, P. J. Wainwrigh, Waste Management Series, Elsevier Science Ltd., Amsterdam, the Netherlands, Vol. 1, 2000, pp. 671–684.
178. Hosokawa Y., Yamada K., Johannesson B. F., Nilsson L. O., Models for chloride ion bindings in hardened cement paste using thermodynamic equilibrium calculations. In *Second International RILEM Symposium on Advances in Concrete through Science and Engineering*, Eds. J. Marchand et al., Vol. 51, RILEM Publications, Quebec City, Canada, 2006.
179. Jones M. R., Macphee D. E., Chudek J. A., Hunter G., Lannegrand R., Talero R., Scrimgeour S. N., Studies using 27Al MAS NMR of Afm and Aft phases and the Freidel's salt, *Cem. Concr. Res.*, 33, 2003, 177–182.
180. Suryavanshi A. K., Scantlebury J. D., Lyon S. B., Mechanism of Friedel's salt formation in cement rich in tri-calcium aluminate, *Cem. Concr. Res.*, 26, 1996, 717–772.
181. Birnin-Yauri U. A., Glasser F. P., Friedel's salt, Ca2Al(OH)6(Cl,OH)·H2O: Its solid solutions and their role in chloride binding, *Cem. Concr. Res.*, 28, 1998, 1713–1723.
182. Torii K., Sasatani T., Kawamura M., Chloride penetration into concrete incorporating mineral admixtures in marine environment. In *Sixth CANMET/ACI International Conference on Fly Ash, Silica Fume, Slag and Natural Pozzolans in Concrete*, Bangkok, Thailand, May 31–June 5, 1998.
183. Collepardi M., Marcialis A., Turriziani, R., Penetration of chloride ions into cement pastes and concretes, *J. Am. Ceram. Soc.*, 55(10), 1972, 534–535.
184. Ishida T., Kawai K., Sato R., Experimental study on decomposition processes of Friedel's salt due to carbonation. In *Proceedings of International RILEM-JCI Seminar on Concrete Durability (ConcreteLife'06)*, Ed. K. Kovler, Ein-Bokek, Israel, 2006, pp. 51–58.
185. Samson E., Marchand J., Modeling the effect of temperature on ionic transport in cementitious materials, *Cem. Concr. Res.*, 37, 2007, 455–468.
186. Tixier R., Mobasher B., Modeling of damage in cement-based materials subjected to external sulfate attack. I: Formulation, *ASCE J. Mater. Civil Eng.*, 15(4), 2003, 305–313.
187. Al-Amoudi O. S. B., Performance of 15 reinforced concrete mixtures in magnesium-sodium sulphate environments, *Constr. Build. Mater.*, 9(3), 1995, 149–158.
188. Lea F. M., *The Chemistry of Cement and Concrete*, 3rd edn., Chemical Publishing Company, Inc., New York, 1971.

189. Mehta P. K., Sulfate attack on concrete: A critical review. In *Materials Science of Concrete*, Ed. J. Skalny, American Ceramic Society, Westerville, OH, Vol. III, 1993, pp. 105–130.

190. Shah V. N., Hookham C. J., Long-term aging of light water reactor concrete containments, *Nucl. Eng. Des.*, 185(1), 1998, 51–81.

191. Harrison W., Sulphate resistance of buried concrete. The third report on a long-term investigation at Northwick Park on similar concretes in sulphate solutions at BRE, BR 164, Building Research Establishment, U.K., 1992.

192. Rasheeduzzafar F. H. D., Al-Gahtani A. S., Saadoun S. S., Bader M. A., Influence of cement composition on the corrosion of reinforcement and sulphate resistance of concrete, *ACI Mater. J.*, 87(2), 1990, 114–122.

193. Ferraris C. F., Stutzman P. E., Snyder K. A., Sulfate resistance of concrete: A new approach, PCA R&D Serial No. 2486, Portland Cement Association, Skokie, IL, 2006.

194. Crammond N. J., Halliwell M. A., The thaumasite form of sulfate attack in concretes containing a source of carbonate ions—A microstructural overview. In *Second CANMET/ACI Symposium on Advanced Concrete Technology*, Las Vegas, NV, 1995.

195. Barker A. P., Hobbs D. W., Performance of Portland limestone cements in mortar prisms immersed in sulfate solutions at 5°C, *Cem. Concr. Compos.*, 21, 2002, 129–137.

196. Bensted J., Thaumasite—Background and nature in deterioration of cements, mortars and concretes, *Cem. Concr. Compos.*, 21, 1999, 117–121.

197. Nixon P. J., Longworth T. I., Matthews J. D., New UK guidance on the use of concrete in aggressive ground, *Cem. Concr. Compos.*, 25, 2003, 1177–1184.

198. The thaumasite form of sulfate attack: Risks, diagnosis, remedial works and guidance on new construction, Report of the Thaumasite Expert Group, Construction Directorate, Department of the Environment, Transport and the Regions (now, the Construction Industry Directorate of the Department of Trade and Industry), London, U.K., 1999.

199. Clifton J. R., Frohnsdorff G., Ferraris C., Standards for evaluating the susceptibility of cement-based materials to external sulfate attack. In *Materials Science of Concrete—Sulfate Attack Mechanisms*, Eds. J. Skalny, J. Marchand, Special Volume, American Ceramic Society, Westerville, OH, 1999, pp. 337–355.

200. Santhanam M., Cohen M. D., Olek J., Mechanism of sulfate attack: A fresh look part 2. Proposed mechanisms, *Cem. Concr. Res.*, 33, 2003, 341–346.

201. DePuy G. W., Chemical resistance of concrete. In *Concrete and Concrete—Making Materials*, ASTM STP 169C, Philadelphia, PA, 1994, pp. 263–281.

202. Tixier R., Mobasher B., Modeling of damage in cement-based materials subjected to external sulfate attack. II: Comparison with experiments, *ASCE J. Mater. Civil Eng.*, 15(4), 2003, 314–322.

203. Atkinson A., Hearne J. A., Mechanistic model for the durability of concrete barriers exposed to sulphate-bearing groundwaters, *Mater. Res. Soc. Symp. Proc.*, 176, 1990, 149–156.

204. Clifton J. R., Pommersheim J. M., Sulfate attack of cementitious materials: Volumetric relationships and expansions, NISTIR 5390, National Institute of Standards and Technology, Gaithersburg, MD, 1994.

205. Snyder K. A., Clifton J. R., 4SIGHT manual: A computer program for modeling degradation of underground low level waste concrete vaults, NISTIR 5612, National Institute of Standards and Technology, Gaithersburg, MD, 1995.

206. Basista M., Weglewski W., Micromechanical modelling of sulphate corrosion in concrete: Influence of ettringite forming reaction, *J. Theor. Appl. Mech.*, 35(1–3), 2008, 29–52.
207. Santhanam M., Cohen M. D., Olek J., Sulfate attack research—Whither now? *Cem. Concr. Res.*, 31, 2001, 845–851.
208. Scalny J., Pierce J., Sulfate attack: An overview. In *Materials Science of Concrete—Sulfate Attack Mechanisms*, Eds. J. Skalny, J. Marchand, Special Volume, American Ceramic Society, Westerville, OH, 1999, pp. 49–64.
209. Cohen M. D., Mather B., Sulfate attack on concrete. Research needs, *ACI Mater. J.*, 88(1), 1991, 62–69.
210. Mehta P. K., Evaluation of sulfate resistance of cements by a new test method, *ACI Mater. J.*, 72(10), 1975, 573–575.
211. Brown P. W., An evaluation of the sulfate resistance of cements in a controlled environment, *Cem. Concr. Res.*, 11(5–6), 1981, 719–727.
212. Atkinson A., Hearne J. A., Mechanistic model for the durability of concrete barriers exposed to sulphate bearing groundwaters. In *Scientific Basis for Nuclear Waste Management XIII*, Eds. V. M. Oversby, P. W. Brown, Materials Research Society, Pittsburgh, PA, Vol. 176, 1989, pp. 149–156.
213. Ferraris C., Stutzman P., Peltz M., Winpigler J., Developing a more rapid test to assess sulphate resistance of hydraulic cements, *J. Res. Natl. Inst. Stand. Technol.*, 110, 2005, 529–540.
214. Mobasher B., Bonakdar A., Anantharaman S., Modeling of sulfate resistance of fly ash blended cement concrete materials. In *Proceeding of the World of Coal Ash (WOCA) Conference*, Covington, KY, May 7–10, 2007.
215. Planel D., Sercombe J., Bescop P. L., Adenot F., Torrenti J. M., Long-term performance of cement paste during combined calcium leaching-sulfate attack: Kinetics and size effect, *Cem. Concr. Res.*, 36, 2006, 137–143.
216. Kunther W., Lothenbach B., Scrivener K., Influence of carbonate in sulfate environments. In *Proceedings of Concrete in Aggressive Aqueous Environments, Performance, Testing and Modeling*, Eds. M. G. Alexander, A. Bertron, RILEM, Toulouse, France, June 3–5, 2009, pp. 498–499.
217. Lothenbach B., Bary B., Bescop P. L., Schmidt T., Leterrier N., Sulfate ingress in Portland cement, *Cem. Concr. Res.*, 40, 2010, 1211–1225.
218. Kalousek G. L., Porter L. C., Benton E. J., Concrete for long-time service in sulfate environment, *Cem. Concr. Res.*, 2(1), 1972, 79–89.
219. ACI 201.2R-08: Guide to durable concrete, American Concrete Institute, Farmington Hills, MI.
220. DIN 4030-2: Beurteilung betonangreifender Wässer, Böden und Gase. Entnahme und Analyse von Wasser-und Bodenproben (Assessment of concrete attacking water, soil and gases; collection and analysis of water and soil samples), 1991.
221. Troli R., Collepardi M., Technical contradictions in the European Norm EN206 for concrete durability. In *Proceedings of the International Symposium on Role of Concrete in Sustainable Development*, University of Dundee, Scotland, U.K., September 3–4, 2003, pp. 665–674.
222. The Concerte Centre, *Concrete in Aggressive Ground*, Special Digest 1, 3rd edn., BRE Construction Division, Garston, U.K., 2005.
223. Bakkar R. F. M., Permeability of blended cement concretes, American Concrete Institute Publication, SP-79-30, Detroit, MI, 1983, pp. 589–605.

224. Fraay A. L. A., Bijen J. M., Haan Y. M., The reaction of fly ash in concrete. A critical examination, *Cem. Concr. Res.* 19, 1989, 235–246.
225. Givi A. N., Rashid S. A., Aziz F. N. A., Salleh M. A. M., Contribution of rice husk ash to the properties of mortar and concrete: A review, *J. Am. Sci.*, 6(3), 2010, 157–165.
226. Mehta P. K., Mechanism of expansion associated with ettringite formation, *Cem. Concr. Res.*, 3(1), 1973, 1–6.
227. Al-Amoudi O. S. B., Maslehuddin M., Saadi M. M., Effect of magnesium sulphate and sodium sulphate on the durability performance of plain and blended cements, *ACI Mater. J.*, 92(1), January–February 1995, pp. 15–24.
228. Lawrence C. D., Influence of binder type on sulphate resistance, *Cem. Concr. Res.*, 22(6), 1992, 1047–1058.
229. Hughes D. C., Sulphate resistance of OPC, OPC/fly ash and SRPC pastes: Pore structure and permeability, *Cem. Concr. Res.*, 15(6), 1985, 1003–1012.
230. Park Y. S., Suh J. K., Lee J. H., Shin Y. S., Strength deterioration of high strength concrete in sulfate environment, *Cem. Concr. Res.*, 29, 1999, 1397–1402.
231. ASTM C1157/C1157M–10, Standard performance specification for hydraulic cement, ASTM International, West Conshohocken, PA, 2010, doi: 10.1520/C1157_C1157 M-10, http://www.astm.org
232. ASTM C150/C150M–09, Standard specification for Portland cement, ASTM International, West Conshohocken, PA, 2009, doi: 10.1520/C0150_C0150M-09, http://www.astm.org
233. Irassar E. F., Gonzalez M., Rahhal V., Sulfate resistance of Type V cements with limestone filler and natural pozzolana, *Cem. Concr. Compos.*, 22, 2000, 361–368.
234. Cao H. T., Bucea L., Ray A., Yozghatlian S., The effect of cement composition and pH of environment on sulfate resistance of Portland cements and blended cements, *Cem. Concr. Compos.*, 19, 1997, 161–171.
235. Monteiro P. J., Kurtis K. E., Time to failure for concrete exposed to severe sulfate attack, *Cem. Concr. Res.*, 33, 2003, 987–993.
236. Shanahan N. G., Influence of C_3S content of cement on concrete sulfate durability, MS thesis, Department of Civil and Environmental Engineering, College of Engineering, University of South Florida, Tampa, FL, 2003.
237. Osborne G. J., Determination of sulphate resistance of blast furnace slag cement using small scale accelerated methods of tests, *Adv. Cem. Res.*, 2, 1989, 21–27.
238. Frearson J. P. H., Sulphate resistance of combination of Portland cement and ground granulated blast furnace slag. In *Proceedings of Second International Congress on Fly Ash, Silica Fume, Slag and Natural Pozzolans in Concrete*, Madrid, Spain, ACI Pub-91, Vol. 2, 1986, pp. 1495–524.
239. Mangat P. S., Khatib J. M., Influence of fly ash, silica fume and slag on sulphate resistance of concrete, *ACI Mater. J.*, 92(5), 1995, 542–552.
240. Dikeou J. T., Fly ash increases resistance of concrete to sulphate attack, United States Department of the Interior, Bureau of Reclamation, A Water Resources Technical Publication, Research Report No. 23, 1970.
241. Verbeck A. J., Field and laboratory studies of the sulphate resistance of concrete. In *Performance of Concrete*, Canadian Building Series, University of Toronto Press, Toronto, Canada, 1968, pp. 113–124.
242. Tsivilis S., Kakali G., Skaropoulou A., Sharp J. H., Swamy R. N., Use of mineral admixtures to prevent thaumasite formation in limestone cement mortar, *Cem. Concr. Compos.*, 25, 2003, 969–976.

243. Mulenga D. M., Stark J., Nobst P., Thaumasite formation in concrete and mortars containing fly ash, *Cem. Concr. Compos.*, 24, 2003, 907–912.
244. Mehta P. K., Sulfate attack on concrete: Separating myths from reality, *Concr. Int.*, 22(8), 2000, 57–61.
245. Neville A., Concrete: 40 years of progress? *Concrete*, 38, 2004, 52–54.
246. Taylor H. F. W., Famy C., Scrivener K. L., Delayed ettringite formation, *Cem. Concr. Res.*, 31, 2001, 683–693.
247. Diamond S., Delayed ettringite formation-processes and problems, *Cem. Concr. Compos.*, 18(3), 1996, 205–215.
248. Collepardi M., Damage by delayed ettringite formation, *Concr. Int.*, 21(1), 1999, 69–74.
249. Taylor H. F. W., Distribution of sulfate between phases in Portland cement clinkers, *Cem. Concr. Res.*, 29(8), 1999, 1173–1179.
250. Hime W. G., Delayed ettringite formation—A concern for precast concrete? *PCI J.*, 41, July–August 1996, 26–30.
251. Graf, L. A., Effect of relative humidity on expansion and microstructure of heat-cured mortars, RD139, Portland Cement Association, Skokie, IL, 2007.
252. Sahu S., Thaulow N., Delayed ettringite formation in Swedish concrete railroad ties, *Cem. Concr. Res.*, 34, 2004, 1675–1681.
253. Fu Y., Beaudoin J. J., On the distinction between delayed and secondary ettringite formation in concrete, *Cem. Concr. Res.*, 26(6), 1996, 979–980.
254. Yang R., Lawrence C. D., Lynsdale C. J., Sharp J. H., Delayed ettringite formation in heat-cured Portland cement mortars, *Cem. Concr. Res.*, 29(1), 1999, 17–25.
255. Grabowski E., Czarnecki B., Gillot J. E., Duggan C. R., Scott J. F., Rapid test of concrete expansivity due to internal sulfate attack, *ACI Mater. J.*, September–October 1992, 469–480.
256. Idorn G. M., Skalny J. P., Discussion of rapid test of concrete expansivity due to internal sulphate attack by Grabowski et al. 1992, *ACI Mater. J.*, 90(4), 383–385.
257. Oberholster R. E., Maree H., Brand J. H. B., Cracked prestressed concrete railway sleepers: Alkali-silica reaction or delayed ettringite formation. In *Proceedings of the Ninth International Conference on Alkali-Aggregate Reaction in Concrete*, Concrete Society, Slough, U.K., Vol. 2, 1992, pp. 739–749.
258. Fu Y., Delayed ettringite formation in Portland cement production, PhD thesis, Department of Civil Engineering, University of Ottawa, Ottawa, Ontario, Canada, 1996.
259. Pavoine A., Divet L., Fenouillet S., A concrete performance test for delayed ettringite formation: Part I optimisation, *Cem. Concr. Res.*, 36, 2006, 2138–2143.
260. Pavoine A., Divet L., Fenouillet S., A concrete performance test for delayed ettringite formation: Part II validation, *Cem. Concr. Res.*, 36(12), 2006, 2144–2151.
261. Ramlochan T., Zakarias P., Thomas M. D. A., Hooton R. D., The effect of pozzolans and slag on the expansion of mortars cured at elevated temperature: Part I: Expansive behaviour, *Cem. Concr. Res.*, 33(6), 2003, 807–814.
262. Ramlochan T., Zakarias P., Thomas M. D. A., Hooton R. D., The effect of pozzolans and slag on the expansion of mortars cured at elevated temperature: Part II: Microstructural and microchemical investigations, *Cem. Concr. Res.*, 34(8), 2004, 1341–1356.

263. Heinz D., Ludwig U., Rudiger I., Delayed ettringite formation in heat treated mortars and concretes, *Concr. Precasting Plant Technol.*, 11, 1989, 56–61.
264. Dow C., Glasser F. P., Calcium carbonate efflorescence on Portland cement and building materials, *Cem. Concr. Res.*, 33, 2003, 147–154.
265. Faucon P., Adenot F., Jorda M., Cabrillac R., Behaviour of crystallised phases of Portland cement upon water attack, *Mater. Struct.*, 30, 1997, 480–485.
266. Faucon P., Adenot F., Jacquinot J. F., Petit J. C., Cabrillac R., Jorda M., Long-term behaviour of cement pastes used for nuclear waste disposal: Review of physico-chemical mechanisms of water degradation, *Cem. Concr. Res.*, 28(6), 1998, 847–857.
267. Haga K., Sutou S., Hironaga M., Tanaka S., Nagasaki S., Effects of porosity on leaching of Ca from hardened ordinary Portland cement paste, *Cem. Concr. Res.*, 35, 2005, 1764–1775.
268. Mainguy M., Tognazzi C., Torrenti J. M., Adenot F., Modelling of leaching in pure cement paste and mortar, *Cem. Concr. Res.*, 30, 2000, 83–90.
269. Carde C., Francois R., Torrenti J. M., Leaching of both calcium hydroxide and C-S-H from cement paste: Modeling the mechanical behavior, *Cem. Concr. Res.*, 26(8), 1996, 1257–1268.
270. Saito H., Deguchi A., Leaching tests on different mortars using accelerated electrochemical method, *Cem. Concr. Res.*, 30, 2000, 1815–1825.
271. Moranville M., Kamali S., Guillon E., Physicochemical equilibria of cement-based materials in aggressive environments—Experiment and modeling, *Cem. Concr. Res.*, 34, 2004, 1569–1578.
272. Carde C., Franfois R., Effect of ITZ leaching on durability of cement-based materials, *Cem. Concr. Res.*, 27(7), 1997, 971–978.
273. Eijk R. J., Brouwers H. J. H., Study of the relation between hydrated Portland cement composition and leaching resistance, *Cem. Concr. Res.*, 28(6), 1998, 815–828.
274. Marion A. M., Laneve M. D., Grauw A. D., Study of the leaching behaviour of paving concretes: Quantification of heavy metal content in leachates issued from tank test using demineralised water, *Cem. Concr. Res.*, 35(5), 2005, 951–957.
275. Marinoni N., Pavese A., Voltolini M., Merlini M., Long-term leaching test in concretes: An x-ray powder diffraction study, *Cem. Concr. Compos.*, 30, 2008, 700–705.
276. Sellier A., Lacarriere L. B., Gonnounia M. E., Bourbon X., Behavior of HPC nuclear waste disposal structures in leaching environment, *Nucl. Eng. Des.*, 241(1), 2010, 402–414, doi: 10.1016/j.nucengdes.2010.11.002.
277. Powers T. C., Freezing effects in concrete, American Concrete Institute, Farmington Hills, MI, SP 47-1, 1975, pp. 1–11.
278. Powers T. C., A working hypothesis for further studies of frost resistance of concrete, *J. Am. Concr. Inst.*, 16(4), 1945, 245–272.
279. Powers T. C., Helmuth R. A., Theory of volume changes in hardened Portland-cement paste during freezing. In *Highway Research Board Proceedings*, Vol. 32, 1953, pp. 285–297.
280. Larson T. D., Cady P. D., Identification of frost-susceptible particles in concrete aggregates, National Cooperative Highway Research Program (NCHRP) Report 66, 1969.
281. Litvan G. G., Phase transitions of adsorbates, IV, mechanism of frost action in hardened cement paste, *J. Am. Ceram. Soc.*, 55(1), 1972, 38–42.

282. Janssen D. J., Snyder M. B., Resistance of concrete to freezing and thawing, Strategic Highway Research Program Report, SHRP-C-391, Washington, DC, 1994.

283. Janssen D. J., Freeze-thaw performance of concrete: Reconciling laboratory-based specifications with field experience, *J. ASTM Int.*, 7(1), 2010.

284. Whiting D., Stark D., Control of air content in concrete, National Cooperative Highway Research Program (NCHRP) Report 258, Transportation Research Board, National Research Council, Washington, DC, 1983.

285. Rixom M. R., Mailvaganam N. P., *Chemical Admixtures for Concrete*, The University Press, Cambridge, U.K., 1986.

286. ASTM C260: Standard specification for air-entraining admixtures for concrete, ASTM International, West Conshohocken, PA, 2006, doi: 10.1520/C0260-06, http://www.astm.org

287. Huo J. F., Ji X. X., Yang H., Experimental study on freeze-thaw resistance durability of high performance concrete, *Adv. Mater. Res.*, 168–170, 2010, 393–397.

288. Schwartz D. R., D-cracking of concrete pavements, National Cooperative Highway Research Program (NCHRP) Synthesis of Highway Practice No. 134, 1987.

289. Doherty J. O., D-cracking of concrete pavements, Materials and Technology Engineering and Science (MATES), Materials and Technology Division of the Michigan Department of Transportation, Iss. 7, May 1987.

290. Kaneuji M., Winslow D. N., Dolch W. L., The relationship between an aggregate's pore size distribution and its freeze thaw durability in concrete, *Cem. Concr. Res.*, 10, 1980, 433–441.

291. Andrade C., Types of models of service life of reinforcement: The case of the resistivity, *Concr. Res. Lett.*, 1(2), 2010, 73–80, http://www.crl.issres.net

292. Bouny V. B., Nguyen T. Q., Dangla P., Assessment and prediction of RC structure service life by means of durability indicators and physical/chemical models, *Cem. Concr. Compos.*, 31, 2009, 522–534.

293. Ballim Y., Alexander M. G., Towards a performance-based specification for concrete durability. In *African Concrete Code Symposium-2005*, Tripoli, Libya, November 28–29, 2005.

294. Alexander M., Specification and design for durability of reinforced concrete: South African developments. In *Anna Maria Workshop VII, Sustainability in the Cement and Concrete Industry*, Holmes Beach, FL, November 15–17, 2006.

295. Alexander M. G., Magee B. J., Durability performance of concrete containing condensed silica fume, *Cem. Concr. Res.*, 29, 1999, 917–922.

296. Alexander M. G., Mackechnie J. R., Ballim, Y., Use of durability indexes to achieve durable cover concrete in reinforced concrete structures. In *Materials Science of Concrete*, Eds. J. P. Skalny, S. Mindess, Vol. VI, American Ceramic Society, Westerville, OH, 2001, pp. 483–511.

297. Andrade C., Martinez I., Use of indices to assess the performance of existing and repaired concrete structures, *Constr. Build. Mater.*, 23, 2009, 3012–3019.

298. Getting the numbers right, cement industry energy and CO_2 performance, The Cement Sustainability Initiative, World Business Council for Sustainable Development (WBCSD), Conches, Switzerland.

299. Bapat J. D., Sabnis S. S., Joshi S. V., Hazaree C. V., History of cement and concrete in India—A paradigm shift, American Concrete Institute (ACI) Technical Session on History of Concrete, April 22–26, 2007, Atlanta, GA.

300. Mehta P. K., High-performance, high-volume fly ash concrete for sustainable development. In *International Workshop on Sustainable Development and Concrete Technology, Part I: Critical Issues of Sustainable Development and Emerging Technology for "Green" Concrete*, National Concrete Pavement Technology Centre, Iowa State University, Beijing, China, May 20–21, 2004.
301. Bapat J. D., Sabnis S. S., Hazaree C. V., Deshchaugule A. D., Eco-friendly concrete with high volume of lagoon ash, *ASCE J. Mater. Civil Eng.*, 18(3), 2006, 453–461.
302. Hazaree C., Bapat J. D., Pozzolanic effect, elastic modulus and abrasion resistance: From classified fly ash to lagoon ash. In *Sixth International Conference on Road and Airfield Pavement Technology (ICPT)*, Sapporo, Japan, July 20–23, 2008.
303. Damtoft J. S., Lukasik J., Herfort D., Sorrentino D., Gartner E. M., Sustainable development and climate change initiatives, *Cem. Concr. Res.*, 38, 2008, 115–127.
304. Mehta P. K., Global concrete industry sustainability, *Concr. Int.*, February 2009, 45–48.
305. Concrete the choice for sustainable design, September 2008, http://www.concretethinker.com/Content/Upload%5C443.pdf
306. Haselbach L., Concrete as a carbon sink, Portland Cement Association, Skokie, IL, 2010, http://www.cement.org
307. Gesoglu M., Ozbay E., Effects of mineral admixtures on fresh and hardened properties of self-compacting concretes: Binary, ternary and quaternary systems, *Mater. Struct.*, 40(9), 2007, 923–937.
308. Isaia G. C., High-performance concrete for sustainable constructions. In *Waste Materials in Construction*, Eds. G. R. Woolley, J. J. J. M. Goumans, P. J. Wainwright, Elsevier Science Ltd., Amsterdam, the Netherlands, 2000.
309. Napier T., Construction waste management, U.S. Army Corps of Engineers, Engineer Research and Development Center/Construction Engineering Research Laboratory, January 2011, http://www.wbdg.org/resources/cwmgmt.php
310. Meyer C., The greening of the concrete industry, *Cem. Concr. Compos.*, 31, 2009, 601–605.
311. Eco-Culture, http://www.cowiprojects.com/ecoculture/Project_Goals.html
312. UNEP-SBCI: United Nations Environment Programme's Sustainable Buildings and Climate Initiative, http://www.unep.org/sbci/
313. Schneider M., Romer M., Tschudin M., Bolio H., Sustainable cement production—Present and future, Cem. *Concr. Res.*, 41, 2011, 642–650.
314. Damineli B. L., Kemeid F. M., Aguiar P. S., John V. M., Measuring the eco-efficiency of cement use, *Cem. Concr. Compos.*, 32, 2010, 555–562.
315. Habert G., Billard C., Rossi P., Chen C., Roussel N., Cement production technology improvement compared to factor 4 objectives, *Cem. Concr. Res.*, 40, 2010, 820–826.

Chapter 8

1. Bakker R. R., Elbersen H. W., Managing ash content and quality in herbaceous biomass: An analysis from plant to product, In *14th European Biomass Conference*, Paris, France, October 17–21, 2005, pp. 210–213.
2. Biricik H., Akoz F., Berktay I., Tulgar A. N., Study of pozzolanic properties of wheat straw ash, *Cem. Concr. Res.*, 29, 1999, 637–643.

3. ASTM C618, Standard specification for coal fly ash and raw or calcined natural pozzolan for use in concrete, ASTM International, West Conshohocken, PA, 2008, doi: 10.1520/C0618-08, http://www.astm.org

4. Ganesan K., Rajagopal K., Thangavel K., Evaluation of bagasse ash as supplementary cementitious material, *Cem. Concr. Compos.*, 29, 2007, 515–524.

5. Chusilp N., Jaturapitakkul C., Kiattikomol K., Utilization of bagasse ash as a pozzolanic material in concrete, *Constr. Build. Mater.*, 23, 2009, 3352–3358.

6. Akram T., Memon S. A., Obaid H., Production of low cost self compacting concrete using bagasse ash, *Constr. Build. Mater.*, 23, 2009, 703–712.

7. Wang S., Miller A., Llamazos E., Fonseca F., Baxter L., Biomass fly ash in concrete: Mixture proportioning and mechanical properties, *Fuel*, 87, 2008, 365–371.

8. Thy P., Jenkins B. M., Grundvig S., Shiraki R., Lesher C. E., High temperature elemental losses and mineralogical changes in common biomass ashes, *Fuel*, 85, 2006, 783–795.

9. Ban C. C., Ramli M., The implementation of wood waste ash as a partial cement replacement material in the production of structural grade concrete and mortar: An overview, *Resour. Conserv. Recycl.*, 55(7), 2011, 669–685, doi: 10.1016/j.resconrec.2011.02.002.

10. Adesanya D. A., Raheem A. A., Development of corn cob ash blended cement, *Constr. Build. Mater.*, 23, 2009, 347–352.

11. Goyal A., Kunio H., Ogata H., Garg M., Anwar A. M., Ashraf M., Mandula, Synergic effect of wheat straw ash and rice-husk ash on strength properties of mortar, *J. Appl. Sci.*, Pub: Asian Network of Scientific Information, 7(21), 2007, 3256–3261.

12. GMR 409: Grain Market Report, International Grains Council, March 24, 2011.

13. Adesanya D. A., Raheem A. A., A study of the workability and compressive strength characteristics of corn cob ash blended cement concrete, *Constr. Build. Mater.*, 23, 2009, 311–317.

14. Adesanya D. A., Evaluation of blended cement mortar, concrete and stabilized earth made from ordinary Portland cement and corn cob ash, *Constr. Build. Mater.*, 10(6), 1996, 451–456.

15. Stichnothe H., Schuchardt F., Greenhouse gas reduction potential due to smart palm oil mill residue treatment, Technology Cooperation and Economic Benefit of Reduction of GHG Emissions in Indonesia, Hamburg, November 1–2, 2010.

16. Foo K. Y., Hameed B. H., Value-added utilization of oil palm ash: A superior recycling of the industrial agricultural waste, *J. Hazard. Mater.*, 172, 2009, 523–531.

17. Tay J. H., Show K. Y., Use of ash derived from oil-palm waste incineration as a cement replacement material, *Resour. Conserv. Recycl.*, 13, 1995, 27–36.

18. Chindaprasirt P., Rukzon S., Sirivivatnanon S., Resistance to chloride penetration of blended Portland cement mortar containing palm oil fuel ash, rice husk ash and fly ash, *Constr. Build. Mater.*, 22, 2008, 932–938.

19. Tangchirapat W., Jaturapitakkul C., Chindaprasirt P., Use of palm oil fuel ash as a supplementary cementitious material for producing high-strength concrete, *Constr. Build. Mater.*, 23, 2009, 2641–2646.

20. Tangchirapat W., Saeting T., Jaturapitakkul C., Kiattikomol K., Siripanichgorn A., Use of waste ash from palm oil industry in concrete, *Waste Manage.*, 27, 2007, 81–88.

21. Rukzon S., Chindaprasirt P., Strength and chloride resistance of blended Portland cement mortar containing palm oil fuel ash and fly ash, *Int. J. Miner. Metall. Mater.*, 16(4), 2009, 475–481.

22. Sata V., Jaturapitakkul C., Rattanashotinunt C., Compressive strength and heat evolution of concretes containing palm oil fuel ash, *J. Mater. Civil Eng.*, 22(10), 2010, 1033–1038.

23. Cordeiro G. C., Filho R. D. T., Tavares L. M., Fairbairn E. M. R., Ultrafine grinding of sugar cane bagasse ash for application as pozzolanic admixture in concrete, *Cem. Concr. Res.*, 39, 2009, 110–115.

24. Paya J., Monz J., Borrachero M. V., Diaz-Pinzon L., Ordonez L. M., Sugar-cane bagasse ash (SCBA): Studies on its properties for reusing in concrete production, *J. Chem. Technol. Biotechnol.*, 77(3), 2002, 321–325.

25. CIMMYT (International Maize and Wheat Improvement Center), http:// apps.cimmyt.org/Research/economics/map/facts_trends/wft9596/htm/ wft9596part2.htm

26. Visvesvaraya H. C., Recycling of agricultural wastes with special emphasis on rice husk ash. In *Use of Vegetable Plants and Fibres as Building Materials, Joint Symposium RILEM/CIB/NCCL*, Baghdad, Iraq, 1986, pp. 1–22.

27. Bensted J., Munn J., A discussion of the paper "Study of pozzolanic properties of wheat straw ash" by H. Biricik, F. Akoèz, I. Berktay, and A. N. Tulgar, *Cem. Concr. Res.*, 30, 2000, 1507–1508.

28. Isaia G. C., Synergic action of fly ash in ternary mixtures of high-performance concrete, American Concrete Institute, Farmington Hills, MI, SP186-28, 1999, pp. 481–502.

29. Isaia G. C., Synergic action of fly ash in ternary mixtures with silica fume and rice husk ash. In *Proceedings of 10th International Congress on the Chemistry of Cement*, Ed. H. Justnes, Gothenburg, Sweden, 1997.

30. Rajamma R., Ball R. J., Tarelho L. A. C., Allen G. C., Labrincha J. A., Ferreira V. M., Characterisation and use of biomass fly ash in cement-based materials, *J. Hazard. Mater.*, 172, 2009, 1049–1060.

31. Wiltsee G., Lessons learned from existing biomass power plants, National Renewable Energy Laboratory (NREL), U.S. Department of Energy, Washington, DC, 2000.

32. Udoeyo F. F., Inyang H., Young D. T., Oparadu E. E., Potential of wood waste ash as an additive in concrete, *J. Mater. Civil Eng.*, 18(4), 2006, 605–612.

33. Wang S., Baxter L., Fonseca F., Biomass fly ash in concrete: SEM, EDX and ESEM analysis, *Fuel*, 87, 2008, 372–379.

34. Naylor M. L., Schmidt E. J., Agricultural use of wood ash as a fertilizer and liming material, *Tappi J.*, 69(10), 1986, 114–119.

35. Etietgni L., Campbell A. G., Physical and chemical characteristics of wood ash, *Bioresour. Technol.*, 37, 1991, 173–178. (iii) Ban C. C., Ramli M, *Conserv. Recycl.*, 2011 as earlier.

36. Udoeyo F. F., Dashibil P. U., Sawdust ash as concrete material, *J. Mater. Civil Eng.*, 14(2), 2002, 173–176.

37. Elinwa A. U., Mahmood Y. A., Ash from timber waste as cement replacement material, *Cem. Concr. Compos.*, 24, 2002, 219–222.

38. Naik T. R., Kraus R. N., Singh S. S., Chanodia P. P., CLSM containing mixtures of coal ash and a new pozzolanic material, University of Wisconsin, Milwaukee, WI, Report No. CBU-2000-34, REP-416, 2000.

39. ACI 229R-99, Controlled low-strength materials, American Concrete Institute, Farmington Hills, MI.
40. Miller S. F., Miller B. G., The occurrence of inorganic elements in various biofuels and its effect on ash chemistry and behavior and use in combustion products, *Fuel Process. Technol.*, 88, 2007, 1155–1164.
41. Obernberger I., Biedermann F., Widmann W., Riedi R., Concentrations of inorganic elements in biomass fuels and recovery in the different ash fractions, *Biomass Bioenergy*, 12(3), 1997, 211–224.
42. Grammelis P., Skodras G., Kakaras E., Karangelos D. J., Petropoulos N. P., Anagnostakis M. J., Hinis E. P., Simopoulos S. E., Effects of biomass co-firing with coal on ash properties. Part II: Leaching, toxicity and radiological behaviour, *Fuel*, 85, 2006, 2316–2322.
43. Grammelis P., Skodras G., Kakaras E., Effects of biomass co-firing with coal on ash properties. Part I: Characterisation and PSD, *Fuel*, 85, 2006, 2310–2315.
44. Wang S., Baxter L., Comprehensive study of biomass fly ash in concrete: Strength, microscopy, kinetics and durability, *Fuel Process. Technol.*, 88, 2007, 1165–1170.
45. Tkaczewska E., Małolepszy J., Hydration of coal-biomass fly ash cement, *Constr. Build. Mater.*, 23, 2009, 2694–2700.
46. European List of Wastes, Commission Decision 2000/532/EC of 3 May 2000, amended by the Commission Decision 2001/118/EC, Commission Decision 2001/119/EC and Council Decision 2001/573/EC.
47. EN450-1: Fly ash for concrete, part 1, definition, specification and conformity criteria, European Committee for Standardization, 2005.
48. Martin S., Paper chase, Ecology Global Network: http://www.ecology.com/features/paperchase/index.html
49. U.S. Environmental Protection Agency, Methodology for Estimating Municipal Solid Waste Recycling Benefits, November 2007. http://www.epa.gov/waste/nonhaz/municipal/pubs/06benefits.pdf (retrieved on June 2010).
50. Banfill P., Frias M., Rheology and conduction calorimetry of cement modified with calcined paper sludge, *Cem. Concr. Res.*, 37, 2007, 184–190.
51. Pera J., Amrouz A., Development of highly reactive metakaolin from paper sludge, *Adv. Cem. Based Mater.*, 7, 1998, 49–56.
52. Sabador E., Frias M., Sanchez de Rojas M. I., Vigil R., Garcia R., San Jose J. T., Characterization and transformation of an industrial by-product (coated paper sludge) into a pozzolanic material, *Materiales de Construccion*, 57(285), 2007, 45–59.
53. Vegas I., Urreta J., Frias M., Garcia R., Freeze-thaw resistance of blended cements containing calcined paper sludge, *Constr. Build. Mater.*, 23, 2009, 2862–2868.
54. Villa R. V., Frias M., Rojas M. I. S., Vegas I., Garcia R., Mineralogical and morphological changes of calcined paper sludge at different temperatures and retention in furnace, *Appl. Clay Sci.*, 36, 2007, 279–286.
55. Frias M., Garcia R., Vigil R., Ferreiro S., Calcination of art paper sludge waste for the use as a supplementary cementing material, *Appl. Clay Sci.*, 42, 2008, 189–193.
56. Pera J., Ambroise J., Chabannat M., Transformation of wastes into complementary cementing materials. In *Proceedings of the Seventh CANMET/ACI Conference on Fly Ash, Silica Fume and Slag in Concrete*, Ed. V. M. Malhotra, Madras, India, ACI SP 199, Vol. 2, 2001, pp. 459–475.

57. Frias M., Rojas M. I. S., Rivera J., Influence of calcining conditions on pozzolanic activity and reaction kinetics in paper sludge-calcium hydroxide mixes. In *Proceedings of the Eighth CANMET/ACI Conference on Fly Ash, Silica Fume, Slag and Natural Pozzolan in Concrete*, Ed. V. M. Malhotra, ACI SP 221, Las Vegas, NV, Vol. 1, 2004, pp. 879–892.

58. Rodriguez O., Frias M., Rojas M. I. S., Garcia R., Vigil R., Effect of thermally activated paper sludge on the mechanical properties and porosity of cement pastes, *Materiales de Construccion*, 59(294), 2009, 41–52.

59. Bai J., Chaipanich A., Kinuthia J. M., O'Farrell M., Sabira B. B., Wilda S., Lewis M. H., Compressive strength and hydration of wastepaper sludge ash-ground granulated blastfurnace slag blended pastes, *Cem. Concr. Res.*, 33, 2003, 1189–1202.

60. Mozaffari E., Kinuthia J. M., Bai J., Wild S., An investigation into the strength development of wastepaper sludge ash blended with ground granulated blast-furnace slag, *Cem. Concr. Res.*, 39, 2009, 942–949.

61. DOE/EE-0229: Energy and environmental profile of the U.S. iron and steel industry, U.S. Department of Energy, Office of Industrial Technologies, August 2000

62. Recupac, http://www.recupac.com/index-2.html

63. Wang J. C., Hepworth M. T., Reid K. J., Recovering Zn, Pb, Cd and Fe from electric furnace dust, *J. Metals*, 42, 1990, 42–45.

64. World steel in figures, International Institute of Steel and Iron (IISI), 2003, http://www.worldsteel.org/media/wsif/wsif2003.pdf (accessed on March, 20, 2005).

65. Maslehuddin M., Awan F. R., Shameem M., Ibrahim M., Ali M. R., Effect of electric arc furnace dust on the properties of OPC and blended cement concretes, *Constr. Build. Mater.*, 25, 2011, 308–312.

66. Vargas A. S., Masuero A. B., Vilela A. C. F., Investigations on the use of electric-arc furnace dust (EAFD) in Pozzolan-modified Portland cement I (MP) pastes, *Cem. Concr. Res.*, 36, 2006, 1833–1841.

67. Souza C. A. C., Machado A. T., Andrade Lima L. R. P., Cardoso R. J. C., Stabilization of electric-arc furnace dust in concrete, *Mater. Res.*, 13(4), 2010, 513–519.

68. Al-Zaid R. Z., Al-Sugair F. H., Al-Negheimish A. I., Investigation of potential uses of electric-arc furnace dust (EAFD) in concrete, *Cem. Concr. Res.*, 27(2), 1997, 267–278.

69. Sturm T., Milacic R., Murko S., Vahcic M., Mladenovic A., Suput J. S., Scancar J., The use of EAF dust in cement composites: Assessment of environmental impact, *J. Hazard. Mater.*, 166, 2009, 277–283.

70. Laforest G., Duchesne J., Stabilization of electric arc furnace dust by the use of cementitious materials: Ionic competition and long-term leachability, *Cem. Concr. Res.*, 36, 2006, 1628–1634.

71. Cyr M., Coutand M., Clastres P., Technological and environmental behavior of sewage sludge ash (SSA) in cement-based materials, *Cem. Concr. Res.*, 37, 2007, 1278–1289.

72. Bhatty J. I., Reid K. J., Compressive strength of municipal sludge ash mortars, *ACI Mater. J.*, 86(4), 1989, 394–400.

73. Monzo J., Paya J., Borrachero M. V., Girbe I., Reuse of sewage sludge ashes (SSA) in cement mixtures: The effect of SSA on the workability of cement mortars, *Waste Manage*, 23, 2003, 373–381.

74. Monzo J., Paya J., Borrachero M. V., Morenilla J. J., Bonilla M., Calderon P., Some strategies for reusing residues from waste water treatment plants: Preparation of building materials. In *Proceedings of International RILEM Conference on the Use of Recycled Materials in Building and Structures*, RILEM Publications SARL, Bagneux, France, 2004, pp. 814–823.

75. Lam H. K., Barford J. P., McKay G., Utilization of incineration waste ash residues as Portland cement clinker, *Chem. Eng. Trans.*, 21, 2010, 757–762, doi: 10.3303/CET1021127.

76. Vassilev S. V., Braekman-Danheux C., Laurent P., Thiemann T., Fontana A., Behaviour, capture and inertization of some trace elements during combustion of refuse-derived char from municipal solid waste, *Fuel*, 78(10), 1999, 1131–1145.

77. Sawell S. E., Hetherington S. A., Chandler A. J., An overview of municipal solid waste management in Canada, *Waste Manage.*, 16(5/6), 1996, 351–359.

78. Huang W. J., Chu S. C., A study on the cementlike properties of municipal waste incineration ashes, *Cem. Concr. Res.*, 33, 2003, 1795–1799.

79. Incineration, Wikipedia, http://en.wikipedia.org/wiki/Incineration

80. Lu L. T., Hsiao T. Y., Shang N. C., Yu Y. H., Ma H. W., MSW management for waste minimization in Taiwan: The last two decades, *Waste Manage.*, 26, 2006, 661–667.

81. Li M., Xiang J., Hu S., Sun L., Su S., Li P., Sun X., Characterization of solid residues from municipal solid waste incinerator, *Fuel*, 83, 2004, 1397–1405.

82. Chandler A. J., *Municipal Solid Waste Incinerator Residues*, The International Ash Working Group, Elsevier Science, Amsterdam, the Netherlands, 1997.

83. Stuart B. J., Kosson D. S., Characterization of municipal waste combustion air pollution control residues as a function of particle size, *Combust. Sci. Technol.*, 101(1 and 6), 1994, 527–548.

84. Alba N., Gasso S., Lacorte T., Baldasano J. M., Characterization of municipal solid waste incineration residues from facilities with different air pollution control systems, *J. Air Waste Manage. Assoc.*, 47, 1997, 1170–1179.

85. Bontempi E., Zacco A., Borgese L., Gianoncelli A., Ardesi R., Depero L. E., A new method for municipal solid waste incinerator (MSWI) fly ash inertization, based on colloidal silica, *J. Environ. Monit.*, 12(11), 2010, 2093–2099.

86. Remond S., Pimienta P., Bentz D. P., Effects of the incorporation of municipal solid waste incineration fly ash in cement pastes and mortars. I Experimental Study, *Cem. Concr. Res.*, 32(2), 2002, 303–311, Through: Garboczi E. J., Bentz D. P., Snyder K. A., Martys N. S., Stutzman P. E., Ferraris C. F., Bullard J. W., An electronic monograph: Modeling and measuring the structure and properties of cement-based materials, National Institute of Standards and Technology, Gaithersburg, MD 20899-8615, February 23, 2011.

87. Forestier L. L., Libourel G., Characterization of fuel gas residues from municipal solid waste combustors, *Environ. Sci. Technol.*, 32(15), 1998, 2250–2256.

88. Puri A., Georgescu M., The immobilization of municipal solid waste incineration ashes in hardened concrete. In *Infrastructure Regeneration and Rehabilitation Improving the Quality of Life through better Construction: A Vision for the Next Millennium*, Sheffield Academic Press, Sheffield, U.K., June 28–July 2, 1999.

89. Gougar M. L. D., Scheetz B. E., Roy D. M., Ettringite and C-S-H Portland cement phases for waste ion immobilization: A review, *Waste Manage.*, 16(4), 1996, 295–303.

90. Quina M. J., Bordado J. C., Quinta-Ferreira R. M., Treatment and use of air pollution control residues from MSW incineration: An overview, *Waste Manage.*, 28(11), 2007, 2097–2121, doi: 10.1016/j.wasman.2007.08.030.
91. Conner J. R., Hoeffner S. L., The history of stabilization/solidification technology, *Crit. Rev. Environ. Sci. Technol.*, 28(4), 1998, 325–396.
92. Conner J. R., Hoeffner S. L., A critical review of stabilization/solidification technology, *Environ. Sci. Technol.*, 28(4), 1998, 397–462.
93. Shih C. J., Lin C. F., Lin Y. C., Re-solidification of the solidified heavy metal-containing fly ashes from municipal incinerator, *J. Chin. Inst. Environ. Eng.*, 13(3), 2003, 167–173.

Index